Biofuels and Biochemicals Production

Special Issue Editor

Thaddeus Ezeji

MDPI • Basel • Beijing • Wuhan • Barcelona • Belgrade

MDPI

Special Issue Editor
Thaddeus Ezeji
The Ohio State University
USA

Editorial Office
MDPI AG
St. Alban-Anlage 66
Basel, Switzerland

This edition is a reprint of the Special Issue published online in the open access journal *Fermentation* (ISSN 2311-5637) from 2016–2017 (available at: http://www.mdpi.com/journal/fermentation/special_issues/biofuels_production).

For citation purposes, cite each article independently as indicated on the article page online and as indicated below:

Author 1; Author 2. Article title. *Journal Name* **Year**, *Article number*, page range.

First Edition 2017

ISBN 978-3-03842-554-0 (Pbk)
ISBN 978-3-03842-555-7 (PDF)

Table of Contents

About the Special Issue Editor

Thaddeus Ezeji is an Associate Professor in the Department of Animal Sciences at Ohio State University (OSU) and Ohio State Agricultural and Development Center (OARDC). Dr. Ezeji received his PhD (Magna cum laude) in Microbiology in 2001 from University of Rostock, Germany. During his 6 years as a post-doctoral and Research Assistant Professor at the University of Illinois Urbana-Champaign, he focused on fermentation technologies and metabolic engineering of Clostridium species. In 2007, Dr. Ezeji joined the OSU and OARDC. His current research focuses on the development of non-food substrates for the production of fuels and chemicals, and the design of advanced fermentation systems. Additionally, Dr. Ezeji's laboratory conducts research activities on genetic and metabolic engineering of bacteria of biotechnological significance. Dr. Ezeji is involved in outreach programs that address emerging issues from Ohio's expanding bioenergy sector. He has published more than 60 scientific papers, 16 book chapters, and co-edited 2 books.

fermentation

MDPI

Editorial

Production of Bio-Derived Fuels and Chemicals

Thaddeus Chukwuemeka Ezeji 🅙

Department of Animal Sciences and Ohio Agricultural Research and Development Center (OARDC),
The Ohio State University, 305 Gerlaugh Hall, 1680 Madison Avenue, Wooster, OH 44691, USA; ezeji.1@osu.edu

Received: 28 August 2017; Accepted: 28 August 2017; Published: 30 August 2017

Keywords: lignocellulose; biomass; pretreatment; butanol; ethanol; butanediol; acetone;
LDMIC; hydrogen

The great demand for, and impending depletion of petroleum reserves, the associated impact of fossil fuel consumption on the environment, and volatility in the energy market have elicited extensive research on alternative sources of traditional petroleum-derived products such as biofuels and bio-chemicals. Fossil oil is largely associated with gasoline, however, approximately 6000 petroleum-derived products currently exist in the market, with diverse applications. Ironically, while biofuels are more popular with the public, the other petroleum-derived products have not attracted similar attention despite the vast economic values for these products. Thus, given the finite nature of petroleum, it is timely to deploy substantial resources and research efforts to the development of renewable chemicals (similar to the efforts devoted to biofuels). Theoretically, bio-production of gasoline-like fuels and the 6000 petroleum-derived products is within the realm of possibility, because aquatic and terrestrial ecosystems harbor an abundance of diverse microorganisms, capable of catalyzing unlimited numbers of chemical reactions. Moreover, the fields of synthetic biology and metabolic engineering have evolved to the point that a wide range of microorganisms can be induced or manipulated to catalyze foreign or vastly improve indigenous biosynthetic reactions. Hence the need for this Special Issue to provide a platform for highlighting recent progress on fuel and chemical production from renewable resources such as lignocellulosic biomass.

This Special Issue, titled Biofuels and Biochemicals Production, consists of 13 articles in which eleven and two are research and review articles, respectively. The Special Issue covers themes on the development of different methodologies for efficient conversion of lignocellulosic biomass, agricultural wastes, carbon dioxide, and carbon monoxide to fuels (ethanol, butanol, hydrogen), chemicals (2,3-butanediol, acetone, acetic acid), and enzymes (cellulase). Some of the articles in this Special Issue provide recent advancements on pretreatment and hydrolysis of lignocellulosic biomass (LB) to lignocellulosic biomass hydrolysates (LBH), challenges associated with LBH utilization, and recommended mitigation strategies.

Consistent with the Biofuels and Biochemicals Production theme, the research groups of Moreno [1] and Rosentrater [2] evaluated different pre-treatment technologies for efficient disruption and separation of lignin from the hemicellulose component of the LB to facilitate enzymatic hydrolysis of the carbohydrate fraction to fermentable sugars. By combining acid-catalyzed steam explosion and alkali-based extrusion process, the protective lignin structure of barley straw was disrupted, which resulted in hydrolysates with significant amounts of glucan and hemicellulose sugars, minimal concentrations of lignocellulose derived microbial inhibitory compounds (LDMICs), and a solid residue with significant amounts of lignin [1]. In addition, the Low-Moisture Anhydrous Ammonia (LMAA) pre-treatment method enhanced enzymatic hydrolysis of the cellulose component of the LB to glucose, thus, the potential is great for LMAA for LB pre-treatment [2]. Consistent with enzymatic hydrolysis of the cellulose component of LB, Bajaj's group contributes an article that highlights the capacity of *Bacillus subtilis* SV1 to use agroindustrial residues (LB) as carbon and nitrogen sources for growth and ionic liquid (IL) stable cellulase production followed by the hydrolysis of IL-pretreated LB to fermentable sugars [3]. Unfortunately, pre-treatment and hydrolysis of LB can result in the formation of a complex mixture of LDMICs that are toxic to fermenting microbes. Examples of LDMICs are furfural, hydroxymethylfurfural (HMF), benzaldehyde, syringaldehyde, and acetic, ferulic, glucuronic, p-coumaric, syringic, levulinic acids, and so on [4]. Overcoming the barriers imposed by LDMICs motivated the study conducted by Marinova's group in which LDMICs of phenol origin in LBH were detoxified using nanofiltration, flocculation, laccase, and combinations thereof [5]. Detoxification of LBH by a combination of flocculation and laccase enzymes before fermentation drastically reduced the concentration of LDMICs in LBH, and significantly improved the fermentation of LBH to butanol [5].

To go beyond conversion of LB to fermentable sugars and produce usable products with lesser carbon footprints, Rorke and Kana [6] evaluated the feasibility of using Monod and modified Gompertz models to study the kinetic behaviour of a bioethanol fermentation process using sorghum leaves and *Saccharomyces cerevisiae* as a substrate and fermentation microorganism, respectively. Interestingly, obtained Monod and modified Gompertz coefficients indicated that waste sorghum leaves can serve as an efficient substrate for bioethanol production. Similarly, Bardi and Cutzu [7] evaluated production of ethanol from agricultural wastes (apple, kiwifruit, peach wastes, and corn threshing residue) using residual thermal energy from ethanol distillation column. Their article recapitulates different concentrations of ethanol obtained from these wastes during ethanolic fermentation with *S. cerevisiae*. With the exception of peach wastes, all the waste substrates assessed had promise for industrial ethanol fermentation, a finding that bodes well with use of non-food crops for biofuel production. Additionally, Krömer's group contributes a technical note that describes simultaneous quantitation of sugars, carboxylates, alcohols and aldehydes in fermentation broth by High Performance Liquid Chromatography (HPLC) [8]. The developed method allows quantitation of 21 compounds in a single process, and could be used in LB pretreatment, hydrolysis, and fermentation of LBH to fuels and chemicals' research.

The two articles from Atiyeh's and Sekoai's groups focus on production of ethanol and acetic acid from synthesis gas by *Clostridium ragsdalei* [9] and optimization of fermentative production of hydrogen using Box–Behnken design [10]. Notably, Atiyeh's article is the first study on continuous operation of syngas fermentations in a trickle-bed reactor (TBR) for ethanol and acetic acid production, and the report highlights operational constraints and challenges of continuous syngas fermentation in TBR, and how the bioreactor operation can be restarted after major accidents such as flooding and power shutdown [9]. Sekoai's study indicates that there can be an improved biohydrogen production yield of 603.5 mL H_2/g total volatile solid (TVS) or more which is achievable at optimized operational set point variables of 39.56 g/L, 82.58 h, 5.56, and 37.9 °C for substrate concentration, fermentation time, pH, and temperature, respectively; a finding that could facilitate the use of large-scale biohydrogen production processes [10].

Fermentative production of chiral compounds is currently receiving remarkable attention because of the numerous industrial applications in the biofuel, synthetic rubber, bioplastics, cosmetics, and flavor industries, and high cost of production from chemical synthetic routes. Recognizing the importance of chiral compounds in the biotechnology industry, our group [11] contributed an article in which process development for enhanced 2,3-butanediol (2,3-BD) production by non-pathogenic bacterium, *Paenibacillus polymyxa* DSM 365, was emphasized. Indeed, while our group was able to increase the concentration of 2-3-BD from 47 g/L (un-optimized) to 68.5 g/L (optimized) under fed-batch fermentation condition, the results underscore an interaction between medium components and fermentation conditions, which tends to influence 2,3-BD and undesirable exopolysaccharides (EPS) production [11]. Although butanol is an achiral compound, it is an important chemical with many applications in the production of solvents, butyl acetates, butylamines, plasticizers, amino resins, etc. [12]. These facts were echoed by Li's group whose article focused on the feasibility of using acidified fibrous immobilization materials (cotton balls, modal fiber and charcoal fiber) to improve production [13]. By pre-treating modal fiber materials with 3.5% HCl for 12 h, the structure of modal

2

Fermentation **2017**, *3*, 42

fibers was etched to decrease mass transfer resistance, increased adsorption of *C. acetobutylicum* to the material, and ultimately, enhanced the kinetics of acetone butanol ethanol (ABE) fermentation [13].

The review articles in this Special Issue provide insights into syngas fermentation [14] and the significance of laccases in the development of LB as an important substrate for the production of renewable fuels and chemicals [15]. The review article contributed by Phillips et al. [14] indicates that integration of thermochemical gasification of LB and wastes to syngas (CO, CO_2 and H_2) and syngas fermentation by autotrophic bacteria is a robust and potentially economical process for the production of fuels and chemicals. Important concepts such as Wood–Ljungdahl biochemical pathway reactions and applications, gas solubility, mass transfer, thermodynamics of enzyme-catalyzed reactions, electrochemistry and cellular electron carriers and fermentation kinetics, were highlighted [14]. The review article contributed by Fillat et al. [15] provides important studies and perspectives on the use of laccases as a delignification and detoxification tool for efficient conversion of LB into value-added products, with emphasis on lignocellulosic ethanol production; highlighting major challenges and opportunities, and plausible ways to integrate the enzymes in the future lignocellulose-allied industries.

In conclusion, it is my hope that this Special Issue will serve as a useful resource for students, teachers, professors, engineers, government personnel, and anyone actively or passively involved in renewable fuels and chemical production and research. In summary, I wish to thank our article contributors, Editorial Board members, Ad Hoc reviewers, and Assistant Editors of this journal, whose contributions made the publication of this Special Issue possible.

Conflicts of Interest: The author declares no conflict of interest.

References

1. Oliva, J.M.; Negro, M.J.; Manzanares, P.; Ballesteros, I.; Chamorro, M.A.; Sáez, F.; Ballesteros, M.; Moreno, A.D. A Sequential Steam Explosion and Reactive Extrusion Pretreatment for Lignocellulosic Biomass Conversion within a Fermentation-Based Biorefinery Perspective. *Fermentation* **2017**, *3*, 15. [CrossRef]
2. Yang, M.; Zhang, W.; Rosentrater, K.A. Anhydrous Ammonia Pretreatment of Corn Stover and Enzymatic Hydrolysis of Glucan from Pretreated Corn Stover. *Fermentation* **2017**, *3*, 9. [CrossRef]
3. Nargotra, P.; Vaid, S.; Bajaj, B.K. Cellulase Production from Bacillus subtilis SV1 and Its Application Potential for Saccharification of Ionic Liquid Pretreated Pine Needle Biomass under One Pot Consolidated Bioprocess. *Fermentation* **2016**, *2*, 19. [CrossRef]
4. Ezeji, T.C.; Qureshi, N.; Blaschek, H.P. Butanol Production from Agricultural Residues: Impact of Degradation Products on Clostridium beijerinckii Growth and Butanol Fermentation. *Biotechnol. Bioeng.* **2007**, *97*, 1460–1469. [CrossRef] [PubMed]
5. Allard-Massicotte, R.; Chadjaa, H.; Marinova, M. Phenols Removal from Hemicelluloses Pre-Hydrolysate by Laccase to Improve Butanol Production. *Fermentation* **2017**, *3*, 31. [CrossRef]
6. Rorke, D.C.S.; Kana, E.B.G. Kinetics of Bioethanol Production from Waste Sorghum Leaves Using *Saccharomyces cerevisiae* BY4743. *Fermentation* **2017**, *3*, 19. [CrossRef]
7. Cutzu, R.; Bardi, L. Production of Bioethanol from Agricultural Wastes Using Residual Thermal Energy of a Cogeneration Plant in the Distillation Phase. *Fermentation* **2017**, *3*, 24. [CrossRef]
8. Lai, B.; Plan, M.R.; Hodson, M.P.; Krömer, J.O. Simultaneous Determination of Sugars, Carboxylates, Alcohols and Aldehydes from Fermentations by High Performance Liquid Chromatography. *Fermentation* **2016**, *2*, 6. [CrossRef]
9. Devarapalli, M.; Lewis, R.S.; Atiyeh, H.K. Continuous Ethanol Production from Synthesis Gas by *Clostridium ragsdalei* in a Trickle-Bed Reactor. *Fermentation* **2017**, *3*, 23. [CrossRef]
10. Sekoai, P. Modelling and Optimization of Operational Setpoint Parameters for Maximum Fermentative Biohydrogen Production Using Box-Behnken Design. *Fermentation* **2016**, *2*, 15. [CrossRef]
11. Okonkwo, C.C.; Ujor, V.C.; Mishra, P.K.; Ezeji, T.C. Process Development for Enhanced 2,3-Butanediol Production by *Paenibacillus polymyxa* DSM 365. *Fermentation* **2017**, *3*, 18. [CrossRef]

12. Ezeji, T.C.; Qureshi, N.; Blaschek, H.P. Microbial Production of a Biofuel (Acetone-Butanol-Ethanol) in a Continuous Bioreactor: Impact of Bleed and Simultaneous Product Removal. *Bioprocess Biosyst. Eng.* **2013**, *36*, 109–116. [CrossRef] [PubMed]

13. Zeng, H.-S.; He, C.-R.; Yen, A.T.-C.; Wu, T.-M.; Li, S.-Y. Assessment of Acidified Fibrous Immobilization Materials for Improving Acetone-Butanol-Ethanol (ABE) Fermentation. *Fermentation* **2017**, *3*, 3. [CrossRef]

14. Phillips, J.R.; Huhnke, R.L.; Atiyeh, H.K. Syngas Fermentation: A Microbial Conversion Process of Gaseous Substrates to Various Products. *Fermentation* **2017**, *3*, 28. [CrossRef]

15. Fillat, U.; Ibarra, D.; Eugenio, M.E.; Moreno, A.D.; Tomás-Pejó, E.; Martín-Sampedro, R. Laccases as a Potential Tool for the Efficient Conversion of Lignocellulosic Biomass: A Review. *Fermentation* **2017**, *3*, 17. [CrossRef]

fermentation

MDPI

Technicalnote

Simultaneous Determination of Sugars, Carboxylates, Alcohols and Aldehydes from Fermentations by High Performance Liquid Chromatography

Bin Lai [1,2,†], Manuel R. Plan [3,4,†], Mark P. Hodson [3,4] and Jens O. Krömer [1,2,*]

1   Centre for Microbial Electrochemical Systems (CEMES), The University of Queensland, Brisbane QLD 4072, Australia; b.lai@uq.edu.au
2   Advanced Water Management Centre (AWMC), The University of Queensland, Brisbane QLD 4072, Australia
3   Australian Institute for Bioengineering and Nanotechnology (AIBN), The University of Queensland, Brisbane QLD 4072, Australia; m.plan@uq.edu.au (M.R.P); m.hodson1@uq.edu.au (M.P.H)
4   Metabolomics Australia (Queensland Node), The University of Queensland, Brisbane QLD 4072, Australia
*   Correspondence: j.kromer@uq.edu.au; Tel.: +61-733-463-222; Fax: +61-733-654-726
†   These authors contributed equally to this work.

Academic Editor: Thaddeus Ezeji
Received: 29 January 2016; Accepted: 24 February 2016; Published: 7 March 2016

Abstract: Despite the rise of 'omics techniques for the study of biological systems, the quantitative description of phenotypes still rests to a large extent on quantitative data produced on chromatography platforms. Here, we describe an improved liquid chromatography method for the determination of sugars, carboxylates, alcohols and aldehydes in microbial fermentation samples and cell extracts. Specific emphasis is given to substrates and products currently pursued in industrial microbiology. The present method allows quantification of 21 compounds in a single run with limits of quantification between 10^{-7} and 10^{-10} mol and limits of detection between 10^{-9} and 10^{-11} mol.

Keywords: high performance liquid chromatography; ion-exchange chromatography; metabolite separation; fermentation product quantification

---

1. Introduction

High performance liquid chromatography (HPLC) has been widely used for quantification of compounds in biological samples [1]. It is precise, quantitative and highly reproducible, but, depending on the analysis, HPLC can be slow and the analysis of different compound classes are best performed with dedicated columns and methods [2]. To date, numerous HPLC-based methods have been developed for analyzing sugars [3], organic acids [4,5] and alcohols [6], respectively. However, running multiple dedicated methods has an impact on sample throughput, unless several instruments are available. In addition, sample throughput can only be increased by reducing chromatographic acquisition time, which may subsequently compromise peak resolution and, thus, data reproducibility. Therefore, a combined method permitting analysis of multiple compound classes is preferable and desirable, permitting the analyst to strike a balance between best possible analysis and throughput. However, the few published combined methods were either operated at high temperatures or achieved lower compound resolutions [7,8].

Despite the progress in column development in other areas of chromatography, such as rapid resolution in reversed phase applications, the method of choice for combined quantification of alcohols, organic acids and sugars is still ion-exchange chromatography [9,10], and due to the use of refractive index detection for sugar and alcohol analysis, this is still mainly based on isocratic elution.

Amongst a wide range of applications for such analyses in the food and chemical industries, one important application is in biotechnology research. In particular, the quantitative analysis of compounds from fermentation samples can serve as an essential tool for the understanding of microbial phenotypes and for the development of improved microbial strains for the production of biofuels, fine chemicals or bulk chemical feedstocks as replacements for petrochemicals.

Here, we present a thoroughly tested method that has broad application in microbiological research, providing quantitative data for a range of common substrates in microbial fermentation including hexoses, pentoses and disaccharides, while at the same time covering a broad range of fermentation products including mono-, di-, tri-alcohols, aldehydes, mono-, di- and tri-carboxylic acids, as well as sugar acids. While still based on cation-exchange, the method provides optimized operation temperature and mobile phase composition for a recently commercialized column. It has been optimized for simultaneous quantification of at least 21 compounds, including carbohydrates to varied alcohol products via central metabolism and has been applied to three very different samples.

2. Experimental Section

2.1. Chemicals

A list of 30 compounds was tested, and all chemicals used in the study were of analytical grade and were purchased from Sigma–Aldrich (Sydney, Australia). Aqueous analyte solutions and mobile phase were prepared using high purity water (18.2 kΩ) generated by an Elga Lab water purification system (Veolia Water Solutions and Technologies, Saint Maurice Cedex, France).

2.2. HPLC Set up

Separation of compounds was performed on an Agilent 1200 HPLC system using an Agilent Hiplex H column (300 × 7.7 mm, PL1170-6830, Santa Clara, CA, USA) with a guard column (SecurityGuard Carbo-H, Phenomenex PN: AJO-4490; Lane Cove West, New South Wales, Australia) for extended column life. Moreover, to extend column life, the column is cleaned with 0.2 mL/min of high purity water (18.2 MΩ) at 60 °C overnight and then regenerated with the same conditions using 25 mM sulfuric acid for a few hours, which is ideally performed after each batch of analysis. With regular column maintenance and careful sample preparation (e.g., samples pre-filtered using 0.22 μm PES syringe filter (Millipore: Cork, Ireland) and pre-diluted microbial fermentation samples) we have been able to make more than 200 injections per batch of analysis without change in column performance (*i.e.*, without significant RT drift or increase in back pressure).

Sugars and alcohols were monitored using a refractive index detector (Agilent RID, G1362A) set on positive polarity and optical unit temperature of 40 °C with mobile phase in the reference cell, while organic acids were monitored using RID and/or ultraviolet detector at 210 nm (Agilent MWD, G1365B).

A sample volume of 30 μL was injected onto the column using an autosampler (Agilent HiP-ALS, G1367B) and the column temperature was maintained at 40 °C using a thermostatically controlled column compartment (Agilent TCC, G1316A). Analytes were eluted isocratically with 14 mM H_2SO_4 at 0.4 mL/min for 38 or 65 min (elution time was dependent upon whether higher alcohols were present in the sample). Chromatograms were integrated using Agilent ChemStation (Rev B.03.02; Santa Clara, CA, USA).

3. Results and Discussions

A series of preliminary experiments were conducted to monitor the interaction of the retention times (RT) of compounds from various classes with column temperature (30, 50 and 65 °C), mobile phase concentration (2, 4, 6, 8, 10, 12 and 14 mM) and flow rate (0.4, 0.5 and 0.6 mL/min) (see supplementary information) and found that a column temperature of 40 °C, aqueous solution of H_2SO_4 (14 mM) and a flow rate of 0.4 mL/min was the best combination to achieve separation of the highest number of target compounds.

With the optimized operating parameters, the method developed in this article is suitable for mapping varied metabolic routes from carbohydrates, via carboxylic acids to alcoholic products (Figure 1) and is, thus, highly relevant for fermentation process development.

Looking at widely used sugar substrates for fermentation [11] and sugar products occurring in bioprocesses [12], our method has the capacity to separate D-trehalose, D-glucose, D-galactose, L-arabinose and D-ribose in the same sample. D-fructose and D-galactose partially overlap, which means they should only be quantified if the other sugar is known to be absent from the sample, the same holds for the disaccharides D-maltose and D-trehalose. Sucrose exhibited partial on/in-column inversion and cannot be analyzed reliably with the presented method, however the same column with water as the mobile phase would be suitable for sucrose quantification (data not shown).

In addition to the fermentation substrates, 10 organic acids related to central metabolism were identifiable and quantifiable in a single injection with this method, including two specific sugar acids, gluconic acid and 2-ketogluconic acid, making this method suitable for microbes that favour the Embden–Meyerhof–Parnas, as well as those using the Entner–Doudoroff pathway for sugar utilization. This extends the applicability of the method amongst others to the group of *Pseudomonads*, which contains a range of new strains for biotechnology that are currently widely studied for biosynthesis of chemicals [13]. It has to be noted that citric acid and 2-ketogluconic acid co-elute with this method and should not occur simultaneously. Previously published methods struggled to separate compounds like formic acid and fumaric acid, 2-ketoglutaric acid and citric acid, pyruvic acid and glucose [14]. These can now be successfully resolved and quantified in the same sample.

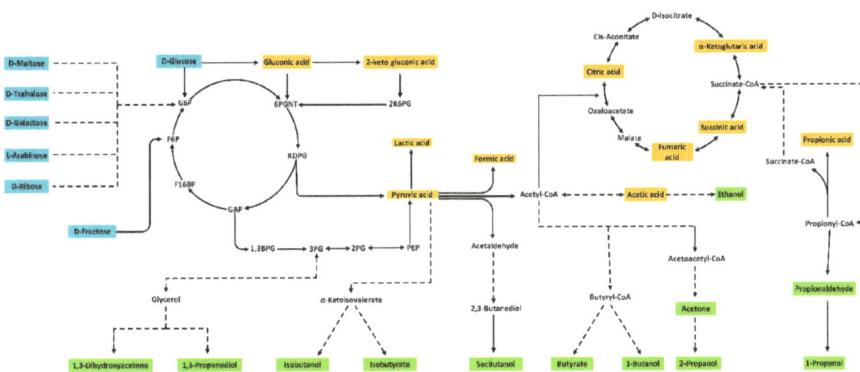

Figure 1. Substrates, intermediates and products of microbial fermentation captured by the presented method in culture broth. Sugars (blue), organic acids (yellow), alcohols, ketones and aldehydes (green).

Looking at target compounds for biotechnology, this method is able to analyze a range of alcohols currently studied as biofuels and chemical feedstock replacements. This includes ethanol, 1-butanol, *sec*-butanol, *iso*-butanol, 1-propanol as well as 2-propanol (Table 1). Acetone and its structural isomer propionaldehyde are the metabolic precursors to 2-propanol and 1-propanol, respectively, and can now be analyzed with their respective end product in the same solution. Butyric acid and *iso*-butyric acid, the main by-products of butanol fermentation, can be quantified simultaneously as well. One problem is the separation of 2-propanol and butyric acid, these will partly overlap with the current chromatographic conditions. In any case, peak identification should be confirmed with alternative means (e.g., mass spectrometry) in complex samples. The calibration curves achieved a good fit and recoveries in the standard matrix were high (Table 1). The achieved peak shape and elution profiles were acceptable for an isocratic HPLC method (Figure 2).

Table 1. Analytes quantified with the presented method in order of retention time. LOQ: limit of quantification; S/N: Signal-to-Noise ratio; LOD: limit of detection. LOQ and LOD are given both as concentration in the sample, as well as amount injected. LOQ, LOD and S/N were detected and calculated based on the specified detector for quantification for each compound. In other sample matrices LOQ, LOD and S/N might vary. Calibration curves were obtained through linear regression (forced through the origin) of five standard points covering the linear detection range. UV/RT: retention time in UV detector; RI/RT: retention time in RI detector, detectors in series.

Peak	Compound	RT (min)		LOQ		S/N_{LOQ}	LOD		S/N_{LOD}	Calibration Curve			Recovery (%)
		UV	RI	mM	nmol		mM	nmol		Detector	Slope	R^2	
1	D-Trehalose	-	12.53	0.098	2.94	10.9	0.012	0.37	2.8	RI	203,384	0.9999	100.6
	D-Maltose	-	12.56	0.117	3.52	12.4	0.029	0.88	2.9	RI	218,036	0.99998	100.3
2	2-Ketogluconic acid	13.34	13.70	0.158	4.73	12.3	0.005	0.16	2.9	UV	374	0.99997	102.3
	Citric acid	13.42	13.78	0.060	1.80	15.0	0.010	0.30	2.9	UV	827	0.99999	100.0
	Gluconic acid	14.13	14.49	0.159	4.77	11.4	0.005	0.16	3.2	UV	343	0.99999	100.6
3	D-Glucose	-	14.50	0.391	11.72	9.4	0.049	1.46	1.9	RI	116,091	0.99977	101.1
	2-Ketoglutaric acid	14.94	15.29	0.025	0.75	10.7	0.003	0.09	2.1	UV	706	0.99996	99.9
4	D-Galactose	-	15.38	0.300	9.00	20.1	0.075	2.24	3.1	RI	119,106	0.99974	96.0
	D-Fructose	-	15.64	0.365	10.94	14.9	0.024	0.73	3.6	RI	109,365	0.99996	101.8
5	Pyruvic acid	16.33	16.68	0.020	0.59	7.2	0.005	0.15	2.3	UV	1976	0.99992	96.5
	L-Arabinose	-	16.68	0.365	10.95	13.8	0.091	2.74	1.8	RI	a	0.99999	100.0
6	D-Ribose	-	17.17	0.266	7.99	11.3	0.018	0.53	1.6	RI	86,258	0.99996	102.7
7	Succinic acid	19.01	19.43	0.196	5.87	9.4	0.024	0.73	1.8	UV	285	0.99996	99.8
	Lactic acid	19.90	20.27	0.235	7.06	10.9	0.029	0.88	1.7	UV	297	0.99997	99.4
8	Glycerol	-	20.45	0.156	4.68	44.1	0.078	2.34	3.0	RI	199,288	0.99999	98.9
9	1,3-Dihydroxyacetone	20.51	20.88	0.363	10.90	9.2	0.060	1.8	2.6	UV	192	0.9995	105.6
11	Formic acid	21.53	21.89	0.469	14.08	12.3	0.059	1.76	2.0	UV	186	0.99999	99.8
11	Acetic acid	23.46	23.83	0.361	10.84	9.2	0.045	1.36	1.7	UV	147	0.99999	100.2
12	Fumaric acid	25.33	-	0.006	0.19	32.6	0.001	0.02	4.2	UV	44,143	0.99967	95.8
13	1,3-Propanediol	-	26.48	0.811	24.33	7.3	0.203	6.08	3.1	RI	37,638	0.99986	98.9
14	Propionic acid	28.22	28.59	0.276	8.27	6.5	0.034	1.03	1.8	UV	177	0.99998	100.2
15	Ethanol	-	32.45	9.609	288.26	10.8	0.601	18.02	2.0	RI	12,451	0.99994	99.7
	iso-Butyric acid	32.76	33.13	0.394	11.82	17.8	0.049	1.48	2.9	UV	274	0.99987	99.5
16	Propionaldehyde	-	34.94	1.634	49.03	9.8	0.204	6.13	1.8	RI	21,298	0.99982	97.5
	Acetone	-	35.16	1.568	47.05	9.7	0.399	11.97	2.3	RI	16,063	0.99999	101.0
	2-Propanol	-	35.96	1.616	48.48	10.1	0.404	12.12	1.7	RI	23,012	0.99998	101.2
17	Butyric acid	36.02	36.38	0.450	13.51	12.9	0.074	2.23	2.3	UV	212	0.99998	102.3
18	1-Propanol	-	41.38	1.566	46.97	15.5	0.196	5.87	2.0	RI	10,988	0.99966	100.7
19	sec-Butanol	-	48.87	1.558	46.73	9.8	0.195	5.84	1.5	RI	33,209	0.99999	99.5
20	iso-Butanol	-	52.07	2.286	68.58	13.9	0.377	11.3	2.3	RI	35,461	0.99999	101.1
21	1-Butanol	-	59.97	1.565	46.96	11.9	0.391	11.74	2.6	RI	34,911	0.99999	99.3

^{a} The line of best fit for arabinose was quadratic ($y = 497.65*x^2 + 64,692.83*x$).

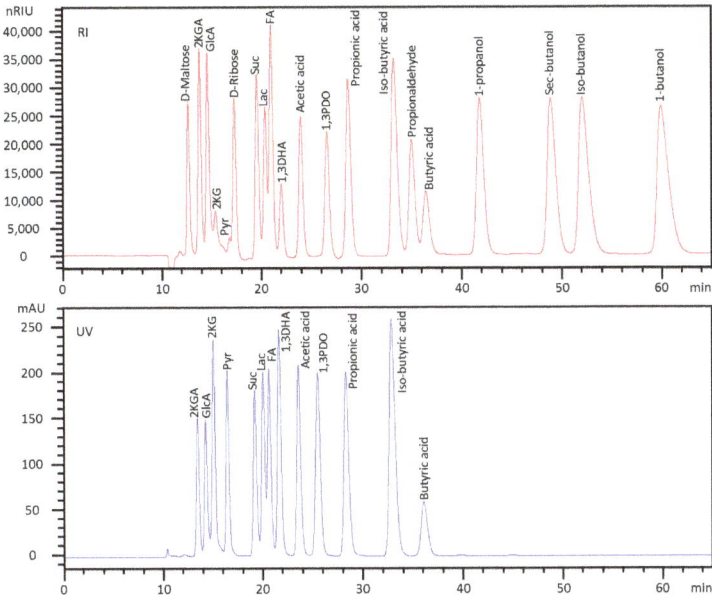

Figure 2. HPLC chromatograms of selected standard mixture that is quantifiable in a single injection. Note: Not all compounds listed in Table 1 are included in the mixture and focus should be on peak shape. Abbreviations: Mal (Maltose), 2KGA (2-Ketogluconic acid), GlcA (Gluconic acid), 2KG (2-Ketoglutaric acid), Pyr (Pyruvic acid), Suc (Succinic acid), Lac (Lactic acid), FA (Formic acid), 1,3DHA (1,3-Dihydroxyacetone), 1,3PDO (1,3-Propanediol).

Finally, the method was tested on three different samples: (i) fermentation broth of a genetically modified *Escherichia coli* fermentation during aerobic growth on glucose in minimal medium; (ii) fermentation broth of *Saccharomyces cerevisiae* growing on the carbon sources glycerol and ethanol in minimal medium and finally; (iii) the culture supernatant of Chinese Hamster Ovary (CHO) cells in complex cell culture medium using glucose and galactose as the main carbon source (Figure 3). In all three samples peak shapes and separation was good and the main substrates and products could successfully be quantified. This demonstrates the robustness and versatility of the described method. Based on the limits of detection, this method should also in principle be applicable to cell extracts. Quantitative data obtained from this method on fermentation samples of *Pseudomonas putida* was of sufficient quality to close the carbon and redox balances [15], underlining the value of this method for a range of applications.

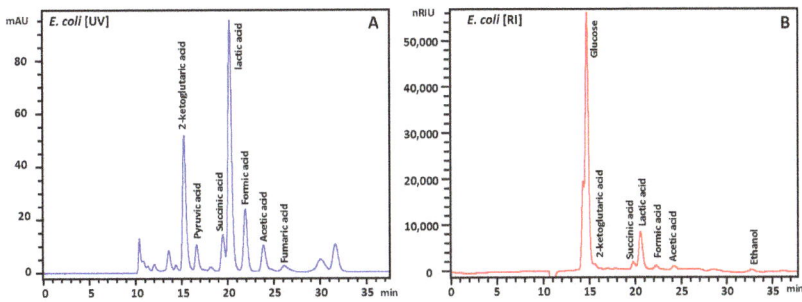

Figure 3. *Cont.*

9

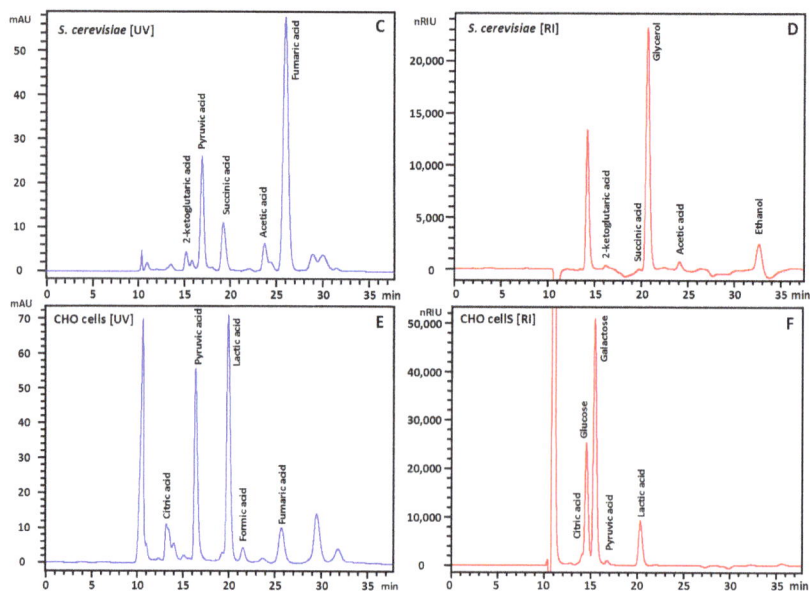

Figure 3. HPLC chromatograms for supernatants of genetically modified *E. coli* (**A,B**) growing on glucose in minimal medium, genetically modified *S. cerevisiae* growing on glycerol and ethanol (**C,D**) in minimal medium and Chinese Hamster Ovary (CHO) cells growing on galactose and glucose in defined cell culture medium. Chromatograms serve the purpose of highlighting resolving power and are not a reference chromatogram for the respective organisms. Detection is performed with UV (**A,C,E**) and RI (**B,D,F**). In these samples, no signals were observed beyond 35 min and chromatograms have been shortened for better visibility of the peaks. Injection signals are visible at around 10 min in UV and 10.4 min RI. Complex medium components in the CHO experiment mainly pass through the column (**E,F**).

4. Conclusions

In summary, a broad range of metabolites could be separated and quantified in one HPLC injection with LOQ and LOD in ranges that will be suitable for a large range of fermentation samples, including microbial culture broth and cell culture media and could be extended to intracellular samples. The data quality allows drawing of carbon and degree of reduction balances. The method can be extended to other compounds, if presence of co-eluting compounds can be ruled out with alternative methods (e.g., by mass spectrometry on pooled samples).

Supplementary Materials: Supplementary materials can be found at http://www.mdpi.com/2311-5637/2/1/6/s1.

Acknowledgments: Bin Lai acknowledges scholarship support from The University of Queensland. Manuel R. Plan and Mark P. Hodson acknowledge financial support from Metabolomics Australia. Jens O. Krömer was supported by the Australian Research Council (DE120101549). We thank Alex Prima, Nils Averesch, Axayacatl Gonzalez Garcia and Veronica Martinez for the supply of samples.

Author Contributions: Bin Lai and Manuel R. Plan designed, performed the experiments and analyzed the data, under the guidance from Mark P. Hodson and Jens O. Krömer. All authors contributed to the drafting and editing of the manuscript.

Conflicts of Interest: The authors declare no conflict of interest.

Fermentation **2016**, *2*, 6

References

1. Pereira da Costa, M.; Conte-Junior, C.A. Chromatographic methods for the determination of carbohydrates and organic acids in foods of animal origin. *Compr. Rev. Food Sci. Saf.* **2015**, *14*, 586–600. [CrossRef]
2. Ball, S.; Lloyd, L. Agilent Hi-Plex Columns for Carbohydrates, Alcohols, and Acids. Available online: http://www.agilent.com/cs/library/applications/5990–8264EN.pdf (accessed on 7 March 2016).
3. Agblevor, F.A.; Hames, B.R.; Schell, D.; Chum, H.L. Analysis of biomass sugars using a novel hplc method. *Appl. Biochem. Biotechnol.* **2007**, *136*, 309–326. [CrossRef] [PubMed]
4. Womersley, C.; Drinkwater, L.; Crowe, J.H. Separation of tricarboxylic acid cycle acids and other related organic acids in insect haemolymph by high-performance liquid chromatography. *J Chromatogr. A* **1985**, *318*, 112–116. [CrossRef]
5. van Hees, P.A.W.; Dahlén, J.; Lundström, U.S.; Borén, H.; Allard, B. Determination of low molecular weight organic acids in soil solution by hplc. *Talanta* **1999**, *48*, 173–179. [CrossRef]
6. Kumar, M.; Saini, S.; Gayen, K. Acetone-butanol-ethanol fermentation analysis using only high performance liquid chromatography. *Anal. Methods* **2014**, *6*, 774–781. [CrossRef]
7. Sluiter, A.; Hames, B.; Ruiz, R.; Scarlata, C.; Sluiter, J.; Templeton, D. *Determination of Sugars, Byproducts, and Degradation Products in Liquid Fraction Process Samples: Laboratory Analytical Procedure (LAP)*; National Renewable Energy Laboratory: Golden, CO, USA, 2008.
8. Castellari, M.; Versari, A.; Spinabelli, U.; Galassi, S.; Amati, A. An improved hplc method for the analysis of organic acids, carbohydrates, and alcohols in grape musts and wines. *J. Liq. Chromatogr. Relat. Technol.* **2000**, *23*, 2047–2056. [CrossRef]
9. López, E.F.; Gómez, E.F. Simultaneous determination of the major organic acids, sugars, glycerol, and ethanol by hplc in grape musts and white wines. *J. Chromatogr. Sci.* **1996**, *34*, 254–257. [CrossRef]
10. Chinnici, F.; Spinabelli, U.; Riponi, C.; Amati, A. Optimization of the determination of organic acids and sugars in fruit juices by ion-exclusion liquid chromatography. *J. Food Compos. Anal.* **2005**, *18*, 121–130. [CrossRef]
11. Vertes, A.A. *Biomass to Biofuels: Strategies for Global Industries*; Wiley: Chichester, UK, 2010; Volume 1.
12. Wang, P.-M.; Zheng, D.-Q.; Chi, X.-Q.; Li, O.; Qian, C.-D.; Liu, T.-Z.; Zhang, X.-Y.; Du, F.-G.; Sun, P.-Y.; Qu, A.-M.; *et al.* Relationship of trehalose accumulation with ethanol fermentation in industrial saccharomyces cerevisiae yeast strains. *Bioresour. Technol.* **2014**, *152*, 371–376. [CrossRef] [PubMed]
13. Nikel, P.I.; Martinez-Garcia, E.; de Lorenzo, V. Biotechnological domestication of pseudomonads using synthetic biology. *Nat. Rev. Microbiol.* **2014**, *12*, 368–379. [CrossRef] [PubMed]
14. Dietmair, S.; Timmins, N.E.; Gray, P.P.; Nielsen, L.K.; Krömer, J.O. Towards quantitative metabolomics of mammalian cells: Development of a metabolite extraction protocol. *Anal. Biochem.* **2010**, *404*, 155–164. [CrossRef] [PubMed]
15. Lai, B.; Yu, S.; Bernhardt, P.V.; Rabaey, K.; Virdis, B.; Krömer, J.O. Anoxic metabolism and biochemical production in *pseudomonas putida* f1 driven by a bioelectrochemical system. *Biotechnol. Biofuels* **2016**, *9*, 39. [CrossRef] [PubMed]

fermentation

MDPI

Article

Modelling and Optimization of Operational Setpoint Parameters for Maximum Fermentative Biohydrogen Production Using Box-Behnken Design

Patrick T. Sekoai

Sustainable Energy & Environment Research Unit, School of Chemical and Metallurgical Engineering, Faculty of Engineering and the Built Environment, University of the Witwatersrand, Private Bag 3, Wits, Johannesburg 2050, South Africa; 679314@students.wits.ac.za; Tel.:+27-807-3783

Academic Editors: Thaddeus Ezeji and Badal C. Saha
Received: 3 June 2016; Accepted: 15 July 2016; Published: 20 July 2016

Abstract: Fermentative biohydrogen production has been flagged as a future alternative energy source due to its various socio-economical benefits. Currently, its production is hindered by the low yield. In this work, modelling and optimization of fermentative biohydrogen producing operational setpoint conditions was carried out. A box-behnken design was used to generate twenty-nine batch experiments. The experimental data were used to produce a quadratic polynomial model which was subjected to analysis of variance (ANOVA) to evaluate its statistical significance. The quadratic polynomial model had a coefficient of determination (R^2) of 0.7895. The optimum setpoint obtained were potato-waste concentration 39.56 g/L, pH 5.56, temperature 37.87 °C, and fermentation time 82.58 h, predicting a biohydrogen production response of 537.5 mL H_2/g TVS. A validation experiment gave 603.5 mL H_2/g TVS resulting to a 12% increase. The R^2 was above 0.7 implying the model was adequate to navigate the optimization space. Therefore, these findings demonstrated the feasibility of conducting optimized biohydrogen fermentation processes using response surface methodology.

Keywords: biohydrogen production; modelling and optimization; box-behnken design

1. Introduction

The adverse effects of climate change coupled with environmental pollution makes it necessary to search for clean and sustainable energy resources [1–3]. Hydrogen is considered as one of the potential alternative fuels because it is a clean energy source and its combustion results in pure water. It can also be used in various applications such as fuel for automobiles, electricity, and thermal energy generation. Moreover, it can be derived from diverse substrates including waste materials.

Amongst the hydrogen producing methods, biological hydrogen production processes are highly recommended in hydrogen research fraternity as compared to thermo-chemical processes because they are environmentally friendly and less-energy intensive, i.e., can be carried out at ambient temperature and pressure. They mainly include photosynthetic and fermentative biohydrogen production. The challenges facing photosynthetic biohydrogen production are low production yields and the requirement for a light source. Meanwhile, fermentative biohydrogen production can produce hydrogen for long periods of time without any light using diverse substrates such as organic wastes and thus has a higher feasibility for industrialization. Moreso, it is more viable and extensively used [4]. Therefore, fermentative biohydrogen production process from waste materials plays a pivotal role because it simultaneously generates hydrogen while curbing environmental pollution.

The optimization of biohydrogen operational setpoint parameters is of critical importance in the research and development of biohydrogen fermentation technology owing to its impact on the economy and practicability of the process. The one dimensional search with successive variation in variables, such as the one-variable-at-a-time (OVAT) method, is still used, albeit it is well understood that it is impractical for the one dimensional search to achieve an appropriate optimum results in a restricted number of experiments [5]. The complexity of combinational interactions of operational setpoint variables and production does not allow for satisfactory detailed modelling [5]. Furthermore, single parameter optimization methods are not only tedious but can lead to misinterpretation of results, especially because the interaction between different factors is overlooked [6,7].

Statistical experimental approaches have been extensively used for many years and it can be implemented at various stages of an optimization strategy, such as screening of experiments or for investigating optimal setpoint parameters on production responses [8]. Fermentation optimization is conducted using a statistically designed experiment in a sequential process [9,10]. This involves a large number of variables that are initially screened and the irrelevant ones are eliminated in order to obtain a fewer and manageable set of parameters. The remaining variables are then optimized by a response surface modelling (RSM) method. Finally, after model building and optimization, the predicted optimum is verified [11,12]. The box-behnken RSM design uses a spherical design with good certainty within the design space. It requires fewer experiments as compared to other RSM designs [13]. In addition, box-behnken design is rotatable regardless of the number of parameters under investigation [14]. This statistical approach has been successfully applied in various fermentative biohydrogen production processes and has been proven to be very efficient in optimizing these processes [15–19].

This study modelled and optimized the operational setpoint parameters of potato-waste concentration, pH, temperature, and fermentation time for maximum biohydrogen production process using box-behnken design. Moreover, the pairwise interactive effect of the above mentioned setpoint parameters was investigated on biohydrogen production response.

2. Materials and Methods

2.1. Inoculum Development

Biohydrogen-producing anaerobic mixed sludge was collected from Olifantvlei Wastewater Treatment Plant, Johannesburg, South Africa. The sludge was boiled at 100 °C for 30 min. This was done in order to deactivate the biohydrogen-consuming methanogenic bacteria and enumerate the biohydrogen spore-forming bacteria. The sludge was supported with a nutrient stock solution (all in g/L): yeast extract 2.0, glucose 10, K_2HPO_4 0.420, $CaCl_2$ 0.375, $MgSO_4$ 0.312, $NaHCO_3$ 8.0, KCl 0.25. It was then transferred into an Erlenmeyer flask (100 mL) which was covered with foil, and cultured for three days at (30 °C) using a water-bath shaker, this was done to boost the population of biohydrogen-producing bacteria. In addition, the inoculum preparation stage is essential because biohydrogen-producing bacteria, such as *Clostridium* species, are fastidious and, therefore, a preliminary stage is carried out in order to revive them and increase their cell concentration. This served as inoculum for the twenty nine experimental designs.

2.2. Experimental Design

The four parameters studied and their search ranges were the concentration of potato-waste 10–40 g/L, pH 3–8, temperature 32–38 °C, and fermentation time 5–120 h. Based on these, box-behnken design was used to generate 29 different experiments by varying the operational setpoint parameters, as shown in Table 1.

Table 1. Biohydrogen production response from the box-behnken design.

Run	PW	FT	pH	Temp	H_2 Yield
1	10	62.5	8	35	89.8
2	10	120	5.5	35	111.3
3	25	5	3	35	0.5
4	40	120	5.5	35	214.2
5	40	5	5.5	35	30.9
6	25	5	8	35	50.4
7	25	120	8	35	58.6
8	25	120	3	35	48.7
9	10	5	5.5	35	10.5
10	25	62.5	3	38	139.5
11	25	120	5.5	38	405.0
12	40	62.5	5.5	38	495.5
13	25	5	5.5	38	0
14	10	62.5	5.5	38	0
15	25	62.5	8	38	528.0
16	40	62.5	8	35	474.5
17	25	62.5	5.5	35	373.0
18	25	62.5	5.5	35	245.5
19	25	62.5	5.5	35	333.0
20	25	62.5	5.5	35	384.5
21	10	62.5	3	35	0
22	40	62.5	8	35	275.0
23	25	62.5	5.5	35	432.5
24	25	62.5	3	32	10.0
25	25	5	5.5	32	0
26	10	62.5	5.5	32	61.0
27	25	62.5	8	32	310.0
28	40	62.5	5.5	32	277.0
29	25	120	5.5	32	0

PW: Potato-waste concentration (g/L), FT: Fermentation time (h), Temp (°C), H_2 yield (mL H_2/g TVS).

2.3. Substrates and Pretreatment

Potato-waste was obtained from various dumping sites in the city of Johannesburg, South Africa. The effluents were oven dried at 60 °C for 24 h, and then grounded into fine particles (0.2–0.5 mm). The total volatile solid (TVS) of potato-waste was determined using Equation (1).

$$TVS = \frac{\text{Weight of dried waste} - \text{Weight of ash}}{\text{Weight of dried waste}} \times 100\% \qquad (1)$$

2.4. Fermentation Process

Substrate concentrations as specified in the design (Table 1) were weighed into 250 mL Erlenmeyer flask, and the volume was raised to 100 mL with distilled water. These were autoclaved prior to the fermentation process. One ml of inoculum was added to each 250 mL flask. The operational setpoint parameters were kept as specified in the design. The fermentation process was conducted in a temperature controlled shaking water-bath. Anaerobic microenvironments were achieved by flushing the fermenter flasks with nitrogen gas for 3 min. The twenty-nine batch fermentation processes were carried out in duplicates.

2.5. Analytical Procedures

Hydrogen was measured and monitored using the hydrogen sensor at 1 h interval (BCP-H_2 Bluesens GmbH, Herten, Germany) connected to a computer measuring software system. The sensor has a measuring range of 0%–100% and use a thermal conductivity detector and infrared technology. The cumulative volume of hydrogen was calculated using Equation (2).

$$V_{H,i} = V_{H,i-1} + C_{H,i}(V_{G,i} - V_{G,i-1}) + V_H(C_{H,i} - C_{H,i-1}) \qquad (2)$$

$V_{H,i}$ and $V_{H,i-1}$ are cumulative hydrogen gas volume at the current (i) and previous (i − 1) time intervals, $V_{G,i}$ and $V_{G,i-1}$ the total biogas volumes in the current and previous time intervals, $C_{H,i}$ and $C_{H,i-1}$ the fraction of hydrogen gas in the headspace of the reactor in the current and previous time intervals, and V_H the total volume of headspace in the reactor [20].

2.6. General Model

The results obtained from the experiments were used to develop a quadratic model that relates hydrogen production to the considered parameters. The general form of the model with four parameters is represented by Equation (3).

$$Y = \alpha_0 + \alpha_1 x_1 + \alpha_2 x_2 + \alpha_3 x_3 + \alpha_4 x_4 + \alpha_{11} x_1^2 + \alpha_{22} x_2^2 + \alpha_{33} x_3^2 + \alpha_{44} x_4^2 + \alpha_{12} x_1 x_2 +$$
$$\alpha_{13} x_1 x_3 + \alpha_{14} x_1 x_4 + \alpha_{23} x_2 x_3 + \alpha_{24} x_2 x_4 + \alpha_{34} x_3 x_4 \qquad (3)$$

where Y is the biohydrogen production response, α_0 is the regression coefficient, $\alpha_1 x_1$ to $\alpha_4 x_4$ are linear terms, $\alpha_{11} x_1^2$ to $\alpha_{44} x_4^2$ are linear coefficient and $\alpha_{12} x_1 x_2$ to $\alpha_{34} x_3 x_4$ shows the interaction between parameters on biohydrogen production. The model fitness was evaluated by the analysis of variance (ANOVA) using Design Expert software (Stat Ease, Inc., Minneapolis, MN, USA).

3. Results and Discussion

3.1. The Linear Interactive Effect of Parameters on Biohydrogen Production

Table 1 shows the linear interaction of operational setpoint parameters on biohydrogen production. The hydrogen yields varied from 0 to 528 mL H_2/g TVS. The highest biohydrogen production yield was observed in runs 12 and 15, i.e., a maximum biohydrogen yield of 495.5 and 528.0 mL H_2/g TVS, respectively, were obtained from these batch experiments. Analysis of individual parameters impact on the biohydrogen production pattern indicated that the fermentation times of 5 and 62.5 h, low pH (3 and 5.5), and low concentration of potato waste (10 and 20 g/L) produce low yields of hydrogen. This is likely attributed to the low pH as confirmed in literature. pH has been identified as one of the most pivotal parameters that influence the growth of biohydrogen-producing bacteria. It also affects the activity of biohydrogen-producing hydrogenase enzymes and its metabolic pathway [4]. Moreover, it was shown that low pH values (below 4) have an inhibitory effect on the activity of biohydrogen-producing bacteria [4].

However, low fermentation time and high pH, moderate temperature, and concentration of potato waste increases the hydrogen yield. Similar findings were reported by Sekoai and Gueguim Kana [4], hence this highlights the importance of operational setpoint parameters on biohydrogen production process modelling and optimization.

3.2. Development of Model for Optimization of Biohydrogen Production

3.2.1. Model Analysis Based on Input Parameters

The experimental data were used to generate a quadratic polynomial equation (Equation (4)). This mathematical model relates hydrogen production to pH, temperature, fermentation time, and substrate concentration. Where Y represents the hydrogen production response; A, B, C, and D represents the operational setpoint parameters of potato-waste concentration, fermentation time, pH, and temperature respectively. Moreover A^2, B^2, C^2, and D^2 represents the quadratic coefficients of the above mentioned setpoint parameters.

$$Y = 707.40 + 248.00A + 123.58B + 107.25C + 152.00D + 41.75AB - 148.50AC +$$
$$140.75AD - 18.00BC + 202.5BD + 44.25CD - 141.66A^2 - 421.28B^2 - 155.28C^2 - 96.66D^2 \qquad (4)$$

The ANOVA was also conducted to test the significance and the fitness of the regression equation. Data from the analysis of variance is presented in Table 2, a high F-value (3.75) and low p-value (0.0094) indicates that the model is significant. The model's coefficients of estimates and their confidence intervals are presented in Table 3. The generated model had a coefficient of determination (R^2) value of 0.7895, this implies that 78.95% of the data can be explained by the model. The results obtained from this study correlate with literature, it has been reported that R^2 values greater than 0.75 show that the model is accurate [21].

Table 2. Analysis of variance (ANOVA) of the box-behnken model.

Source	SS	df	MS	F-Value	p-Value	R²
Model	2,890,000	14	207,000	3.75	0.0094	0.7895
A	738,000	1	738,000	13.4	0.0026	
B	183,000	1	183,000	3.33	0.0895	
C	138,000	1	138,000	2.51	0.1357	
D	277,000	1	277,000	5.03	0.0415	
AB	6972.25	1		0.13	0.7273	
AC	88,209	1		1.6	0.2263	
AD	79,242.25	1		1.44	0.2502	
BC	1296	1		0.024	0.8803	
BD	164,000	1		2.98	0.1064	
CD	7832.25	1	7832.25	0.7117		
A²	130,000	1		2.36	0.1465	
B²	1,150,000	1		20.9	0.0004	
C²	156,000	1		2.84	0.1141	
D²	60,602.16	1		1.1	0.3119	

A: Potato-waste concentration, B: Fermentation time, C: pH, D: Temperature, AB: interaction between potato-waste concentration and fermentation time, AC: interaction between potato-waste concentration and pH, AD: interaction between potato-waste concentration and temperature, BC: interaction between fermentation time and pH, BD: interaction between fermentation time and temperature, CD: interaction between pH and temperature, A^2: quadratic value for potato-waste concentration, B^2: quadratic value for fermentation time, C^2: quadratic value for pH, D^2: quadratic value for temperature, SS: Sum of squares, MS: Mean of squares, df: degrees of freedom, F-value: Fisher-Snedecor distribution value, p-value: Probability value, R^2: Coefficient of determination.

Table 3. Coefficients of estimates and their confidence intervals for box-behnken design.

Factor	CE	df	SE	95% CIL	95% CIH	VIF
Intercept	707.4	1	104.95	482.31	932.49	
A	248	1	67.74	102.7	393.3	1
B	123.58	1	67.74	−21.71	268.88	1
C	107.25	1	67.74	−38.05	252.55	1
D	152	1	67.74	6.7	297.3	1
AB	41.75	1	117.34	−209.91	293.41	1
AC	−148.5	1	117.34	−400.16	103.16	1
AD	140.75	1	117.34	−110.91	392.41	1
BC	−18	1	117.34	−269.66	233.66	1
BD	202.5	1	117.34	−49.16	454.16	1
CD	44.25	1	117.34	−207.41	295.91	1
A²	−141.66	1	92.14	−339.28	55.97	1.08
B²	−421.28	1	92.14	−618.91	−223.66	1.08
C²	−155.28	1	92.14	−352.91	42.34	1.08
D²	−96.66	1	92.14	−294.28	100.97	1.08

A: Potato-waste concentration, B: Fermentation time, C: pH, D: Temperature, AB: interaction between potato-waste concentration and fermentation time, AC: interaction between potato waste concentration and pH, AD: interaction between potato-waste concentration and temperature, BC: interaction between fermentation time and pH, BD: interaction between fermentation time and temperature, CD: interaction between pH and temperature, A^2: quadratic value for potato-waste concentration, B^2: quadratic value for fermentation time, C^2: quadratic value for pH, D^2: quadratic value for temperature, CE: Coefficient of estimate, df: degrees of freedom, SE: Standard error, 95% CIL: 95% Confidence Intervals (Low limit), 95% CIH: 95% Confidence Intervals (High limit), VIF: Variance Inflation Factor.

3.2.2. Effect of Parameter Interaction on Biohydrogen Production Response

The three dimensional response surface curves showing the production of biohydrogen as a function of parameters interaction are shown in Figures 1–6. The interactive effect of fermentation time and substrate concentration is illustrated in Figure 1; it was observed that an increase in fermentation time (55–80 h) and concentration of potato-waste (22–30 g/L) maximized the production of biohydrogen. It has been reported that an increase in substrate concentration enhances the activity of biohydrogen-producing bacterial species especially during their exponential growth phase [17]. This implies that a large-scale biohydrogen production process can be achieved within this range. Moreover, from these findings it can be deduced that increasing the concentration of potato-waste has a positive effect on biohydrogen production, but higher substrate concentration may have an inhibitory effect on its production [17,22,23].

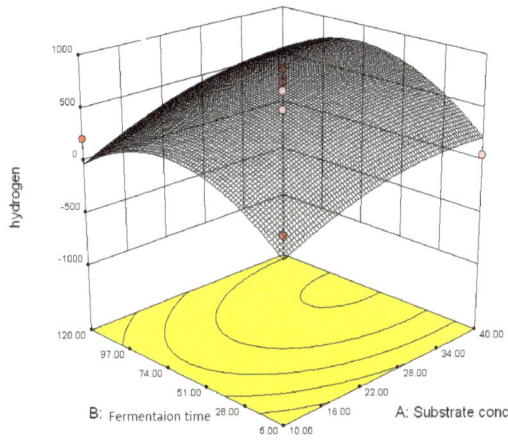

Figure 1. Response surface graph showing the interactive effect of fermentation time (h) and potato-waste concentration (conc, g/L) on hydrogen yield (mL H_2/g TVS).

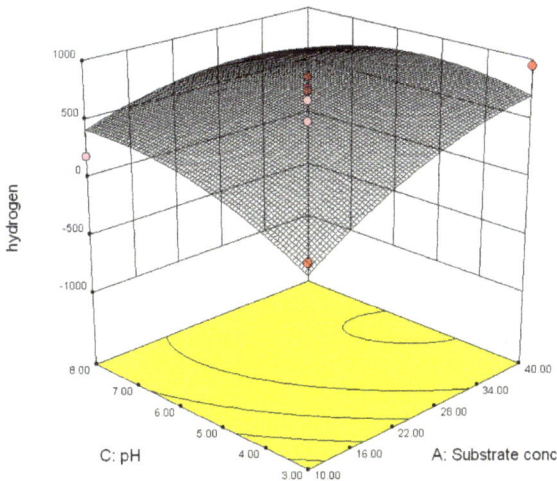

Figure 2. Response surface graph showing the interactive effect of pH and potato-waste concentration (conc, g/L) on hydrogen yield (mL H_2/g TVS).

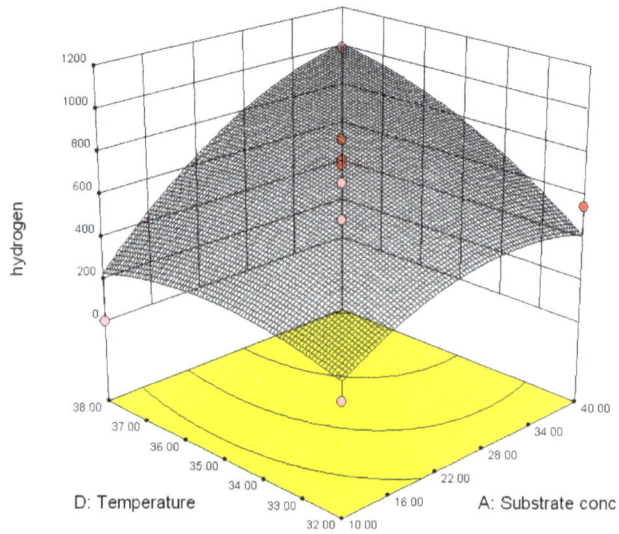

Figure 3. Response surface graph showing the interactive effect of temperature (°C) and potato-waste concentration (conc, g/L) on hydrogen yield (mL H_2/g TVS).

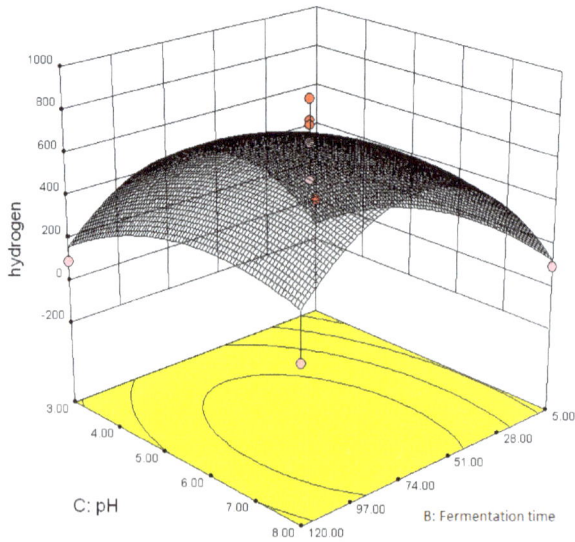

Figure 4. Response surface graph showing the interactive effect of fermentation time (h) and pH on hydrogen yield (mL H_2/g TVS).

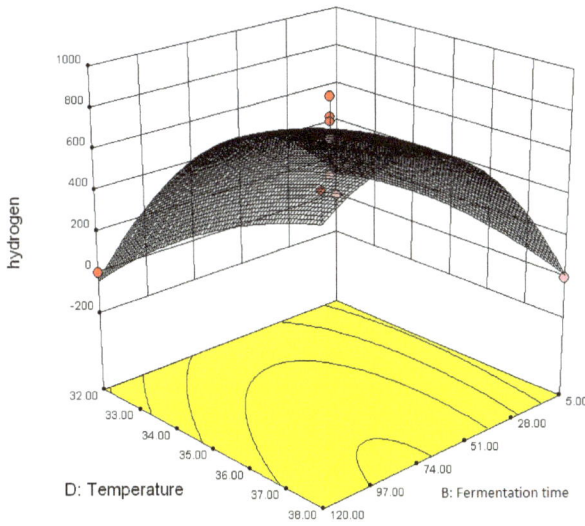

Figure 5. Response surface graph showing the interaction of fermentation time (h) and temperature (°C) on hydrogen yield (mL H_2/g TVS).

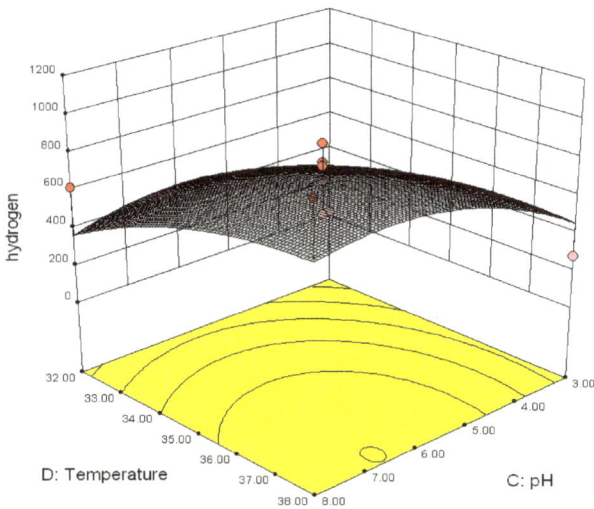

Figure 6. Response surface graph showing the interactive effect of temperature (°C) and pH on hydrogen yield (mL H_2/g TVS).

The interaction between pH and potato-waste concentration is shown in Figure 2, a simultaneous increase in pH (above 5) and potato-waste concentration (above 22 g/L), increases biohydrogen production. It has been confirmed that at an appropriate range, increasing pH could potentially increase the metabolic activities of biohydrogen-producing bacteria during dark fermentative process, but extreme pH values may inhibit their metabolic pathways [24]. For instance, Sekoai and Gueguim Kana [4] reported an optimal pH value of 7.9. In contrast, low concentrations of potato-waste generate low yields of biohydrogen (Figure 2). It has also been confirmed in various studies of biohydrogen production that increasing substrate concentration within the experimental range enhances its production. Earlier studies by Mu et al. [25] and Wang et al. [26] reported optimal concentrations above 25.0 g/L from organic effluents, whereas Sekoai and Gueguim Kana [19] reported an optimal concentration of 40.45 g/L from organic fraction of solid municipal waste.

The synergistic effect of temperature and potato-waste concentration showed than an increase in both temperature (35 °C) and potato-waste concentration (above 22 g/L) resulted in maximum biohydrogen production (Figure 3). Several studies of biohydrogen fermentation process have shown that mesophilic and thermophilic temperature have the ability to increase the population of biohydrogen-producing bacteria; however some extreme temperatures may inhibit their metabolic activities as reported in literature [1]. An increase in potato-waste concentration enhanced the biohydrogen yield but the ability of biohydrogen-producing bacteria to produce hydrogen decreased rapidly with increasing potato-waste concentration from 100 to 300 g/L (Figure 3). Thus, it is reasonable to predict that when the potato-waste concentration continues to increase to 520 g/L, the activity of biohydrogen-producing bacteria will be inhibited completely by the substrate at such high concentration, and the fermentative biohydrogen production by mixed cultures will stop accordingly [22]. With regards to the interactive effect of pH and fermentation time (Figure 4), it was observed that low pH (below 5) and short fermentation time (below 51 h) minimizes the production of biohydrogen. Khanal et al. [27] indicated that low pH values of 4.0–4.5 cause longer lag periods. On the other hand, high initial pH values such as 9.0 decrease lag time, but have a lower yield of biohydrogen production [28]. An optimum retention time between 8.0 and 14 h was reported to yield maximum H_2 without activating methanogenic process [29,30].

Considering the effect of temperature and fermentation time (Figure 5), decreasing both temperature (below 35 °C) and fermentation time (below 51 h) generated low biohydrogen production. Similarly, Wang and Wan [1] observed that the concentration of hydrogen in batch tests increased with increasing temperature from 20 to 35 °C, however it decreased with further increase from 35 to 55 °C. A plausible explanation for such results might be due to the fact that the inoculum consisted of high population of mesophilic biohydrogen-producing bacteria. Conflicting results were reported by Hussy et al. [30]; they observed that reducing fermentation time from 18 to 12 h improved the biohydrogen yield without affecting starch removal efficiency when wheat starch was used as substrate. This might be attributed to various factors such as inoculum type, mode of fermentation, and operational setpoint parameters, i.e., organic loading rate.

In Figure 6, it is seen that low temperature (below 35 °C) coupled with low pH (below 5), decreases the overall production of biohydrogen. Therefore, temperature is one of the most critical parameters in biohydrogen process optimization because its affects the specific growth rate, substrate utilization rate, and the metabolic pathway of microorganisms [31–34]. pH is also highlighted as one of the most vital process parameters in biohydrogen production studies. It affects hydrogenase activity, metabolic activity, and substrate hydrolysis [35–37].

3.3. Modelling and Optimization of Setpoint Parameters Using Box-Behnken Design

Optimization studies revealed that a maximum hydrogen production of 537.5 mL H_2/g TVS can be obtained with potato-waste concentration of 39.56 g/L, fermentation time 82.58 h, pH 5.56, and temperature 37.87 °C. Model validation gave 603.5 mL H_2/g TVS resulting to a 12% increase. Thus, the models accurately optimized the biohydrogen production.

4. Conclusions

This study modelled and optimized the production of biohydrogen using box-behnken response surface methodology. It was shown that an enhanced biohydrogen production yield of 603.5 mL H_2/g TVS is achievable at optimized operational setpoint variables of 39.56 g/L, 82.58 h, 5.56, and 37.87 °C for substrate concentration, fermentation time, pH, and temperature, respectively. Therefore, these findings could pave a way for large-scale biohydrogen production process by offering reliable fermentation data and, thus, make this technology economically viable. The scaling-up of biohydrogen production process will accelerate its commercialization and contribute in the global sustainable energy supply. Moreso, it is pivotal to conduct similar findings on large-scale processes to fully understand the process complexities of biohydrogen-producing fermentation processes from these setpoint conditions.

Acknowledgments: The Author would like to acknowledge financial support from the National Research Foundation (NRF-DST, grant no. 95061) and University of the Witwatersrand (Johannesburg, South Africa).

Author Contributions: Patrick T. Sekoai conducted the experimental design, and wrote the manuscript for publication.

Conflicts of Interest: The author declares no conflict of interest.

References

1. Wang, J.L.; Wan, W. Effect of temperature on fermentative hydrogen production by mixed cultures. *Int. J. Hydrog. Energy* **2008**, *33*, 5392–5397. [CrossRef]
2. Xing, D.F.; Ren, N.Q.; Wang, A.J.; Li, Q.B.; Feng, Y.J.; Ma, F. Continuous hydrogen production of auto-aggregative *Ethanoligenens harbinense* YUAN-3 under non-sterile condition. *Int. J. Hydrog. Energy* **2008**, *33*, 2137–2146. [CrossRef]
3. Yang, H.J.; Shen, J.Q. Effect of ferrous iron concentration on anaerobic bio-hydrogen production from soluble starch. *Int. J. Hydrog. Energy* **2006**, *31*, 2137–2146. [CrossRef]
4. Sekoai, P.T.; Gueguim Kana, E.B. Semi-pilot scale production of hydrogen from organic fraction of solid municipal waste and electricity generation from process effluents. *Biomass Bioenergy* **2014**, *60*, 156–163. [CrossRef]
5. Lotfy, W.; Ghanem, K.M.; Helow, E.R. Citric acid by a novel *Aspergillus niger* isolate: II. Optimization of process parameters through statistical experimental designs. *Bioresour. Technol.* **2007**, *98*, 3470–3477. [CrossRef] [PubMed]
6. He, G.Q.; Chen, Q.H.; Ju, X.J.; Shi, N.D. Improved elastase production by *Bacillus* sp. EL3140-further optimization and kinetics studies of culture medium for batch fermentation. *J. Zhejiang Univ. Sci.* **2004**, *5*, 149–156. [CrossRef] [PubMed]
7. Haltrich, D.; Preiss, M.; Steiner, W. Optimization of a culture medium for increased xylanase production by a wild strain of Schizophyllum commune. *Enzyme Microbiol. Technol.* **1993**, *15*, 854–860. [CrossRef]
8. Box, G.E.P.; Behnken, D.W. Some new three level designs for the study of quantitative variables. *Technometrics* **1960**, *2*, 455–475. [CrossRef]
9. Haaland, P.D. *Experimental Design in Biotechnology*; Marcel Dekker: New York, NY, USA, 1989.
10. Box, G.E.; Hunter, W.G.; Hunter, J.S. *Statistical for Experimenters*; John Wiley & Sons: Hoboken, NJ, USA, 1978.
11. Strobel, R.J.; Nakatsukasa, W.M. Response surface for optimization *Saccharopolyspora spinosa*, a novel macrolide producer. *J. Indian Microbiol.* **1993**, *11*, 121–127. [CrossRef]
12. Swanson, T.R.; Carroll, J.O.; Britto, R.A.; Durhart, D.J. Development and field confirmation of a mathematical model for amyloglucosidase/pullulanase saccharafication. *Starch* **1986**, *38*, 382–387. [CrossRef]
13. Dhillon, G.S.; Brar, S.K.; Verma, M.; Tygi, R.D. Apple pomace ultrafiltration sludge—A novel substrate for fungal bioproduction of citric acid: Optimisation studies. *Food Chem.* **2011**, *128*, 864–871. [CrossRef]
14. Nath, K.; Muthukumar, M.; Kumar, A.; Das, D. Kinetics of two stage fermentation process for the production of hydrogen. *Int. J. Hydrog. Energy* **2008**, *33*, 1195–1203. [CrossRef]
15. Argun, H.; Kargi, F.; Kapdan, I.K.; Oztekin, R. Bio hydrogen production by dark fermentation of wheat powder solution: Effects of C/N and C/P ratio on hydrogen yield and formation rate. *Int. J. Hydrog. Energy* **2008**, *33*, 1813–1819. [CrossRef]
16. Rorke, D.; Gueguim Kana, E.B. Biohydrogen process development on waste sorghum (*Sorghum bicolor*) leaves: Optimization of saccharafication, hydrogen production and preliminary scale up. *Int. J. Hydrog. Energy* **2016**, *41*, 12941–12952. [CrossRef]
17. Mafuleka, S.; Gueguim Kana, E.B. Modelling and optimization of xylose and glucose production from Napier grass using hybrid pre-treatment techniques. *Biomass Bioenergy* **2015**, *77*, 200–208. [CrossRef]

18. O-Thong, S.; Prasertsan, P.; Intrasungkha, N.; Dhamwichukorn, S.; Birkeland, N.K. Optimization of simultaneous thermophilic fermentative hydrogen production and COD reduction from palm oil mill effluent by *Thermoanaerobacterium*-rich sludge. *Int. J. Hydrog. Energy* **2008**, *33*, 1221–1231. [CrossRef]
19. Sekoai, P.T.; Gueguim Kana, E.B. A two-stage modelling and optimization of biohydrogen production from a mixture of agro-municipal waste. *Int. J. Hydrog. Energy* **2013**, *38*, 8657–8663. [CrossRef]
20. Chong, M.L.; Rahim, R.A.; Shirai, Y.; Hassan, M.A. Biohydrogen production by *Clostridium butyricum* EB6 from palm oil mill effluent. *Int. J. Hydrog. Energy* **2009**, *34*, 746–771. [CrossRef]
21. Myers, R.H.; Montgomery, D.C. *Response Surface Methodology: Process and Product Optimization Using Designed Experiments*; John Wiley & Sons: Hoboken, NJ, USA, 1995.
22. Moodley, P.; Gueguim Kana, E.B. Optimization of xylose and glucose production from sugarcane leaves (*Saccharum officinarum*) using hybrid pretreatment techniques and assessment for hydrogen generation at semi-pilot scale. *Int. J. Hydrog. Energy* **2015**, *40*, 3859–3867. [CrossRef]
23. Wu, J.H.; Lin, C.Y. Biohydrogen production by mesophilic fermentation of food wastewater. *Water Sci. Technol.* **2004**, *49*, 223–228. [PubMed]
24. Sinha, P.; Pandey, A. An evaluative report and challenges for fermentative biohydrogen production. *Int. J. Hydrog. Energy* **2011**, *36*, 7460–7478. [CrossRef]
25. Mu, Y.; Wang, G.; Yu, H.Q. Response surface methodological analysis on bio hydrogen production by enriched anaerobic cultures. *Enzyme Microbiol. Technol.* **2006**, *38*, 905–913. [CrossRef]
26. Wang, G.; Mu, Y.; Yu, H.Q. Response surface analysis to evaluate the influence of pH, temperature and substrate concentration on the acidogenesis of sucrose-rich wastewater. *Biochem. Eng. J.* **2005**, *23*, 175–184. [CrossRef]
27. Khanal, S.K.; Chen, W.H.; Li, L.; Sung, S. Biological hydrogen production: Effects of pH and intermediate products. *Int. J. Hydrog. Energy* **2004**, *29*, 1123–1131. [CrossRef]
28. Zhang, T.; Liu, H.; Fang, H.H.P. Biohydrogen production from starch in wastewater under thermophilic conditions. *J. Environ. Manag.* **2003**, *69*, 49–56. [CrossRef]
29. Chen, C.C.; Lin, C.Y.; Lin, M.C. Acid-base enrichment enhances anaerobic hydrogen production process. *Appl. Microbiol. Biotechnol.* **2002**, *58*, 224–228. [PubMed]
30. Hussy, I.; Hawkes, F.R.; Dinsdale, R.; Hawkes, D.L. Continuous fermentative hydrogen production from a wheat starch coproduct by mixed microflora. *Biotechnol. Bioeng.* **2003**, *84*, 619–629. [CrossRef] [PubMed]
31. Cheng, X.Y.; Liu, C.Z. Hydrogen production via thermophilic fermentation of cornstalk by *Clostridium thermocellum*. *Energy Fuels* **2011**, *25*, 1714–1720. [CrossRef]
32. Lay, J.J. Modelling and optimization of anaerobic digested sludge converting starch to hydrogen. *Biotechnol. Bioeng.* **2000**, *68*, 269–278. [CrossRef]
33. Li, C.L.; Fang, H.H.P. Fermentative hydrogen production from wastewater and solid wastes by mixed cultures. *Crit. Rev. Environ. Sci. Technol.* **2007**, *37*, 1–39. [CrossRef]
34. Van Ginkel, S.W.; Oh, S.E.; Logan, B.E. Bio hydrogen gas production from food processing and domestic wastewaters. *Int. J. Hydrog. Energy* **2005**, *30*, 1535–1542. [CrossRef]
35. De Gioannis, G.; Muntoni, A.; Polettini, A.; Pomi, R. A review of dark fermentative hydrogen production from biodegradable municipal waste fractions. *Waste Manag.* **2013**, *33*, 1345–1361. [CrossRef] [PubMed]
36. Sewsynker, Y.; Gueguim Kana, E.G. Modelling of biohydrogen generation in microbial electrolysis cells (MECs) using a committee of artificial neural networks (ANNs). *Biotechnol. Biotechnol. Equip.* **2015**, *29*, 1208–1215. [CrossRef]
37. Faloye, F.D.; Gueguim Kana, E.B.; Schmidt, S. Optimization of biohydrogen inoculum development via a hybrid pH and microwave treatment technique—Semi pilot scale production assessment. *Int. J. Hydrog. Energy* **2014**, *39*, 5607–5616. [CrossRef]

fermentation

MDPI

Article

Cellulase Production from *Bacillus subtilis* SV1 and Its Application Potential for Saccharification of Ionic Liquid Pretreated Pine Needle Biomass under One Pot Consolidated Bioprocess

Parushi Nargotra, Surbhi Vaid and Bijender Kumar Bajaj *

School of Biotechnology, University of Jammu, Bawe Wali Rakh, Jammu 180006, India;
parushi11nargotra@gmail.com (P.N.); sur.vaid@gmail.com (S.V.)
* Correspondence: bajajbijenderk@gmail.com; Tel.: +91-941-9102-201; Fax: +91-191-2456534

Academic Editor: Thaddeus Ezeji
Received: 29 August 2016; Accepted: 8 November 2016; Published: 23 November 2016

Abstract: Pretreatment is the requisite step for the bioconversion of lignocellulosics. Since most of the pretreatment strategies are cost/energy intensive and environmentally hazardous, there is a need for the development of an environment-friendly pretreatment process. An ionic liquid (IL) based pretreatment approach has recently emerged as the most appropriate one as it can be accomplished under ambient process conditions. However, IL-pretreated biomass needs extensive washing prior to enzymatic saccharification as the enzymes may be inhibited by the residual IL. This necessitated the exploration of IL-stable saccharification enzymes (cellulases). Current study aims at optimizing the bioprocess variables viz. carbon/nitrogen sources, medium pH and fermentation time, by using a Design of Experiments approach for achieving enhanced production of ionic liquid tolerant cellulase from a bacterial isolate *Bacillus subtilis* SV1. The cellulase production was increased by 1.41-fold as compared to that under unoptimized conditions. IL-stable cellulase was employed for saccharification of IL (1-ethyl-3-methylimidazolium methanesulfonate) pretreated pine needle biomass in a newly designed bioprocess named as "one pot consolidated bioprocess" (OPCB), and a saccharification efficiency of 65.9% was obtained. Consolidated bioprocesses, i.e., OPCB, offer numerous techno-economic advantages over conventional multistep processes, and may potentially pave the way for successful biorefining of biomass to biofuel, and other commercial products.

Keywords: ionic liquid stable cellulase; *Bacillus subtilis* SV1; response surface methodology; pine needle biomass; ionic liquid pretreatment; one pot consolidated bioprocess

1. Introduction

Ever-increasing world demand of energy, fast depleting fossil fuel reserves, and climate change issues have motivated investigations for potential renewable sources of energy [1]. Among different alternatives, lignocellulose biomass (LB) may be one of the most appropriate renewable resources for the production of energy/biofuels [2]. LB is composed of cellulose, hemicellulose and lignin in a densely compact form which in fact poses a major hurdle in its conversion into simple fermentable sugars that, in turn, can be used for production of biofuels/chemicals and other commercial products [3]. Extensive pretreatments are required for disrupting the recalcitrance of LB and to make the cellulose accessible to saccharifying enzymes [4,5]. However, most of the pretreatment approaches are expensive, tedious, energy intensive, and need harsh conditions like high temperature, pressure, extreme pH, usage of hazardous chemicals, can cause sugar loss and may produce microbial inhibitors [6] that may be detrimental to the fermentation microorganisms [7]. Therefore, there is a need for development of environmentally-benign pretreatment methods that may be executed under ambient process conditions.

Pretreatment of LB with ionic liquid(s) may represent a relatively novel and efficient approach for reducing recalcitrance of biomass; this approach proposes several merits over the traditional pretreatment methods such as it can be accomplished under ambient process conditions, does not require high temperature/pressure/extremes of pH or harmful chemicals, does not produce inhibitors, does not cause sugar loss, and, finally, it is cost and energy efficient [7,8]. However, the IL-pretreated biomass needs ample washing prior to enzymatic hydrolysis for removal of residual IL as the latter is considered as a strong inhibitor of enzymes that are used for saccharification. However, extensive washing of IL-pretreated biomass leads to loss of sugar, wastage of water and consequential escalated generation of effluent, and this overall undermines the efficacy of ILs as LB pretreatment agents [5]. The washing step could be obviated should the ionic liquid tolerant saccharification enzymes viz. cellulases and others are available. Considering that the techno-economic sustainability and success of the IL based pretreatment approach is substantially determined by the availability of IL-stable cellulases, extensive research attempts are being undertaken for IL-stable cellulases [9]. Several microorganisms have been reported to produce cellulases that are tolerant towards ionic liquids, viz. *Bacillus subtilis* [4], *Paenibacillus tarimensis* [10], *Pseudoalteromonas* sp. [11], and from metagenomic sources [12]. Availability of ionic liquid tolerant cellulases may potentially be used for developing novel consolidated bioprocesses such as "one pot consolidated bioprocess" (OPCB). In OPCB, unit operations like pretreatment and saccharification, or pretreatment, saccharification, and fermentation are executed in a single reaction vessel. OPCB offers several advantages over a conventional multi-operational strategy such as cost effectiveness, high product recovery, and no or minimal sugar loss, among others. [13]. Consolidated bioprocesses may be operated at different levels like pretreatment and enzymatic saccharification of LB or pretreatment, enzymatic saccharification and sugar fermentation in a single vessel, either using single microorganism or microbial consortium [14–16].

The high cost of saccharifying enzymes viz. cellulases, xylanases and others is another hurdle for the success of LB-biofuel technology. Microbial sources are generally used for saccharifying enzymes production [17]. High cost is incurred due to usage of expensive carbon/nitrogen sources for growth of microorganisms for enzyme production [5]. The enzyme production cost may be reduced by usage of agro-residues as carbon/nitrogen sources for microbial growth and enzyme production [18,19]. Furthermore, the bioprocess optimization may enhance enzyme yield and overall process economy [20]. Design of Experiments (DoE) based optimization offers several advantages over the conventional one-variable-at-a-time (OVAT) optimization strategy [19]. One of the most commonly used DoE approaches is response surface methodology (RSM). RSM represents an effective and proficient tool for elucidation of processes involving multiple variables [21], and has been used extensively for cellulase production [19,20]. Nonetheless, limited studies have been done on DoE mediated optimization of process variables for producing ionic liquid tolerant cellulases [4,5].

Exploration of new LB resources for potential production of biofuels/chemicals has been a continuous process [6], and pine needle biomass (PNB) may represent an important feedstock [22]. In coniferous forests, accumulation of leaves of pine trees (pine needles) on soil causes multifaceted problems viz. destroys the nutrient dynamics of soil, affects the decomposition/mineralization of organic matter and the flora/fauna of soil; tannins released from pine needles may inhibit the growth of various beneficial soil microbes [5,22,23], and, finally, the dried heaps of pine needles may risk forest fires [22,23]. Pine needles are mainly composed of polysaccharides (cellulose, hemicelluloses) which can be hydrolysed into simple sugars, that in turn may be microbially fermented into valuable products of commercial importance like biofuel, biomaterials, energy, and other products (biorefining) [5,22,23]. Thus, PNB may be exploited as a resource that might help not only mitigate the problems associated with pine needle accumulation but might also realize the "valorization of waste". Rare reports are available on usage of PNB as feedstock for production of biofuels /chemicals [5].

The current study aimed at DoE-based optimization for the production of IL-stable cellulase from *Bacillus subtilis* SV1 using agroindustrial residues as substrates, and its prospective for saccharification of PNB through OPCB.

2. Results and Discussion

2.1. Cellulolytic Bacteria

Primary screening showed that all the five bacterial isolates D1, J2, L10, SV1 and SV29 exhibited cellulolytic activity (Figure 1a). Secondary (quantitative) screening indicated that bacterial isolate SV1 produced maximum cellulase (CMCase) titre (2.201 IU/mL ± 0.06) after 72 h of fermentation, and was followed by isolates SV29 (1.574 IU/mL ± 0.035), L10 (1.115 IU/mL ± 0.06), D1 (1.102 IU/mL ± 0.05) and J2 (0.862 IU/mL ± 0.06) (Figure 1b). Of all the five bacterial isolates, D1, L10 and SV29 displayed maximum growth after 72 h, and showed a direct relationship between growth and cellulase production. However, partial association between growth and cellulase production was exhibited by the isolates J2 and SV1 [5] (Figure 1c). The cellulase from all the isolates was also analyzed for its FPase activity from 24 to 72 h. Bacterial isolate SV29 exhibited maximum FPase activity after 72 h (0.233 IU/mL ± 0.0035) (Figure 1d). Microbial cellulases have got vast application potential in various industries [19]. Diverse ecological habitats have been explored for the isolation of several cellulolytic bacteria like *Bacillus subtilis* MS 54 [19], *Bacillus licheniformis* K-3 [20], *Bacillus subtilis* G₂ [5], *Paenibacillus terrae* ME27-1 [24].

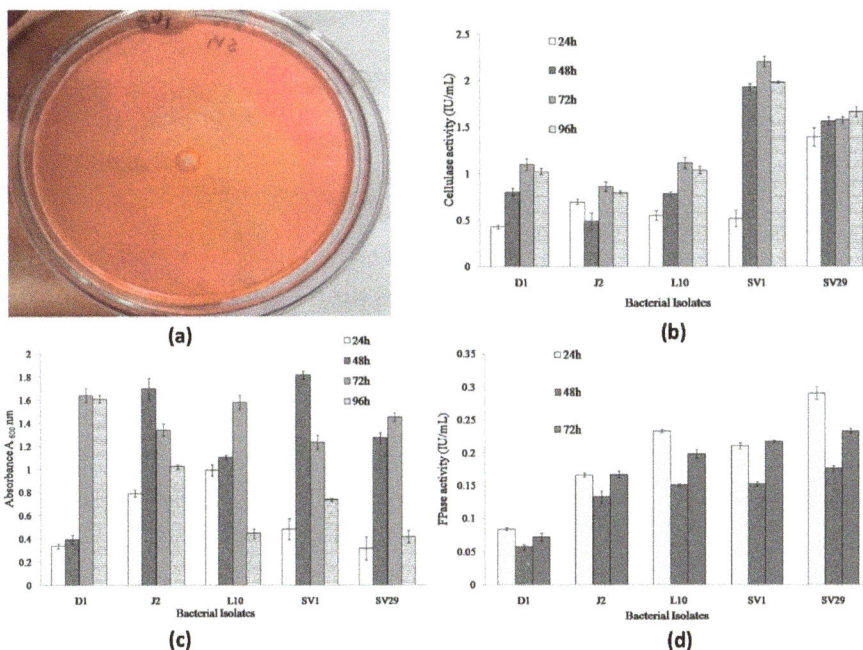

Figure 1. Cellulolytic activity of bacterial isolate SV1 (**a**); CMCase activity (**b**); Growth profile (**c**); and FPase activity (**d**) of the bacterial isolates under submerged fermentation.

2.2. IL Stability of Cellulase

Crude cellulase produced from all the bacterial isolates was examined for its stability/tolerance towards IL 1-ethyl-3-methylimidazolium methanesulfonate (EMIMS, 5%, *v*/*v*). Cellulase from SV1 exhibited maximum stability (residual activity, 161.6%) and retained highest activity after

72 h of prolonged exposure to the IL, EMIMS (5%, v/v) followed by SV29 (115%), J2 (54.8%), L10 (53.1%) and D1 (18%) (Figure 2a). The IL-stability of cellulase from bacterial isolate SV1 was further examined with higher concentrations of EMIMS (10%–50%, v/v). The cellulase from bacterial isolate SV1 exhibited 98.12% residual activity in 50% EMIMS after 4 h of incubation while it showed substantial residual activity of 72.9%, 69.52%, 65.71%, and 88.93% in 10%, 20%, 30% and 40% EMIMS, respectively. The enzyme retained 70.14% residual activity even after 48 h of incubation with EMIMS. However, after 72 h of incubation, the activity decreased considerably (Figure 2b). Thus, cellulase of bacterial isolate SV1 exhibited excellent stability towards IL.

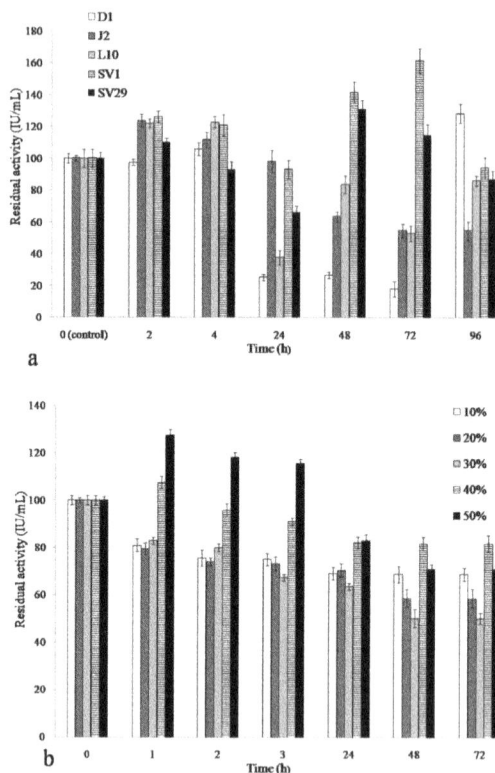

Figure 2. Stability of cellulase from the bacterial isolates against ionic liquid, 1-ethyl-3-methylimidazolium methanosulfonate (**a**); stability of cellulase of bacterial isolate SV1 at different concentrations of 1-ethyl-3-methylimidazolium methanosulfonate (**b**).

The potential of IL-stable cellulases may be exploited for biorefining of LB. IL-stable cellulases have been reported from several microorganisms. *B. subtilis* I-2 cellulase exhibited high stability (activity retention 93%–98%) after 72 h with 1-ethyl-3-methylimidazolium methanesulfonate (EMIMS) [4]. IL tolerant cellulase from *B. subtilis* G_2 showed 95%–100% stability at 20%–50% 1-ethyl-3-methylimidazolium methanosulfonate after 72 h of exposure [5]. Similarly, activity of cellulase Hu-CBH1 from heat tolerant haloalkaliphilic archaeon *Halorhabdus utahensis* remained unchanged or even slightly stimulated in the presence of 20% [EMIM]Ac [9]. *Fusarium oxysporum* cellulase BN showed quite high and long-term stability in the presence of [Emim][DMP] and [Emim][MtSO4] [15]. Thus, ionic liquid stability of cellulases from different microorganisms varies with different types of ionic liquids (Table 1).

Table 1. Ionic liquid stability of cellulases from different microorganisms.

Microorganism	Ionic Liquid (IL)	IL Concentration (%)	Stability (Residual/Relative Activity, %)	Reference
Bacillus subtilis I-2	1-ethyl-3-methylimidazolium methanesulfonate	10	93–98	4
	1-ethyl-3-methylimidazolium methanesulfonate (EMIMS)			
	1-butyl-3-methylimidazolium chloride			
Bacillus subtilis G$_2$	1-ethyl-3-methylimidazolium bromide	20–50	95–100	5
	1-ethyl-3-methylimidazolium acetate			
	1-butyl-3-methylimidazolium trifluoro methanesulfonate			
	1-Ethyl-3-methylimidazolium acetate ([Emim]Ac)		100 (Remained unchanged)	
Halorhabdus utahensis	1-ethyl-3-methylimidazolium chloride ([Emim]Cl)	20	Slightly increased	9
	1-butyl-3-methylimidazolium chloride ([Bmim]Cl)		Slightly increased	
	1-allyl-3-methylimidazolium chloride		100 (remained unchanged)	
	1-ethyl-3-methylimidazolium Methanesulfonate		59	
	1-ethyl-3-methylimidazolium bromide		67	
Pseudoalteromonas sp.	1-ethyl-3-methylimidazolium acetate	20	93.47	11
	1-butyl-1-methylpyrrolidinium trifluromethanesulfonate		80.2	
	1-butyl-3-methylimidazolium trifluoromethanesulfonate		74.69	
	1-butyl-3-methylimidazolium trifluoromethanesulfonate		73.2	
	1-ethyl-3-methyl-imidazolium dimethylphosphate		93	
Fusarium oxysporum BN	1-ethyl-3-methyl-imidazolium methylphosphonate	10	More than 84	15
	1-ethyl-3-methylimidazolium Phosphinate		More than 74	
***Bacillus subtilis* SV1**	**1-ethyl-3-methylimidazolium methanesulfonate**	**10–50**	**72.9–98.12**	**Present study**

2.3. Identification of IL-stable Cellulase Producing Bacterium

The bacterial isolate SV1 that produced cellulase which exhibited substantial IL-stability was examined on the basis of morphological, microscopic, and 16S rDNA sequence analysis. Bacterial isolate SV1 showed rapid growth on nutrient as well as CMC agar plates, and formed slimy, off-white, irregular colonies (Figure 3a). The bacterial isolate SV1 was Gram-positive, rod shaped (bacillus), and had spore-forming ability. The bacterial isolate SV1 possessed potential capability of hydrolyzing starch, xylan, gelatin, casein and triglycerides (Figure 3b). The phylogenetic study of isolate SV1 based on 16S rDNA sequence analysis showed its highest homology with several other *Bacillus subtilis* strains available in the GenBank database (Figure 3c,d). Hence, this isolate is one of the strains of *Bacillus subtilis*, and designated as *Bacillus subtilis* SV1. The sequence was submitted to GenBank under accession number *KU871117*.

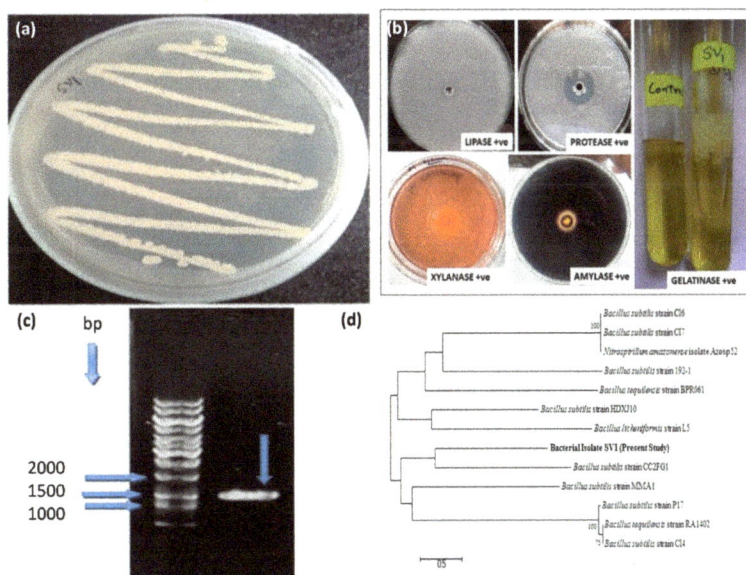

Figure 3. Identification of bacterial isolate SV1. Plate culture of isolate SV1 on nutrient agar (**a**); hydrolytic potential of bacterial isolate SV1 (**b**); PCR-amplified amplicon of 16S rDNA sequence from isolate SV1 (**c**); phylogenetic homology analysis of 16S rDNA sequence from isolate SV1 (**d**).

2.4. DoE Based Optimization of Cultural and Environmental Variables for Cellulase Production

DoE based optimization approach not only surmounts the limitations of the OVAT approach but represents an efficient tool for enhancing product yield by optimization of process variables [17,19]. In the present study, cultural and environmental variables were optimized using response surface methodology (RSM) to boost the production of ionic liquid stable cellulase from *B. subtilis* SV1. Central composite design (CCD) of RSM was applied for determining the optimum levels of the selected independent variables, i.e., wheat bran (A), spirulina powder (B) and medium pH (C) and incubation time (D), and the interactions between the variables. Based on the design, 30 experimental runs were executed, and the corresponding responses are presented in Table 2. ANOVA was performed (Table 3) and a polynomial equation was obtained (Equation 1) in which cellulase yield (Y, response) is presented as a function of various variables. The polynomial equation obtained after multiple regression analysis was as follows:

$$\text{Response Y (Cellulase production)} = 1.77 + 0.32A + 0.17B + 0.25C + 0.093D + 0.12A^2$$

$$- 0.025\ B^2 + 9.688E - 004\ C^2 - 0.012\ D^2 + 0.031\ AB - 0.13\ AC + 0.079\ AD - 0.14\ BC + \qquad (1)$$

$$0.081\ BD + 0.17\ C$$

The equation shows the variation of response (cellulase yield) as a function of various variables, i.e., wheat bran (A), spirulina powder (B) and medium pH (C) and incubation time (D).

Table 2. Experimental and predicted response for CMCase production from *B. subtilis* SV1 based on RSM-designed experiments for optimization of medium and environmental variables.

Runs	Experimental Variables *				Response (Enzyme Activity, IU/mL)	
Run number	A	B	C	D	Experimental	Predicted
1	1.5	1.5	5	72	0.66	0.64
2	1.5	3.0	9	72	1.97	2.18
3	2.25	2.25	7	48	1.89	1.77
4	1.5	1.5	9	72	2.31	2.02
5	2.25	2.25	7	48	1.82	1.77
6	1.5	3.0	9	24	1.80	1.66
7	2.25	2.25	7	96	1.85	1.91
8	1.5	1.5	5	24	1.26	1.11
9	2.25	2.25	7	48	1.19	1.77
10	3.0	1.5	9	24	2.05	1.98
11	2.25	3.75	7	48	1.99	2.02
12	3.75	2.25	7	48	2.29	2.91
13	3.0	3.0	9	72	3.04	2.79
14	1.5	3.0	5	72	1.70	1.36
15	2.25	0.75	7	48	1.08	1.33
16	1.5	1.5	9	24	1.98	1.81
17	2.25	2.25	11	48	2.25	2.28
18	1.5	3.0	5	24	1.63	1.51
19	3.0	1.5	5	72	1.90	1.64
20	3.0	3.0	5	72	2.21	2.49
21	3.0	3.0	5	24	2.43	2.32
22	3.0	3.0	9	24	1.81	1.95
23	3.0	1.5	9	72	2.25	2.50
24	2.25	2.25	7	0	1.32	1.54
25	2.25	2.25	7	48	1.90	1.77
26	2.25	2.25	3	48	1.02	1.28
27	2.25	2.25	7	48	1.97	1.77
28	3.0	1.5	5	24	1.89	1.80
29	0.75	2.25	7	48	1.26	1.62
30	2.25	2.25	7	48	1.87	1.77

*A—Wheat bran (%, *w/v*), *B—Spirulina powder (%, *w/v*), *C—pH, *D—Incubation time (h).

Table 3. Results of ANOVA for cellulase production by *B. subtilis* SV1 based on RSM designed experiments for medium and environmental variables *.

Source	Sum of Squares	DF	Mean Squares	F Value	Prob > F	Significance
Model	6.64	14	0.47	5.18	0.0015	Significant
A	2.49	1	2.49	27.23	0.0001	Significant
B	0.71	1	0.71	7.78	0.0138	Significant
C	1.50	1	1.50	16.35	0.0011	Significant
D	0.21	1	0.21	2.26	0.1533	-
A^2	0.41	1	0.41	4.51	0.0507	-
B^2	0.017	1	0.017	0.19	0.6694	-
C^2	2.574×10^{-5}	1	2.574×10^{-5}	2.813×10^{-4}	0.9868	-
D^2	3.727×10^{-3}	1	3.727×10^{-3}	0.041	0.8428	-
AB	0.016	1	0.016	0.17	0.6859	-
AC	0.27	1	0.27	2.97	0.1051	-
AD	0.099	1	0.099	1.08	0.3146	-
BC	0.30	1	0.30	3.33	0.0879	-
BD	0.11	1	0.11	1.15	0.3007	-
CD	0.45	1	0.45	4.93	0.0422	Significant
Residual	1.37	15	0.092			-
Lack of fit	0.95	10	0.095	1.14	0.4714	Not significant
Pure Error	0.42	5	0.084			-
Cor Total	8.01	29				-

*A—Wheat bran (%, w/v); B—Spirulina powder (%, w/v); C—pH; D—Incubation time (h).

The model F-value of 5.18 implies the model is significant. The chance of getting this high model F-value due to noise is quite low (0.15%). The significance of the model terms is shown by probability > F < 0.05. Probability > F < 0.4714 shows that lack of fit is not significant which in turn implies the strength and sturdiness of the model. Based on p value ($p < 0.05$), A, B, C, and CD were found to be the significant model terms. A low value of standard deviation (0.30) and high coefficient of determination, R-square (0.8286), point towards robustness of the model that has a reasonably good predictability. Adequate precision of 10.587 indicates an adequate signal for the present results.

The interactive effects of independent variables were investigated by analyzing the 3-D response surface plots. The interaction between wheat bran and pH (AC), and spirulina powder and pH (BC) had a negative impact on the cellulase production (Figure 4a,b) whereas the interaction between medium pH and incubation time (CD) was found to have a positive significance for enzyme production as depicted in the 3-D response plot (Figure 4c). It is obvious from the graph that increasing pH as well as incubation time led to enhanced response (cellulase yield). Figure 4d shows the perturbation plot. Perturbation plot explains the increase or decrease in the response when the value of each variable is changed keeping other variables constant with respect to the chosen reference point. When the variable A (wheat bran) was changed from the reference point, it had the maximum positive effect on the response among all the variables whereas variable D (incubation time) had the least effect on the response. This is also substantiated from the equation generated by the design (Equation I). It showed the maximum value of A (+0.32 A) and the least value of D (+0.093).

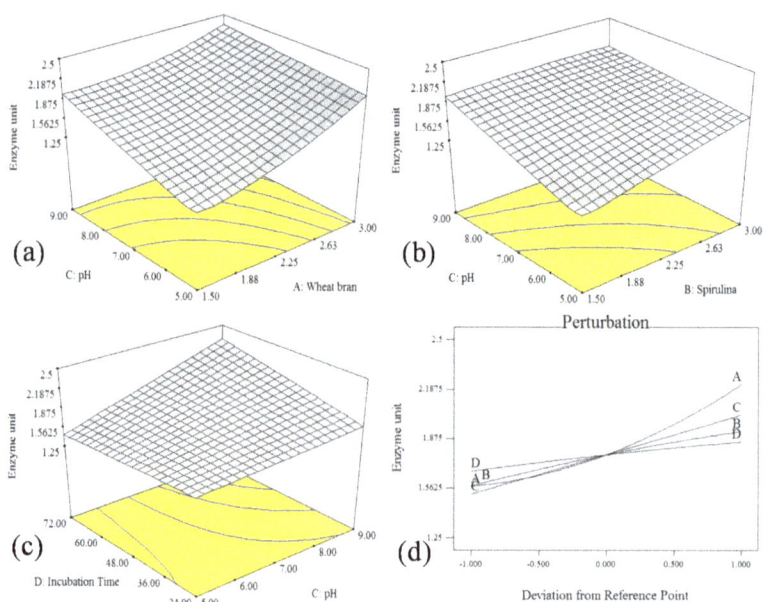

Figure 4. Response surface plots showing the interactions between different variables for cellulase production from *B. subtilis* SV1, wheat bran and medium pH (**a**); spirulina powder and pH (**b**); and pH and incubation time (**c**); perturbation plot showing effect of individual variables (**d**).

The validation of the statistical model was done by using point prediction tool of RSM in which the optimum value of all the four variables, i.e., wheat bran, 3.0 g; spirulina powder, 3 g; medium pH 9 and incubation time, 72 h was determined and the experiment was conducted. The close proximity of observed (CMCase yield of 3.11 IU/mL) and predicted responses (3.10 IU/mL) validated the model. DoE based optimization enhanced cellulase yield by 1.41-fold as compared to that under unoptimized conditions (2.201 IU/mL). One of the major targets for bioprocess development is to achieve enhanced product yield which in turn may contribute substantially towards improving the overall economy of the process; optimization of process variables by DoE, especially the response surface methodology, has quite often been used [18,19]. Though RSM based-optimization of cellulase production has been attempted by several researchers, there are only a few reports on the optimization for production of IL-stable cellulase using RSM. IL-stable cellulase production from *Bacillus subtilis* I-2 [4] was increased by 4.1 and from *B. subtilis* G_2 [5] was increased by 2.66 fold by sequential RSM based optimization of medium and environmental variables. Similarly, many *Bacillus* spp. have been reported to yield enhanced cellulase due to RSM mediated optimization of variables [18–21].

2.5. Some properties of B. subtilis SV1

The *B. subtilis* SV1 cellulase exhibited activity over a broad temperature range (4–90 °C) with an optimal temperature of 45 °C. The cellulase activity at different pH (4–10) showed its optimal activity at pH 10 indicating its alkaline behavior (Figure 5a,b). Similar to the current results, cellulase from *B. aquimaris* [25] and *Pseudomonas fluorescens, B. subtilis, E. coli* and *Serratia marscens* [26] showed alkaline behavior and showed optimum activity at 40–45 °C [25,26]. The cellulase from *B. subtilis* G_2 showed its optimum activity at 45 °C and pH 7 [5]. The optimal temperature of cellulase from *B. subtilis* YJ1 was 60 °C and but exhibited a slightly acidic behavior in contrast to the present study [27]. Thermostability of enzyme depends on molecular interactions which impart a high degree of stabilization due to various forces like hydrophobic and electrostatic interactions, hydrogen, disulphide or other covalent bonding [5,19,20]. Deviations from optimum pH may change the native 3-dimensional structure of enzyme which may lead to alterations in substrate/cofactor/coenzyme binding, and hence decrease or cause total loss of activity [5].

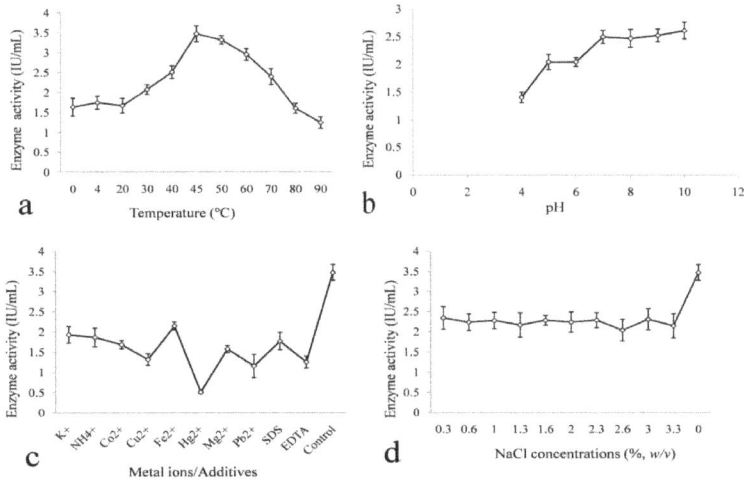

Figure 5. Biochemical properties of cellulase from *B. subtilis* SV1, effect of temperature (**a**); pH (**b**); metal ions/additives (**c**); and sodium chloride on activity of cellulase (**d**).

Each of the metal ions/additives showed an inhibitory effect on cellulase activity. Among all the metal ions/additives, Fe^{2+} had least inhibitive effect while Hg^{2+} had a maximum inhibitory effect on enzyme activity (Figure 5c). Similar to the present study, the activity of cellulase from *B. subitlis* YJ1 was inhibited by Hg^{+}, Cd^{2+}, Fe^{2+}, Fe^{3+} and SDS [27]. *B. vallismortis* RG-07 cellulase was slightly inhibited by Cu^{2+} and Zn^{2+} but Hg^{2+} and Mn^{2+} strongly inhibited cellulase [28]. The Cu^{2+} and Co^{2+} may inhibit cellulase by competing with other cations that might be associated with enzyme resulting in decreased activity. The published reports suggest that Hg^{+} might cause inhibition of activity due to its interactions/binding with –SH or –COOH group of amino acids, or interactions with tryptophan [28]. Metal ions can have a profound effect on the enzyme activity, and can stimulate or hamper activity by multiple mechanisms [5,19].

Salt tolerance of cellulase has been reported to be very much related with IL-stability of the enzyme [11]. Salt tolerance of cellulase was examined by including different concentrations of NaCl (0.3%–3.3%) in enzyme assay reaction mixture. Though cellulase activity decreased in the presence of NaCl, over a range of NaCl concentrations, enzyme activity remained almost constant (66.4%). The results show that cellulase from *B. subtilis* SV1 has considerable salt tolerance (halotolerance) (Figure 5d). Contrary to the present study, cellulase from *Marinimicrobium* sp. LS-A18 retained more than 88% residual activity at 0%–25% NaCl concentrations [29] while cellulase from *B. flexus* NT showed 70% residual activity at 15% NaCl concentration [30]. Halotolerant cellulases are hypothesized to be good candidates that may exhibit good IL-stability due to their adaptation to high salinity [11,12]. Halotolerant enzymes show structural modification like the prevention of protein aggregates' formation through electrostatic repulsion due to the presence of too much charged acidic amino acids on the surface imparting stability in ionic liquid [5,11].

2.6. Pretreatment and Enzymatic Saccharification under a One Pot Consolidated Bioprocess

OPCB allows the execution of multiple unit operations in a single vessel, thus enhancing the overall economy of the process [11]. In the present study, OPCB involved the pretreatment of pine needle biomass with IL (EMIMS) followed by in situ enzymatic saccharification of PNB by using IL-stable cellulase from *B. subtilis* SV1 in a single pot. It is apparent that IL-stability of saccharifying enzymes (cellulase) is mandatory for OPCB. After executing consolidated IL-pretreatment and enzymatic saccharification in one vessel i.e., OPCB, reducing sugar yield was determined to assess the efficacy of the process. Pine needle biomass was pretreated with different IL-concentrations and an increase in sugar yield was observed from 0.149 g/g to 0.200 g/g with the increasing IL concentration from 10% to 50%. The sugar yield in control I was 0.125 g/g (non-pretreated biomass) (Figure 6).

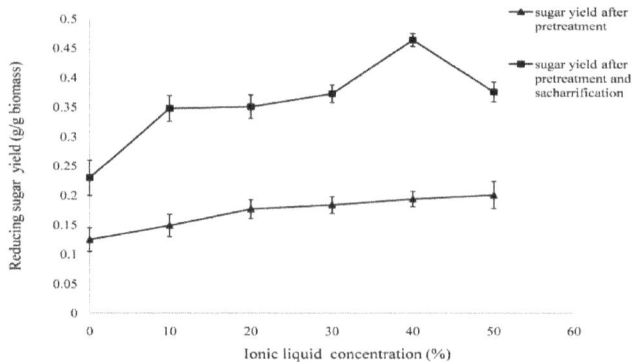

Figure 6. Reducing sugar yield obtained after IL pretreatment and enzymatic saccharification of pine needle biomass using cellulase from *B. subtilis* SV1 (3.1 IU/mL) under one pot consolidated bioprocess. Control shows sugar yield obtained by direct cellulase treatment, i.e., without IL pretreatment.

After pretreatment, enzymatic saccharification was carried out using cellulase from *B. subtilis* SV1 in the same vessel and reducing sugar yield was estimated. A maximum of 0.464 g/g pine needle biomass of reducing sugar yield was observed which was 2.01-fold higher than control II (0.230 g/g pine needle biomass) (Figure 6) at 40% EMIMS concentration. Thus, saccharification efficiency of 65.9% was obtained for PNB under OPCB. From the results, it is inferred that pretreatment resulted in a higher degree of delignification and made cellulose accessible to cellulase and hence yielded higher sugar as compared to the control. OPCB may be commercially vital due to its huge economic benefits [14], but a consolidated process involving application of ionic liquids for LB pretreatment and in situ enzymatic saccharification has scarcely been reported [31]. One pot IL (1-ethyl-3-methylimidazolium acetate, [C2mim][OAc]) pretreatment following the enzymatic saccharification with IL stable cellulase was carried out for switch grass to attain 80% glucose and xylose liberation [31]. Ionic liquid coupled with HCl led to synergistic effects on sugar release from corn stover [32]. Pretreatment of sugarcane bagasse has been effectively done with a combination of ionic liquid and surfactant [33]. Similarly, an enhanced sugar release was observed from eucalyptus, rice straw and grass, but not from pine [34]. Ionic liquid pretreatment of sugarcane bagasse with pure 1,3-dimethylimidazolium dimethyl phosphate gave better results as compared to that obtained with aqueous solution of IL [35]. Pretreatment of rice straw with 20% cholinium lysine IL-water mixtures with subsequent hydrolysis liberated 81% and 48% glucose and xylose yield, respectively [36].

3. Materials and Methods

3.1. Chemicals and Media

Various media, reagents, and chemicals employed in the experiments for current research investigation were of high grade standard and purity, and were procured from suppliers like HiMedia Laboratories, Merck and Co., Ranbaxy Fine Chemicals and Sigma-Aldrich. Ionic liquid

1-ethyl-3-methylimidazolium methanesulfonate used in the current study was purchased from Sigma-Aldrich (St. Louis, MO, USA).

3.2. Cellulase Producing Bacteria

Bacterial isolates used in this study were procured from culture collection of Fermentation Biotechnology Laboratory, School of Biotechnology, University of Jammu (Jammu, India). The isolates were examined primarily for cellulolytic activity by plate assay by using Congo red staining method [5,20]. Congo red stain binds specifically with β-1,4 linked glycosidic linkage in cellulose. Bacterial isolates which produce cellulase cleave β-1,4-glycosidic bonds in carboxymethyl cellulose (CMC) agar, and form a zone of clearance around the colonies. Bacterial isolates exhibiting considerable cellulase activity were subjected to secondary screening.

3.3. Submerged Fermentation for Cellulase Production

Cellulolytic bacterial isolates (D1, J2, L10, SV1 and SV29) selected on the basis of primary screening were subjected to submerged fermentation in shake flasks for production and quantification of cellulase activity. The bacterial isolates were grown under shaking (180 rpm) at 37 °C for 18 h in carboxymethyl cellulose-peptone-yeast extract (CPYE) broth to attain the required cell concentration (A600, 0.9), and then inoculated (2%, v/v) into the CPYE production medium [19]. Submerged fermentation was carried out at 37 °C under shaking (180 rpm). The samples withdrawn at varying time intervals (24, 48, 72 and 96 h) were centrifuged (Eppendorf centrifuge 5804R) at $10,000 \times g$ for 10 min. The supernatant was considered as crude cellulase and assayed for activity. Carboxymethyl cellulase (CMCase, cellulase) and filter paperase (FPase) activities were determined using substrate carboxymethyl cellulose (CMC) and Whatman No. 1 filter paper as substrates, respectively [19]. The amount of reducing sugar released was measured spectrophotometrically (UV-1800, Shimadzu, Japan) by using dinitrosalicylic acid (DNSA) method [37]. One unit (IU) of CMCase and FPase was defined as the amount of enzyme which produced one μmole of glucose equivalent per mL per min under assay conditions. The growth profile of the bacterial isolates was measured spectrophotometrically (A_{600}).

3.4. IL Stability of Bacterial Cellulases

Cellulases produced from bacterial isolates were pre-incubated with IL 1-ethyl-3-methylimidazolium methanesulfonate (EMIMS, 5%, v/v) at room temperature. Samples were withdrawn at various time intervals (1, 2, 3, 4, 24, 48, 72 and 96 h) and examined for residual activity. Cellulase from bacterial isolate that exhibited maximum stability against EMIMS for a prolonged time period was further investigated for its tolerance at higher concentrations of EMIMS i.e., 10%–50%.

3.5. Identification of the Selected Bacterium

Identification of the selected bacterial isolate SV1 that was capable of producing IL stable cellulase was done by studying the cultural, morphological, and microscopic characteristics. The identity of the bacterium was further confirmed by analyzing its 16S rDNA sequence and comparing it with that other sequences available in GenBank (National Center for Biotechnology Information) [5,20]. CMC agar plates were used for the examination of colony morphology of bacterial isolate SV1. Gram staining and endospore staining analysis was done for microscopic examination. The bacterial isolate SV1 was observed for its ability to produce various hydrolytic enzymes viz. lipase, amylase, xylanase, protease, and gelatinase [5,20]. Genomic DNA was extracted (Wizard Genomic DNA Preparation Kit, Promega Co., Madison, WI, USA) and PCR-amplified using universal 16S rDNA primers (forward primer 5′-AGTGTTTGATCCTGGCTCAG-3′, reverse primer 5′-CGGCTACCTTGTTACGACTTT-3′) for 16S rDNA sequence analysis [19]. The amplified product was eluted (Axygen DNA gel extraction kit, Union City, CA, USA) and sequenced (SciGenom Labs Pvt. Ltd., Chennai, India). BLAST analysis of the DNA sequence data was performed for closest homology. The neighbor-joining phylogenetic analysis was carried out with MEGA 6 software. Phylogenetic tree was constructed using MEGA 6 (http://www.megasoftware.net).

3.6. Optimization of Cultural and Environmental Variables for Cellulase Production

Cellulase production from SV1 was enhanced by optimizing the medium components and environmental variables by employing a central composite design of (CCD) of response surface methodology (RSM). The medium variables selected for optimization were crude carbon/nitrogen source viz. wheat bran (A) and spirulina powder (B), and environmental variables were medium pH (C) and incubation time (D). The maximum and minimum range of independent variables was selected based on the already published papers [5]. The variables investigated and full experimental plan are presented in Table 4. A total of 30 experiments were conducted and the results were analyzed using design of expert software version 6.0 (Stat-Ease, Inc., Minneapolis, MN, USA) (Table 5). The three-dimensional (3-D) response surface plots were used to understand the interaction between the variables and to analyze the optimum value of each parameter to maximize cellulase production. The regression equation gave an empirical model that related the measured response to the independent variables of the experiments. The statistical model was then validated in shake-flask experiments for cellulase production under the conditions predicted based on point prediction tool. The response values (Y) were measured as the average of triplicate experiments.

Table 4. Experimental range and levels of the medium and environmental variables used in RSM for cellulase production from *B. subtilis* SV1.

Study Type: Response Surface			Experiments: 30	
Initial Design: Central Composite			Design Model: Quadratic	
Response	Name	Units		
Y	Enzyme activity	IU/mL	Experimental values	
Factors	Name	Units	Lower	Higher
A	Wheat Bran	%, w/v	1.5	3
B	Spirulina powder	%, w/v	1.5	3
C	Medium pH	-	5	9
D	Incubation time	H	24	72

Table 5. RSM-designed experiments for medium and environmental variables for cellulase production from *B. subtilis* SV1.

Runs	Experimental variables *			
Run number	A	B	C	D
1	1.5	1.5	5	72
2	1.5	3.0	9	72
3	2.25	2.25	7	48
4	1.5	1.5	9	72
5	2.25	2.25	7	48
6	1.5	3.0	9	24
7	2.25	2.25	7	96
8	1.5	1.5	5	24
9	2.25	2.25	7	48
10	3.0	1.5	9	24
11	2.25	3.75	7	48
12	3.75	2.25	7	48
13	3.0	3.0	9	72
14	1.5	3.0	5	72
15	2.25	0.75	7	48
16	1.5	1.5	9	24
17	2.25	2.25	11	48
18	1.5	3.0	5	24
19	3.0	1.5	5	72
20	3.0	3.0	5	72
21	3.0	3.0	5	24
22	3.0	3.0	9	24
23	3.0	1.5	9	72
24	2.25	2.25	7	0
25	2.25	2.25	7	48
26	2.25	2.25	3	48
27	2.25	2.25	7	48
28	3.0	1.5	5	24
29	0.75	2.25	7	48
30	2.25	2.25	7	48

*A—Wheat bran (%, w/v), *B—Spirulina powder (%, w/v), *C—pH, *D—Incubation time (h).

3.7. Some Properties of IL-Stable Cellulase

The cellulase produced under optimized process was used for studying the effect of temperature and pH on the activity. Cellulase activity was assayed at different temperatures (4–90 °C) for deducing the effect of temperature on activity. Similarly, the effect of pH was realized by executing an activity assay at different pH by using buffers (50 mM) of appropriate pH: acetate buffer (pH 4–5), glycine-NaOH buffer (pH 6–8) and phosphate buffer (pH 9–10) [4].

For determining halotolerance of cellulase, an activity assay was executed in the presence of varying concentrations of sodium chloride (0.3%–3.0%).

The cellulase activity was assayed in the presence of several metal ions/additives viz. potassium chloride, ammonium chloride, cobalt chloride, copper chloride, ferrous sulphate, mercuric chloride, magnesium sulphate, lead acetate, sodium dodecylsulphate (SDS), and ethylenediaminetetraacetic acid (EDTA) at a final concentration of 1.66 mM. The activity without metal ion/additive was considered as control.

3.8. Pretreatment and Enzymatic Hydrolysis of PNB Using One Pot Consolidated Bioprocess (OPCB)

3.8.1. Pine Needle Biomass (PNB)

Pine needles used in the study were procured from forest area near Udhampur (Jammu, India). Pine needles were thoroughly washed with tap water and then air dried at 50 °C, and dry matter content between 91% and 94% was obtained. The dried material was ground, and the fraction passing through a 4–5 mm sieve was collected and used for further experiments [23]. The powdered PNB was composed of (dry weight basis) holocellulose (64.12%), pentosan (14.12%) and lignin (27.79%) [23].

3.8.2. One Pot Consolidated Bioprocess

Pretreatment of PNB with 1-ethyl-3-methylimidazolium methanesulfonate (EMIMS) was carried out in order to partially remove lignin and disrupt the crystalline structure of cellulose. An appropriate quantity of dried and ground pine needles biomass was immersed in EMIMS at different concentrations (10%–50%, w/v). The contents were incubated at 70 °C under shaking at 180 rpm for 18 h, and then subjected to enzymatic hydrolysis in situ (same pot) using IL-stable cellulase preparation (at 311 IU/g of PNB). After addition of cellulase preparation the contents were incubated at 37 °C under shaking (180 rpm) for 24 h, following which reducing sugar was assayed. The untreated PNB was considered as control I, and enzymatically hydrolyzed untreated PNB was used as control II.

4. Conclusions

Bacillus subtilis SV1 is capable of utilizing agroindustrial residues as carbon and nitrogen sources for growth and IL-stable cellulase production. DoE based optimization of process variables appreciably enhanced cellulase production (1.41-fold). Furthermore, integration of IL based pretreatment and enzymatic saccharification in a single unit (OPCB) gave excellent results as indicated by saccharification efficiency of PNB. Further research on the molecular basis of IL stability of cellulase, and functional mechanisms of IL mediated reduction of LB recalcitrance, is underway in our laboratory. Process-scale-up and other parameters need further investigation for harnessing the full potential of OPCB to ultimately realize a sustainable, economically viable and highly efficient biorefinery process, i.e., conversion of biomass to biofuel (ethanol, butanol, etc.) and other products of commercial importance.

Acknowledgments: Authors thank Department of Science and Technology (DST, Govt. of India), and Department of Biotechnology (DBT, Govt. of India) for financial support (Research Project Ref. SR/SO/BB-66/2007). Bijender Kumar Bajaj gratefully acknowledges the Commonwealth Scholarship Commission, UK, for providing Commonwealth Fellowship (INCF-2013-45) for 'Research Stay' at Institute of Biological, Environmental and Rural Sciences (IBERS), Aberystwyth University, Aberystwyth, UK. Authors thank the Director, School of Biotechnology, University of Jammu, Jammu, for necessary laboratory facilities.

Author Contributions: Bijender Kumar Bajaj conceptualized the research problem; Surbhi Vaid designed the experiments; Parushi Nargotra performed the experiments, and wrote the draft MS; Bijender Kumar Bajaj and Surbhi Vaid corrected the MS, analyzed and interpreted data.

Conflicts of Interest: The authors declare no conflict of interest.

References

1. Mehmood, N.; Husson, E.; Jacquard, C.; Wewetzer, S.; Büchs, J.; Sarazin, C.; Gosselin, I. Impact of two ionic liquids, 1-ethyl-3-methylimidazolium acetate and 1-ethyl-3-methylimidazolium methylphosphonate, on *Saccharomyces cerevisiae*: Metabolic, physiologic, and morphological investigations. *Biotechnol. Biofuels* **2015**, *8*, 8–17.

2. Kumar, R.; Tabatabaei, M.; Karimi, K.; Sárvári-Horváth, I. Recent updates on lignocellulosic biomass derived ethanol: A review. *Biofuel. Res. J.* **2016**, *3*, 347–356. [CrossRef]

3. Moreno, A.D.; Ibarra, D.; Mialon, A.; Ballesteros, M. A bacterial laccase for enhancing saccharification and ethanol fermentation of steam-pretreated biomass. *Fermentation* **2016**, *2*. [CrossRef]

4. Singh, S.; Sambyal, M.; Vaid, S.; Singh, P.; Bajaj, B.K. Process optimization for production of ionic liquid resistant, thermostable and broad range pH-stable cellulase from *Bacillus subtilis* I-2. *Biocatal. Biotransfor.* **2015**, *33*, 224–233. [CrossRef]

5. Vaid, S.; Bajaj, B.K. Production of ionic liquid tolerant cellulase from *Bacillus subtilis* G2 using agroindustrial residues with application potential for saccharification of biomass under one pot consolidated bioprocess. *Waste Biomass Valor.* **2016**. [CrossRef]

6. Badgujar, K.C.; Bhanage, B.M. Factors governing dissolution process of lignocellulosic biomass in ionic liquid: Current status, overview and challenges. *Bioresour. Technol.* **2015**, *178*, 2–18. [CrossRef] [PubMed]

7. Konda, N.M.; Shi, J.; Singh, S.; Blanch, H.W.; Simmons, B.A.; Klein-Marcuschamer, D. Understanding cost drivers and economic potential of two variants of ionic liquid pretreatment for cellulosic biofuel production. *Biotechnol. Biofuels* **2014**, *7*. [CrossRef] [PubMed]

8. Socha, A.M.; Parthasarathi, R.; Shi, J.; Pattathil, S.; Whyte, D.; Bergeron, M.; George, A.; Tran, K.; Stavila, V.; Venkatachalam, S.; et al. Efficient biomass pretreatment using ionic liquids derived from lignin and hemicellulose. *Proc. Natl. Acad. Sci. USA* **2014**, *111*, E3587–E3595. [CrossRef] [PubMed]

9. Zhang, T.; Datta, S.; Eichler, J.; Lvanova, N.; Axen, S.D.; Kerfeld, C.A.; Chen, F.; Kyrpides, N.; Hugenholtz, P.; Cheng, J.F.; et al. Identification of a haloalkaliphilic and thermostable cellulase with improved ionic liquid tolerance. *Green Chem.* **2011**, *13*, 2083–2090. [CrossRef]

10. Raddadi, N.; Cherif, A.; Daffonchio, D.; Fava, F. Halo-alkalitolerant and thermostable cellulases with improved tolerance to ionic liquids and organic solvents from *Paenibacillus tarimensis* isolated from the Chott El Fejej, Sahara desert, Tunisia. *Bioresour. Technol.* **2013**, *150*, 121–128. [CrossRef] [PubMed]

11. Trivedi, N.; Gupta, V.; Reddy, C.R.K.; Jha, B. Detection of ionic liquid stable cellulase produced by the marine bacterium *Pseudoalteromonas* sp. isolated from brown alga *Sargassum polycystum* C. Agardh. *Bioresour. Technol.* **2013**, *132*, 313–319. [CrossRef] [PubMed]

12. Ilmberger, N.; Meske, D.; Juergensen, J.; Schulte, M.; Barthen, P.; Rabausch, U.; Angelov, A.; Mientus, M.; Liebl, W.; Schmitz, R.A. Metagenomic cellulases highly tolerant towards the presence of ionic liquids-linking thermostability and halotolerance. *Appl. Microbiol. Biotechnol.* **2012**, *95*, 135–146. [CrossRef] [PubMed]

13. Park, J.I.; Steen, E.J.; Burd, H.; Evans, S.S.; Redding-Johnson, A.M.; Batth, T.; Benke, P.I.; D'haeseleer, P.; Sun, N.; Sale, K.L.; et al. A thermophilic ionic liquid-tolerant cellulase cocktail for the production of cellulosic biofuels. *PLoS ONE* **2012**, *7*, e37010. [CrossRef] [PubMed]

14. Salehi Jouzani, G.; Taherzadeh, M.J. Advances in consolidated bioprocessing systems for bioethanol and butanol production from biomass: A comprehensive review. *Biofuel Res. J.* **2015**, *5*, 152–195. [CrossRef]

15. Xu, J.; Wang, X.; Hu, L.; Xia, J.; Wu, Z.; Xu, N.; Dai, B.; Wu, B. A novel ionic liquid-tolerant *Fusarium oxysporum* BN secreting ionic liquid-stable cellulase: Consolidated bioprocessing of pretreated lignocellulose containing residual ionic liquid. *Bioresour. Technol.* **2015**, *181*, 18–25. [CrossRef] [PubMed]

16. Brethauer, S.; Studer, M.H. Consolidated bioprocessing of lignocellulose by a microbial consortium. *Energy Environ. Sci.* **2014**, *7*, 1446–1453. [CrossRef]

17. Premalatha, N.; Gopal, N.O.; Jose, P.A.; Anandham, R.; Kwon, S.W. Optimization of cellulase production by *Enhydrobacter* sp. ACCA2 and its application in biomass saccharification. *Front. Microbiol.* **2015**, *6*. [CrossRef] [PubMed]
18. Sharma, A.; Tewari, R.; Soni, S.K. Application of statistical approach for optimizing cmcase production by *Bacillus tequilensis*s 28 strain via submerged fermentation using wheat bran as carbon source. *Int. J. Biol. Food Vet. Agric. Eng.* **2015**, *9*, 76–86.
19. Sharma, M.; Bajaj, B.K. Cellulase production from *Bacillus subtilis* MS 54 and its potential for saccharification of biphasic acid-pretreated rice straw. *J. Biobased Mater. Bioenerg.* **2014**, *8*, 449–456. [CrossRef]
20. Gupta, M.; Sharma, M.; Singh, S.; Gupta, P.; Bajaj, B.K. Enahnced production of cellulase from *Bacillus Licheniformis* K-3 with potential saccharification of rice straw. *Energy Technol.* **2015**, *3*, 216–224. [CrossRef]
21. Singh, S.; Moholkar, V.S.; Goyal, A. Optimization of carboxymethylcellulose production from *Bacillus amyloliquefaciens* SS35. *3 Biotech.* **2013**, *4*, 411–424. [CrossRef]
22. Vats, S.; Maurya, D.P.; Jain, A.; Mall, V.; Negi, S. Mathematical model-based optimization of physico-enzymatic hydrolysis of *Pinus roxburghii* needles for the production of reducing sugars. *Indian J. Exp. Biol.* **2013**, *51*, 944–953. [PubMed]
23. Singh, S.; Anu, S.V.; Singh, P.; Bajaj, B.K. Physicochemical pretreatment of pine needle biomass by design of experiments approach for efficient enzymatic saccharification. *J. Mater. Environ. Sci.* **2016**, *7*, 2034–2041.
24. Liang, Y.L.; Zhang, Z.; Wu, M.; Wu, Y.; Feng, X.J. Isolation, screening, and identification of cellulolytic bacteria from natural reserves in the subtropical region of China and optimization of cellulase production by *Paenibacillus terrae* ME27–1. *Bio. Med. Res. Int.* **2014**. [CrossRef]
25. Trivedi, N.; Gupta, V.; Kumar, M.; Kumari, P.; Reddy, C.R.K.; Jha, B. Solvent tolerant marine bacterium *Bacillus aquimaris* secreting organic solvent stable alkaline cellulase. *Chemosphere* **2011**, *83*, 706–712. [CrossRef] [PubMed]
26. Sethi, S.; Datta, A.; Gupta, B.L.; Gupta, S. Optimization of cellulase production from bacteria isolated from soil. *ISRN Biotechnol.* **2013**. [CrossRef] [PubMed]
27. Yin, L.J.; Lin, H.H.; Xiao, Z.R. Purification and characterization of a cellulase from *Bacillus subtilis* YJ1. *J. Mar. Sci. Technol.* **2010**, *18*, 466–471.
28. Gaur, R.; Tiwari, S. Isolation, production, purification and characterization of an organic-solvent-thermostable-alkalophilic cellulase from *Bacillus vallismortis* RG-07. *BMC Biotechnol.* **2015**, *15*, 19–31. [CrossRef] [PubMed]
29. Zhao, K.; Guo, L.Z.; Lu, W.D. Extracellular production of novel halotolerant, thermostable, and alkali-stable carboxymethyl cellulase by marine bacterium *Marinimicrobium* sp. LS-A18. *Appl. Biochem. Biotechnol.* **2012**, *168*, 550–567. [CrossRef] [PubMed]
30. Trivedi, N.; Gupta, V.; Kumar, M.; Kumari, P.; Reddy, C.R.K.; Jha, B. An alkali-halotolerant cellulase from *Bacillus flexus* isolated from green seaweed *Ulva lactuca. Carbohydr. Polym.* **2011**, *82*, 891–897. [CrossRef]
31. Shi, J.; Gladden, J.M.; Sathitsuksanoh, N.; Kambam, P.; Sandoval, L.; Mitra, D.; Zhang, S.; George, A.; Singer, S.W.; Simmonsa, B.A.; et al. One-pot ionic liquid pretreatment and saccharification of switchgrass. *Green Chem.* **2013**, *15*, 2579–2589. [CrossRef]
32. Qing, Q.; Hu, R.; He, Y.; Zhang, Y.; Wang, L. Investigation of a novel acid-catalyzed ionic liquid pretreatment method to improve biomass enzymatic hydrolysis conversion. *Appl. Microbiol. Biotechnol.* **2014**, *98*, 5275–5286. [CrossRef] [PubMed]
33. Nasirpour, N.; Mousavi, S.M.; Shojaosadati, S.A. A novel surfactant-assisted ionic liquid pretreatment of sugarcane bagasse for enhanced enzymatic hydrolysis. *Bioresour. Technol.* **2014**, *169*, 33–37. [CrossRef] [PubMed]
34. An, Y.X.; Zong, M.H.; Wu, H.; Li, N. Pretreatment of lignocellulosic biomass with renewable cholinium ionic liquids: biomass fractionation, enzymatic digestion and ionic liquid reuse. *Bioresour. Technol.* **2015**, *192*, 165–171. [CrossRef] [PubMed]
35. Bahrani, S.; Raeissi, S.; Sarshar, M. Experimental investigation of ionic liquid pretreatment of sugarcane bagasse with 1,3-dimethylimadazolium dimethyl phosphate. *Bioresour. Technol.* **2015**, *185*, 411–415. [CrossRef] [PubMed]

36. Hou, X.D.; Li, N.; Zong, M.H. Significantly enhancing enzymatic hydrolysis of rice straw after pretreatment using renewable ionic liquid–water mixtures. *Bioresour. Technol.* **2013**, *136*, 469–474. [CrossRef] [PubMed]
37. Miller, G.L. Use of dinitrosalicylic acid reagent for determination of reducing sugar. *Anal. Chem.* **1959**, *31*, 426–428. [CrossRef]

fermentation

MDPI

Article

Assessment of Acidified Fibrous Immobilization Materials for Improving Acetone-Butanol-Ethanol (ABE) Fermentation

Hong-Sheng Zeng [1], Chi-Ruei He [1], Andy Tien-Chu Yen [1], Tzong-Ming Wu [2] and Si-Yu Li [1,*]

[1]   Department of Chemical Engineering, National Chung Hsing University, Taichung 402, Taiwan;
      jack5880500@gmail.com (H.-S.Z.); rexchre@hotmail.com (C.-R.H.); andyyen42@gmail.com (A.T.-C.Y.)
[2]   Department of Materials Science and Engineering, National Chung Hsing University, Taichung 402,
      Taiwan; tmwu@dragon.nchu.edu.tw
*   Correspondence: syli@dragon.nchu.edu.tw; Tel.: +886-4-2284-0510 (ext. 509)

Academic Editor: Thaddeus Ezeji
Received: 13 October 2016; Accepted: 23 December 2016; Published: 30 December 2016

Abstract: Acetone-butanol-ethanol (ABE) fermentation using *Clostridium acetobutylicum* is a process that can be used to produce butanol, which can be utilized as an alternative to petroleum-based fuels. Immobilization of the bacteria using three different fibrous materials was studied in order to see how to improve the ABE fermentation process. The results were compared to those of non-immobilized bacteria. Modal and charcoal fibers had OD levels below one at 72 h with the butanol concentration reaching 11.0 ± 0.5 and 10.7 ± 0.6 g/L, respectively, each of which were close to the free cell concentration at 11.1 ± 0.4 g/L. This suggests that bacteria can be efficiently immobilized in these fibrous materials. Although an extended lag phase was found in the fermentation time course, this can be easily solved by pre-treating fibrous materials with 3.5% HCl for 12 h. From comparisons with previous studies, data in this study suggests that a hydrophilic surface facilitates the adsorption of *C. acetobutylicum*.

Keywords: immobilization; modal fibers; acid treatment; *Clostridium acetobutylicum*; Acetone-Butanol-Ethanol (ABE) fermentation

---

1. Introduction

After the industrial revolution, the consumption of fossil fuels such as natural gas, coal and gasoline increased dramatically. The industrial revolution also created lots of greenhouse gases such as CO_2 that brought about global warming and climate change. In order to reduce the consumption of petrol chemical fuels, people looked to develop alternatives to fossil fuels.

Acetone-butanol-ethanol (ABE) fermentation, developed during World War II, is an anaerobic fermentation process used to produce butanol. The strain of bacteria used in ABE fermentation is *Clostridium acetobutylicum*. ABE fermentation was replaced in favor of petrochemical production in the late 1960s due to having a lower productivity and yield. With the pollution generated by petrochemicals growing every day, people looked to ABE fermentation as an alternative and tried to find a way to improve the process. The key to increasing butanol production is improving the survivability of the bacteria. Thus, researchers have utilized extract separation and immobilization to maintain the bacterial count and improve ABE fermentation. Separation of butanol can decrease its toxicity to bacteria whereas immobilization can enhance the tolerance of toxicity via improved cell density [1–3]. The immobilization technique also facilities the downstream butanol separation while it reduces the amount of carbon needed for biomass formation. Multiple immobilization materials for increasing ABE fermentation, such as brick [2,4], polyvinyl alcohol [5], metals [6], agriculture wastes [7], and weaving fibers [8], were used in earlier studies. Weaving fibers, such as vegetable fibers, are highly malleable hydrophilic materials that can be mixed with other materials to improve various characteristics. Animal fibers made from protein can even act as nitrogen sources. Various kinds of fibrous materials are acquired from weaving industry wastes, which would result in lower costs. With these advantages, weaving fibers have the potential to make ABE fermentation more economical and environmentally friendly.

This study aims to observe the performance of cotton balls, modal fibers, and charcoal fibers. Cotton is a readily available fibrous material, and so can easily be looked into as an immobilization material for the bacteria. Modal fiber is a cellulose-based fiber originating from renewable wood pulp. Its high water absorbance, strength, and availability make it ideal for immobilization testing. Charcoal fiber consists primarily of bamboo fiber, which is harvested from bamboo plants and then burned in an oven. As a result, charcoal is made almost entirely of carbon, giving it high tensile strength. Like modal, its ability to absorb water and its renewability (due to the ease of harvesting from bamboo plants) make it ideal as an immobilization material. Some of these fibers received an acid pretreatment to generate more surface area. Overall butanol production, kinetic performance, and in vitro performance will be analyzed for the cotton balls, modal fibers, and bamboo charcoal fibers. The morphology of the *C. acetobutylicum* adsorbed on these three materials was scanned by field-emission scanning electron microscope (FESEM) and the specific surface area was determined by the Brunauer–Emmett–Teller (BET) method [9].

2. Experimental Section

2.1. Microorganism and Culture Environment

C. acetobutylicum ATCC 824 was the bacteria used in this study for ABE fermentation. A detailed description of *C. acetobutylicum* ATCC 824 cultivation can be found in this previous study [10]. While preculturing the microorganism, the bacterial stock was injected into Reinforced Clostridial Medium (RCM) at a concentration of 5% and heat-shocked at 80 °C for 5 min. The bacteria were then inoculated into a batch bottle containing 100 mL LB-s medium, which consisted of NaCl at 10 g/L, tryptone at 10 g/L, yeast extract at 5 g/L, $MgSO_4 \cdot 7H_2O$ at 0.6 g/L, $FeSO_4 \cdot 7H_2O$ at 0.11 g/L, and $CaCl_2$ at 0.008 g/L. Glucose was added to the bottle at an initial concentration of 60 g/L. The bottles were incubated at 37 °C and 200 rpm, and underwent ABE fermentation for 96 h. Note that the cultivation of *C. acetobutylicum* was achieved under anaerobic condition as previously described [10]. Then 3 g of each immobilization material were used when applied.

2.2. Immobilization Materials and Pretreating Process

The immobilization materials used in this study were cotton balls (CSD Ltd., Changhua, Taiwan), modal fiber (Shing-Long Ltd., Yunlin, Taiwan), and charcoal fiber (composed of 20% of bamboo charcoal fiber and 80% of cotton, Shing-Long Ltd.). For the acid pretreating process, all fibrous materials were soaked in 3.5% $HCl_{(aq)}$ for 12 h and washed with deionized (DI) water.

2.3. Analytical Methods

An UV-VIS spectrophotometer (GENESYS 10S, Thermo Scientific, Waltham, MA, USA) was used to determine the optical density of the bacterial culture at 600 nm. The ABE concentrations for the three materials was determined by gas chromatography (Hewlett Packard HP 5890 Series II, Agilent Technologies, Santa Clara, CA, USA) [11,12]. The DNS method was used to determine the concentration of the glucose remaining [13]. Samples were diluted so that the above measurements were in the dynamic range of calibration curve.

The surface of the three materials was characterized by a field-emission scanning electron microscope (FESEM, model JSM-6700F, JEOL Ltd., Tokyo, Japan). Before pictures were taken, the materials were washed with DI water and put into a drying oven to eliminate moisture. The specific surface area of each immobilization material was measured by the BET method. A micromeritics ASAP 2010 porosimeter was used to absorb nitrogen and the de-gassed condition was controlled at 80 °C.

2.4. In Vitro Performance of the Materials

Bacterial solutions containing *C. acetobutylicum* was prepared by first filling test tubes with 9 mL of the Reinforced Clostridial Medium (RCM, OXOID, Hampshire, UK), capping them with rubber septa, and then sealing them with aluminum. The head space of the test tubes was flushed for 10 min with nitrogen filtered in a 0.2 μm syringe filter. Test tubes containing an oxygen-free headspace were sterilized by autoclaving at 121 °C for 20 min. After cooling the test tubes to room temperature, a mineral stock solution (60 g/L $MgSO_4 \cdot 7H_2O$, 11 g/L $FeSO_4 \cdot 7H_2O$, 0.8 g/L $CaCl_2$) and the glucose solution of 600 g/L were added to them to create a primary medium for growing the bacteria. 1 mL of the bacteria was transferred to each test tube and cultivated at 37 °C and 200 rpm for 72–96 h. Different amounts of modal and charcoal supports were added into test tubes containing 5-mL bacterial solutions to start the in vitro immobilization experiments at 100 rpm. These experiments were performed in an aerobic environment at room temperature to prevent excessive bacterial growth. The performance of in vitro immobilization was determined by calculating the difference between the initial and final OD_{600} of the liquid phase. The OD_{600} was converted into cell dry weight (g/L) by a calibration curve with the conversion factor of $0.748 \text{ g} \cdot \text{L}^{-1} \cdot OD_{600}^{-1}$.

3. Results

3.1. The Performance of Immobilized Materials for ABE Fermentation

The growth curves and butanol production of non-immobilized bacteria and immobilized bacteria were compared with each other. It can be seen in Figure 1 that at a fermentation time of 48 h, the OD_{600} of non-immobilized bacteria reached 7.7 ± 2.9 with a butanol concentration of 8.9 ± 2.5 g/L. On the other hand, the OD_{600} of bacterial cultures in cotton reached 6.0 ± 1.3 with a butanol concentration of 7.0 ± 4.4 g/L. The OD_{600} of cultures in modal and charcoal fiber reached lower numbers of 1.2 ± 0.4 and 3.0 ± 1.8 while the butanol concentrations were still at high values of 4.2 ± 2.5 and 2.5 ± 3.4 g/L, respectively. Overall, OD_{600} of the immobilized bacteria was lower than that of the non-immobilized bacteria.

Modal and charcoal fibers had OD levels below one during the period of 72 h and beyond. The highest butanol concentration was 11.8 ± 0.6, 11.0 ± 0.5, and 10.7 ± 0.6 g/L for cotton, modal, and charcoal materials, respectively, each of which were close to the non-immobilized bacterial concentration at 11.1 ± 0.4 g/L. This suggests that bacteria can be effectively immobilized in the fibrous materials tested in this study, with modal and charcoal fibers displaying the best results. Additionally, a slow onset of butanol production for bacterial cultures in these three materials was observed in Figure 1b, indicating a noticeable mass transfer resistance. This, on the other hand, reflects the effectiveness of the immobilization capability of the fibrous materials tested.

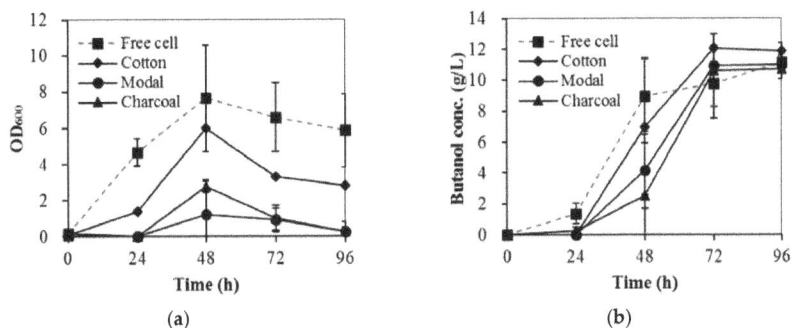

Figure 1. (**a**) Growth curve and (**b**) butanol production of *C. acetobutylicum* on different immobilized materials with 60 g/L glucose. The standard deviation was used for the error with *n* = 3.

3.2. The Effects of Acid Pretreatment

Samples of the three fibrous materials used were treated with hydrochloric acid (HCl). ABE fermentation performance values using these acidified fibrous materials can be seen in Figure 2. At a fermentation time of 48 h, the acidified cotton, modal fibers, and charcoal fibers had OD_{600} values of 6.1 ± 1.7, 1.5 ± 0.3, and 5.0 ± 1.7, respectively, while the normal cotton, modal fibers and charcoal fibers had OD_{600} values of 6.0 ± 1.3, 1.2 ± 0.4, and 2.7 ± 1.8 (Figure 3). This indicates that the immobilization capabilities of three acidified fibrous materials were kept intact. The butanol concentration increased with the acidified modal fibers (with a *p*-value of 0.01). The acidified modal fibers had a butanol concentration level of 10.4 ± 0.0 g/L, respectively, whereas the non-acidified modal fibers had a concentration level of 4.2 ± 2.5 g/L. The elevated butanol concentrations, even higher than those of non-immobilized bacteria at 48 h as shown in Figure 1, indicate that the mass transfer resistance was significantly reduced while keeping the immobilization capabilities. On the other hand, acidified cotton and charcoal fibers had no significant impact on the butanol concentration compared to non-acidified fibers. In summary, evidence demonstrated in terms of kinetics points to acidifying modal fiber as a method for improving butanol productivity, whereas no statistical evidence points to acidifying cotton and charcoal fibers having the same effect.

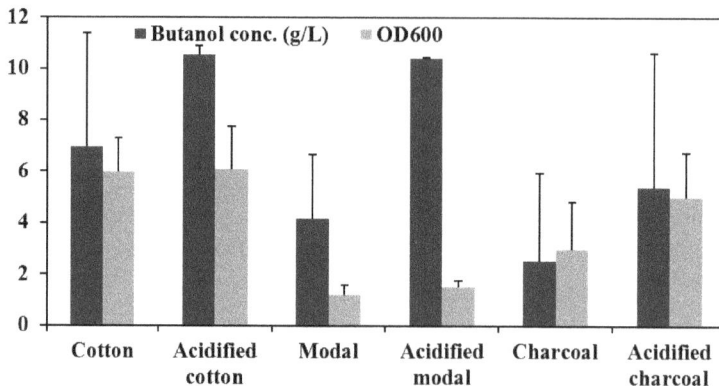

Figure 2. Butanol concentration and cell growth of *C. acetobutylicum* on normal and acidified immobilization materials with 60 g/L glucose at 48 h. The standard deviation was used for the error with *n* = 3.

Figure 3a,b show FESEM images of the modal and charcoal fibers. These fibrous materials had diameters of 5–10 μm. The surface porosity, defined as the distance between each fiber, was greater than 50 μm. Figure 4a,b reveal FESEM images of bacteria on cotton and acidified cotton, respectively, with a more homogeneous distribution of the immobilized bacteria seen in Figure 4b. This may reflect that the stability of butanol production can be increased by using acidified cotton where the standard deviation of the butanol concentration was decreased (Figure 2). It can be seen in Figure 4c that only some areas on the modal fibers were occupied by bacterial colonies. This drawback can be solved by using acid pretreatment so the whole surface area on an acidified fiber can be fully employed for bacterial immobilization (Figure 4d). Figure 4e,f show no significant difference in how much bacteria were immobilized. The effects of acid pretreatment on modal fiber can be perceived by measuring its specific surface area. It can be seen in Figure 5 that there was a nine-fold increase in the specific area for the acidified modal material. This increase in the specific area provides excellent immobilization while minimizing the mass transfer resistance.

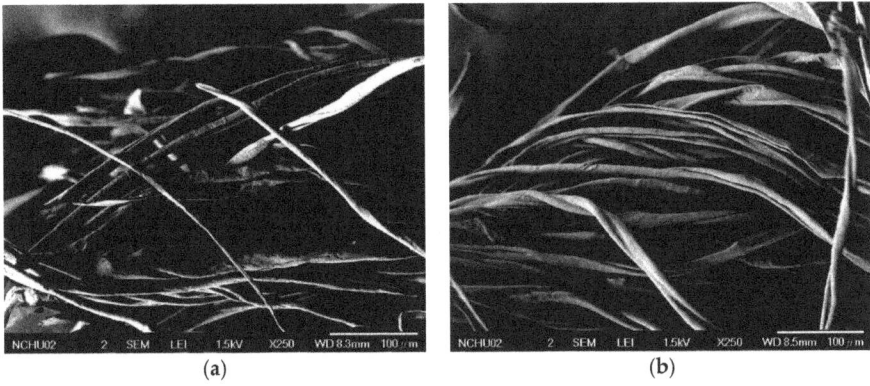

Figure 3. Field-emission scanning electron microscope images of (**a**) modal and (**b**) charcoal fibers.

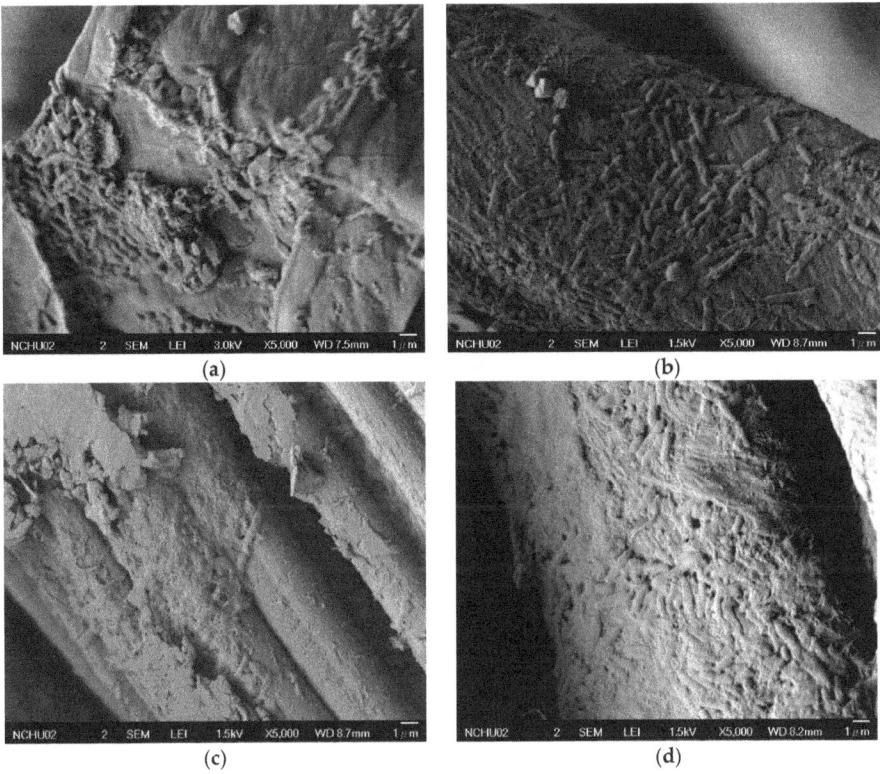

Figure 4. *Cont.*

(e)

(f)

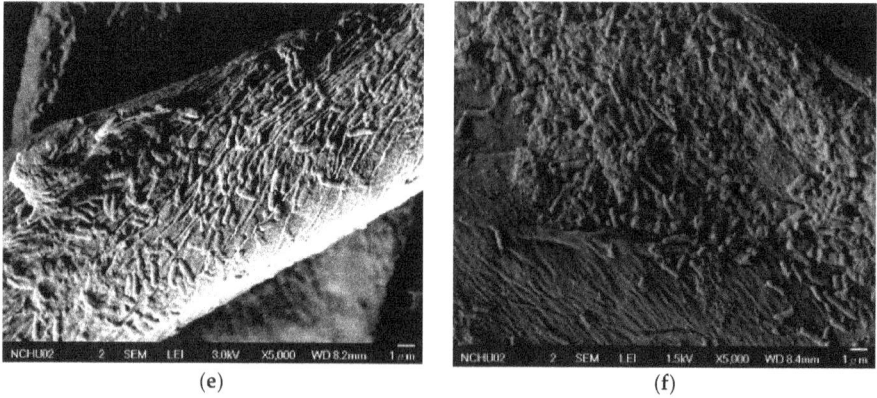

Figure 4. FESEM images of *C. acetobutylicum* bacteria immobilized on supports. (**a**) Bacteria on cotton fiber; (**b**) bacteria on acidified cotton fiber; (**c**) bacteria on normal modal fiber; (**d**) bacteria on acidified modal fiber; (**e**) bacteria on charcoal fiber; (**f**) bacteria on acidified charcoal fiber.

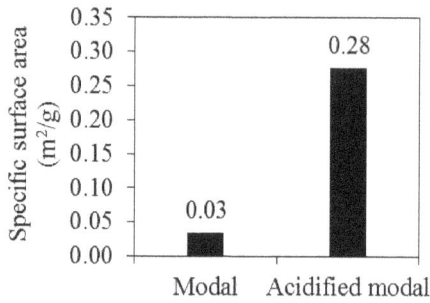

Figure 5. Specific surface area of modal and acidified modal materials.

3.3. In Vitro Performance of the Materials

Figure 6 indicates that modal fiber has an immobilization capability of 17.1 ± 0.3 mg-biomass/g-modal while charcoal fiber has a capability of m. Note that one unit of OD_{600} is equivalent to 0.748 mg-biomass/L.

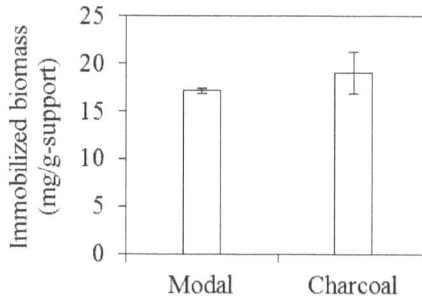

Figure 6. In vitro performance of the modal fiber and charcoal fiber. The standard deviation was used for the error with $n = 3$.

4. Discussion

In this study, experiments using the materials cotton, modal fiber, and charcoal fiber for bacterial immobilization were conducted. It was found that modal and charcoal fibers were suitable bacterial immobilization materials for ABE fermentation. We propose that a hydrophilic surface facilitates more adsorption of *C. acetobutylicum*. This can be argued as follows. When poly(3-hydroxybutyrate) (PHB) materials were fabricated into a fibrous structure, it could in vitro immobilize up to 90 mg-biomass/g-PHB with a surface area of 18.3 m^2/g-support (unpublished data). The immobilization capabilities for modal fiber or charcoal fiber were significantly lower than that of the PHB-based material because of the low specific surface area. The modal fibers have a specific surface area of 0.03 m^2/g-modal while the PHB-based material has a specific surface area of 18.3 m^2/g-PHB (unpublished data). Thus, modal fibers in vitro can immobilize 570 mg-biomass/m^2 while PHB-based supports in vitro can immobilize 4.9 mg-biomass/m^2. This difference in two orders of magnitude can be attributed to surface hydrophobicity, where modal fibers with a hydrophilic surface favor adsorption of the bacteria. The good adsorption of bacteria on the modal or charcoal fibers can be seen from FESEM images.

By pre-treating the modal fiber materials with 3.5% HCl for 12 h, the kinetics of ABE fermentation with the acidified modal fibers were significantly enhanced. It can be argued that the structure of modal fibers was etched by the acids so that the mass transfer resistance was decreased. Therefore, the long lag phase of batch ABE fermentation cultivated with modal fibers was solved. Furthermore, the acid treatment not only presumably increased the adsorption of *C. acetobutylicum* by increasing the specific surface area of acidified modal fibers (Figure 5), but also provided a correct scale of geometry for the formation of homogenous biofilms (Figure 4c,d). This facilitated the entrapment of *C. acetobutylicum*. Therefore, an improved ABE fermentation in terms of kinetics was shown in Figure 2. To have a maximum bacterial immobilization capacity, both adsorption and entrapment should be considered. We propose that a hydrophilic surface on the immobilization material facilitates the adsorption of *C. acetobutylicum*. To further improve the immobilization capability, we need to consider a support structure, which in turn takes into account entrapment.

Acknowledgments: This work was funded by the Ministry of Science and Technology Taiwan, MOST-103-2221-E-005-072-MY3 and MOST-104-2621-M-005-004-MY3.

Author Contributions: Hong-Sheng Zeng, Chi-Ruei He, and Si-Yu Li conceived and designed the experiments; Hong-Sheng Zeng, Chi-Ruei He, and Andy Tien-Chu Yen performed the experiments; Hong-Sheng Zeng, Chi-Ruei He, Andy Tien-Chu Yen, Tzong-Ming Wu, and Si-Yu Li analyzed the data. Hong-Sheng Zeng, Andy Tien-Chu Yen, and Si-Yu Li wrote the paper.

Conflicts of Interest: The authors declare no conflict of interest.

References

1. Chang, Z.; Cai, D.; Wang, Y.; Chen, C.; Fu, C.; Wang, G.; Qin, P.; Wang, Z.; Tan, T. Effective multiple stages continuous acetone-butanol-ethanol fermentation by immobilized bioreactors: Making full use of fresh corn stalk. *Bioresour. Technol.* **2016**, *205*, 82–89. [CrossRef]
2. Wang, Y.-R.; Chiang, Y.-S.; Chuang, P.-J.; Chao, Y.-P.; Li, S.-Y. Direct in situ butanol recovery inside the packed bed during continuous acetone-butanol-ethanol (ABE) fermentation. *Appl. Microbiol. Biotechnol.* **2016**, *100*, 7449–7456. [CrossRef] [PubMed]
3. Li, S.-Y.; Chiang, C.-J.; Tseng, I.T.; He, C.-R.; Chao, Y.-P. Bioreactors and in situ product recovery techniques for acetone-butanol-ethanol fermentation. *FEMS Microbiol. Lett.* **2016**, *363*, fnw107. [CrossRef] [PubMed]
4. Yen, H.W.; Li, R.J.; Ma, T.W. The development process for a continuous acetone-butanol-ethanol (ABE) fermentation by immobilized *Clostridium acetobutylicum*. *J. Taiwan Inst. Chem. Eng.* **2011**, *42*, 902–907. [CrossRef]
5. Kheyrandish, M.; Asadollahi, M.A.; Jeihanipour, A.; Doostmohammadi, M.; Rismani-Yazdi, H.; Karimi, K. Direct production of acetone-butanol-ethanol from waste starch by free and immobilized *Clostridium acetobutylicum*. *Fuel* **2015**, *142*, 129–133. [CrossRef]

6.   Shamsudin, S.; Kalil, M.; Yusoff, W. Production of acetone, butanol and ethanol (ABE) by *Clostridium saccharoperbutylacetonicum* N1–4 with different immobilization systems. *Pak. J. Biol. Sci.* **2006**, 9, 1923–1928.

7.   Qureshi, N.; Li, X.-L.; Hughes, S.; Saha, B.C.; Cotta, M.A. Butanol production from corn fiber xylan using *Clostridium acetobutylicum. Biotechnol. Prog.* **2006**, 22, 673–680. [CrossRef] [PubMed]

8.   Kittithanesuan, N.; Phisalaphong, M. Enhanced acetone-butanol production from sugarcane juice by immobilized *Clostridium acetobutylicum* (ATCC 824) on thin-shell silk cocoons. *Biotechnol. Bioprocess Eng.* **2015**, 20, 599–607. [CrossRef]

9.   Brunauer, S.; Emmett, P.H.; Teller, E. Adsorption of gases in multimolecular layers. *J. Am. Chem. Soc.* **1938**, 60, 309–319. [CrossRef]

10.   Lu, K.-M.; Chiang, Y.-S.; Wang, Y.-R.; Chein, R.-Y.; Li, S.-Y. Performance of fed-batch acetone-butanol-ethanol (ABE) fermentation coupled with the integrated in situ extraction-gas stripping process and the fractional condensation. *J. Taiwan Inst. Chem. Eng.* **2016**, 60, 119–123. [CrossRef]

11.   Lu, K.-M.; Li, S.-Y. An integrated in situ extraction-gas stripping process for acetone-butanol-ethanol (ABE) fermentation. *J. Taiwan Inst. Chem. Eng.* **2014**, 45, 2106–2110. [CrossRef]

12.   Chen, S.-K.; Chin, W.-C.; Tsuge, K.; Huang, C.-C.; Li, S.-Y. Fermentation approach for enhancing 1-butanol production using engineered butanologenic *Escherichia coli. Bioresour. Technol.* **2013**, 145, 204–209. [CrossRef] [PubMed]

13.   Miller, G.L. Use of dinitrosalicylic acid reagent for determination of reducing sugar. *Anal. Chem.* **1959**, 31, 426–428. [CrossRef]

fermentation

MDPI

Article

Anhydrous Ammonia Pretreatment of Corn Stover and Enzymatic Hydrolysis of Glucan from Pretreated Corn Stover

Minliang Yang, Weitao Zhang and Kurt A. Rosentrater *

Department of Agricultural and Biosystems Engineering, Iowa State University, 3327 Elings Hall, Ames, IA 50011, USA; minlyang@iastate.edu (M.Y.); wtzhang1@iastate.edu (W.Z.)
* Correspondence: karosent@iastate.edu; Tel.: +1-515-294-4019

Academic Editors: Thaddeus Ezeji, Mohammad J. Taherzadeh and Badal C. Saha
Received: 27 October 2016; Accepted: 15 February 2017; Published: 17 February 2017

Abstract: As a promising alternative of fossil fuel, ethanol has been widely used. In recent years, much attention has been devoted to bioethanol production from lignocellulosic biomass. In previous research, it is found that the pretreatment method named low-moisture anhydrous ammonia (LMAA) has the advantage of high conversion efficiency and less washing requirements. The purpose of this study was to explore the optimal conditions by employing the LMAA pretreatment method. Corn stover was treated under three levels of moisture content: 20, 50, 80 w.b.% (wet basis), and three levels of particle size: <0.09, 0.09–2, >2 mm; it was also ammoniated with a loading rate of 0.1g NH_3/g biomass (dry matter). Ammoniated corn stover was then subjected to different pretreatment times (24, 96, 168 h) and temperatures (20, 75, 130 °C). After pretreatment, compositional analysis and enzymatic digestibility were conducted to determine the highest glucose yield. As a result, the highest glucose yield was obtained under the condition of 96 h and 75 °C with 50 w.b.% and 0.09–2 mm of corn stover. The main findings of this study could improve the efficiency of bioethanol production processing in the near future.

Keywords: anhydrous ammonia; corn stover; cellulosic ethanol; low-moisture anhydrous ammonia (LMAA); pretreatment

---

1. Introduction

Due to concerns about environmental, long-term economic and national security, there has been increasing interest in renewable and domestic sources of fuels to replace fossil fuels in recent decades. [1]. Bioethanol, produced from renewable materials, is regarded as an alternative to gasoline. There are multiple raw materials to produce bioethanol; one of the most widely adopted is sugar- or starch- based material, such as corn. Bioethanol produced from corn is called first generation biofuel. It has been commercialized in several places and is considered quite efficient. However, a problem arose because of land use and competition with food crops, the so-called food versus fuel debate [2]. Bioethanol can also be produced from lignocellulosic biomass, which is known as second generation biofuel [3]. In general, four major processes are involved in converting lignocellulosic biomass to bioethanol: pretreatment, hydrolysis, fermentation, and ethanol recovery [4]. Among the four steps, pretreatment is critical because of the difficulties in removing the lignin-carbohydrate complex (LCC) structure in lignocellulosic biomass. With the assistance of pretreatment, the LCC structure could be removed, and the exposed cellulose could be broken down into monosaccharides, then the resulting glucose can be fermented into ethanol [1].

Numerous efforts have been invested in exploring various pretreatment methods on various biomass to enhance enzymatic digestibility. Additionally, various pretreatment reagents have been studied, such as carbon dioxide, dilute acid, hot water, ammonia and alkaline. Based on the results of extensive research, each different reagent exhibited its unique characteristics. Several reagents are compared as following.

Carbon dioxide (CO_2) has many advantages, as it is environmentally friendly, inexpensive, and easy to recover after use. The pretreatment method based on CO_2 is supercritical carbon dioxide (SC-CO_2). It has been applied to a few lignocellulosic biomass, such as aspen and south yellow pine [5], wheat straw [6], guayule [7], and corn stover [8]. As for corn stover, the maximum glucose yield obtained under 3500 psi and 150 °C was 30 g/100 g dry corn stover [8]. However, the need for high-pressure equipment by using the SC-CO_2 pretreatment method may result in high capital cost; besides, the low efficiency of this treatment may be a barrier as well to large-scale production [5].

Hot water has also been used as a reagent in pretreatment studies. Hot water has been studied in materials like aspen [9], soybean straw [10], corn stover [9,10], alfalfa [11], and cattails [12]. As a convenient pretreatment method, liquid hot water is effective for soybean straw with the combination of fungal degradation pretreatment, but the combination of these two pretreatment methods is not efficient for corn stover, when compared with fungal degradation pretreatment alone [10].

Another reagent, ammonia, is also broadly explored in this field. Pretreatment methods of ammonia have attracted much attention due to its effectiveness in delignification. For example, ammonia fiber explosion [13–16], ammonia fiber expansion [17–20], and aqueous ammonia soaking [21–23] have been developed. In addition, the improvement in glucose yield is clearly observed. However, water consumption, environmental concerns, and high cost are problematic for ammonia-based pretreatment methods.

Yoo et al. [24] developed the low moisture anhydrous ammonia (LMAA) pretreatment method to eliminate the washing step and reduce capital costs in the ammonia-based pretreatment method. In their study, corn stover pretreated with 3% glucan loading at 80 °C for 84 h resulted in the highest ethanol yield, that is: 89% of theoretical ethanol yield. However, the reactor used in the research conducted by Yoo et al. [24] was a 2.9-inch (8.1 cm) internal diameter with a 6.5-inch (18.5 cm) length (690 mL internal volume). The small sealed reactor may not be capable of providing optimal conditions for bioethanol production at industrial scales. Yang and Rosentrater [25] and Cayetano and Kim [26] have expanded on this initial study. Yang and Rosentrater [25] investigated the effectiveness of LMAA as a method to both pretreat and preserve corn stover prior to fermentation, and found that LMAA is beneficial to preserving sugar yields during storage, with sealed containers being more effective at ammonia treatment.

The main objective of this study was to investigate the LMAA pretreatment process with a larger-scale reactor; four pretreatment conditions (moisture content, particle size, pretreatment temperature, and pretreatment time) were considered in this study. Furthermore, optimal conditions for higher ethanol yield were explored.

2. Materials and Methods

2.1. Biomass

In this study, freshly-harvested, air-dried corn stover was collected from central Iowa in 2012 and stored at ambient temperature. Prior to pretreatment, the corn stover was ground and sieved into three size fractions (<0.09, 0.09–2.0, and >2.0 mm). Then, the sieved corn stover was stored at room temperature (~21 °C) until use.

2.2. Equipment

The reactor (Figure 1), which was purchased from Pall Corporation, Ann Arbor, MI, USA, was used in the ammoniation process. Compared to Yoo's [24] study, this sealed reactor was about 4.35 times larger (the internal capacity is 3 L). It is anticipated that the potential errors caused by different ammonia loadings and reaction times could be eliminated by the use of a larger reactor. High Performance

Liquid Chromatography (HPLC) with a Bio-Rad Aminex HPX-87P column (Aminex HPX-87P, Bio-Rad Laboratories, Hercules, CA, USA) and a refractive index detector (Varian 356-LC, Varian, Inc., Palo Alto, CA, USA) were used to measure sugar contents. Acid soluble lignin (ASL) content was determined by UV-Visible spectrophotometer (UV-2100 Spectrophotometer, Unico, United Products & Instruments, Inc., Dayton, NY, USA).

Figure 1. Ammoniation reactor with internal volume of 3 L.

2.3. Enzymes

In this study, GC 220 cellulase, purchased from Genencor International, Inc. (Rochester, NY, USA), was a mixture of endogluconases and cellobiohydrolases. The cellulase activity was expressed in filter paper units (FPU); the average activity of GC 220 was determined to be 45 FPU/mL. The β-glucosidase enzyme (Novozyme 188), provided by Sigma-Aldrich, Inc. (St. Louis, MO, USA), was used to convert cellobiose to glucose. The activity of Novozyme 188 was 750 cellobiase units (CBU)/mL.

2.4. LMAA Pretreatment Process

The original moisture content was measured before ammoniation, then certain amounts of water were added to the corn stover in order to achieve the target moisture content (20, 50, and 80 w.b.%). Moisturized corn stover was equilibrated for 24 h afterwards.

The moisturized corn stover was placed in the sealed reactor, and ammonia gas was introduced. On top of the reactor, a pressure gauge and a temperature gauge were equipped to monitor the pressure and temperature change during the whole process. However, temperature change was not controlled during this study. The pressure of the anhydrous ammonia was maintained at 0.1 g NH_3/g DM biomass for 30 min in order to achieve a complete reaction. After the ammoniation process, the reactor was cooled down for 5 min and the lid was removed in the fume hood. Then the ammoniated corn stover was transferred into several glass bottles (250 mL) with screw caps. A pipe was connected between the top of the reactor and the fume hood to ventilate surplus ammonia.

The bottles packed with ammoniated corn stover were placed in various heating ovens at varying pretreatment temperatures (20, 75, and 130 °C) for 24, 96, and 168 h. As soon as the pretreatment process was complete, the lid of the glass bottles was removed in the fume hood and surplus ammonia was evaporated for 12 h before compositional analysis.

2.5. Experimental Design

In this study, four independent variables that may have influence on the reaction severity were investigated. Biomass moisture contents were 20, 50 and 80 wet basis (w.b.) %; the pretreatment times were 24, 96, and 168 h; the pretreatment temperatures were 20, 75, and 130 °C; and the particle sizes were <0.9, 0.9–2.0 and >2.0 mm, respectively. By combining different levels of these four independent variables, 17 treatments were designed in this study, i.e., $2 \times 2 \times 2 \times 2 + 1$ center point. As dependent variables, moisture content, lignin, glucan, xylan, galactan, arabinan, mannan and ash content were measured and compared in the experiment. The experimental design for this study is shown in Table 1.

Table 1. Experimental design in this study. *

Treatment	Moisture Content (w.b. %)	Time (h)	Temperature (°C)	Particle Size (mm)
1	20	24	20	<0.9
2	20	24	20	>2.0
3	20	24	130	<0.9
4	20	24	130	>2.0
5	20	168	20	<0.9
6	20	168	20	>2.0
7	20	168	130	<0.9
8	20	168	130	>2.0
9	80	24	20	<0.9
10	80	24	20	>2.0
11	80	24	130	<0.9
12	80	24	130	>2.0
13	80	168	20	<0.9
14	80	168	20	>2.0
15	80	168	130	<0.9
16	80	168	130	>2.0
CP	50	96	75	0.9–2.0

* CP denotes center point of the design.

2.6. Compositional Analysis

Carbohydrates and lignin (both acid-soluble lignin and acid-insoluble lignin) contents were determined by NREL LAP [27]. Each sample was analyzed in duplicate. The glucan and xylan content in the corn stover were analyzed by HPLC, following the NREL standards. Acid soluble lignin was measured by UV-Visible Spectrophotometer. Moisture content was determined by an oven drying method [27].

2.7. Enzymatic Digestibility Test

The enzymatic digestibility test was done in duplicate under conditions of pH 4.8 (0.1 M sodium citrate buffer) with 40 mg/L tetracycline and 30 mg/L cyclohexamide in 250 mL Erlenmeyer flasks according to NREL LAP [28]. The initial glucan concentration was 1% (w/v). Cellulase enzyme (GC 220) loading was 15 FPU/g of glucan, and β-glucosidase enzyme (Novozyme 188) loading was equal to 30 CBU/g of glucan. Flasks were incubated at 50 ± 1 °C and 150 rpm in an incubator shaker (Excella E24 Incubator Shaker Series, New Brunswick Scientific, Edison, NJ, USA). Time for enzymatic digestibility test ranged from 0 to 168 h for sugar analysis.

Total glucose detected from HPLC was used to calculate the glucan digestibility following Equation (1) below. The conversion factor for glucose to equivalent glucan was 0.9 based on the calculation. The quantification of glucose in HPLC is based on the separation of the solvent into its constituent parts due to the different affinities of different molecules for the mobile phase and stationary phase. All the statistical results were anaylzed by SAS 9.4 (SAS Institute Inc., Cary, NC, USA).

$$\text{Glucan digestibility } [\%] = \frac{\text{Total released glucose} \times 0.9}{\text{Initial glucan loading}} \times 100\% \qquad (1)$$

3. Results and Discussion

3.1. Effects of LMAA Pretreatment on Biomass Composition

In this study, the employment of low-moisture anhydrous ammonia (LMAA) pretreatment method didn't result in significant changes in lignin, glucan, xylan, arabinose, mannan or ash contents. Table 2 exhibits the main effect. As can be seen by the letter, temperature had an effect mainly on lignin and ash. With higher temperature, the ash content increased as well as the lignin content. Time also had effect on lignin and glucan as well; longer time resulted in higher glucan. The effect of size is primarily on ash content; larger size resulted in lower ash content. The forth factor, moisture content, did not have much influence on the compositions. According to Table 3, the majority of the *p*-values of interactions among these four independent variables were higher than 0.05, which indicates that little evidence of significant interactions among independent variables was observed. Similar findings were found in treatment effect (Table 4).

The reason for insignificant compositional analyses results in this study was because the ammonia used in the LMAA pretreatment process was meant to break the LCC structure for later enzymatic saccharification and ethanol fermentation process, not to change composition per se. This has also been studied by Cayetano and Kim [26]. Their work showed that the LMAA pretreatment method did not result in significant changes to the chemical composition.

Table 2. Main effects of compositional analysis on corn stover. *

Factor	Levels	Lignin (%)	AIL (%)	ASL (%)	Glucan (%)	Xylan (%)	Galactan (%)	Arabinose (%)	Mannan (%)	Ash (%)
Temperature (°C)	20	20.86a (0.73)	16.86a (0.74)	3.99a (0.44)	35.73a (2.97)	21.35a (2.96)	0.67a (0.34)	3.70a (0.47)	0.05a (0.05)	1.67a (0.69)
	75	21.20ab (0.26)	16.23a (0.48)	4.97b (0.74)	38.89a (2.75)	25.59b (3.07)	0.55a (0.06)	4.31a (0.64)	0.02b (0.01)	1.96ab (0.30)
	130	21.36b (0.80)	17.83b (0.87)	3.54c (0.57)	37.08a (2.86)	22.47ab (1.77)	0.83a (0.40)	3.88a (0.53)	0.04b (0.02)	2.20b (0.55)
Time (h)	24	20.89a (0.95)	17.27a (1.28)	3.62a (0.58)	35.38a (3.25)	21.89a (2.93)	0.75a (0.41)	3.75a (0.64)	0.05a (0.05)	1.94a (0.69)
	96	21.20ab (0.26)	16.23a (0.48)	4.97b (0.74)	38.89ab (2.75)	25.59b (3.07)	0.55a (0.06)	4.31a (0.64)	0.02b (0.01)	1.96a (0.30)
	168	21.33b (0.55)	17.42a (0.72)	3.91a (0.50)	37.43b (2.26)	21.92ab (2.00)	0.75a (0.34)	3.83a (0.32)	0.04b (0.02)	1.93a (0.66)
Moisture Content (w.b.%)	20	21.12a (0.95)	17.32a (1.08)	3.80a (0.45)	35.54a (2.76)	22.02ab (2.66)	0.82a (0.45)	3.88a (0.62)	0.06a (0.04)	1.79a (0.65)
	50	21.20a (0.26)	16.23a (0.48)	4.97b (0.74)	38.89a (2.75)	25.59a (3.07)	0.55a (0.06)	4.31a (0.64)	0.02b (0.01)	1.96a (0.30)
	80	21.10a (0.64)	17.36a (0.81)	3.73a (0.65)	37.27a (2.95)	21.79b (2.33)	0.69a (0.27)	3.70a (0.35)	0.03b (0.02)	2.08a (0.68)
Size	S	21.31a (0.92)	17.56a (1.12)	3.75a (0.60)	35.6a (2.77)	20.67a (2.38)	0.79a (0.36)	3.65a (0.46)	0.04a (0.04)	2.32a (0.59)
	M	21.2a (0.26)	16.23b (0.48)	4.97b (0.74)	38.89a (2.75)	25.59b (3.07)	0.55a (0.06)	4.31a (0.64)	0.02b (0.01)	1.96ab (0.30)
	L	20.91a (0.62)	17.12ab (0.67)	3.78a (0.51)	37.21a (2.98)	23.14b (1.91)	0.71a (0.39)	3.94a (0.52)	0.04a (0.03)	1.56b (0.51)

* Similar letters after means in each level of the main factor indicates insignificant difference for that dependent variable at $\alpha = 0.05$, LSD. Values in parentheses are standard deviation (S.D.). S denotes size less than 0.9 mm, M denotes size between 0.9 and 2.0 mm, and L denotes size larger than 2.0 mm. AIL stands for Acid-Insoluble Lignin; ASL stands for Acid-Soluble Lignin.

Table 3. Interaction effects of compositional analysis (*p*-values) on corn stover. *

Factor	Lignin (%)	AIL (%)	ASL (%)	Glucan (%)	Xylan (%)	Galactan (%)	Arabinose (%)	Mannan (%)	Ash (%)
Temp	0.004	0.004	0.029	0.160	0.188	0.120	0.245	0.014	0.004
Time	0.978	0.612	0.150	0.038	0.967	0.955	0.580	0.003	0.978
MC	0.089	0.885	0.738	0.075	0.788	0.193	0.239	<0.0001	0.089
Size	<0.0001	0.150	0.865	0.097	0.008	0.418	0.072	0.807	<0.0001
Temp * Time	0.437	0.793	0.772	0.546	0.283	0.618	0.010	<0.0001	0.437
Temp * MC	0.285	0.110	0.065	0.426	0.466	0.672	0.036	<0.0001	0.285
Temp * Size	0.922	0.282	0.678	0.205	0.190	0.056	0.927	1.000	0.922
Time * MC	0.083	0.244	0.240	0.178	0.308	0.426	0.765	0.000	0.083
Time * Size	0.377	0.410	0.753	0.722	0.507	0.003	0.053	0.807	0.377
MC * Size	0.507	0.946	0.423	0.714	0.308	0.236	0.233	0.807	0.507
Temp * Time * MC	0.097	0.219	0.975	0.073	0.344	0.077	0.188	0.005	0.097
Temp * Time * Size	0.272	0.939	0.865	0.407	0.457	0.358	0.552	0.155	0.272
Temp * MC * Size	0.070	0.361	0.738	0.836	0.650	0.015	0.765	0.335	0.070
Time * MC * Size	0.512	0.852	0.701	0.315	0.635	0.654	0.510	0.100	0.512
Temp * Time * MC * Size	0.806	0.340	0.356	0.956	0.502	0.100	0.685	0.064	0.806

* Temp = Temperature; MC = Moisture Content; AIL = Acid-Insoluble Lignin; ASL = Acid-Soluble Lignin.

Table 4. Treatment effects of compositional analysis on corn stover. *

Treatment	Lignin (%)	AIL (%)	ASL (%)	Glucan (%)	Xylan (%)	Galactan (%)	Arabinose (%)	Mannan (%)	Ash (%)
1	20.01de	16.01d	4.00a–c	30.04c	16.37c	0.47c	2.90c	0.13a	2.23a–d
2	19.78e	16.22cd	3.56bc	34.44bc	23.34ab	1.20ab	3.69bc	0.12a	0.57e
3	21.46a–c	17.86a–c	3.60bc	36.69ab	20.61bc	0.55c	3.51bc	0.03cd	2.35a–c
4	21.02b–e	16.88b–d	4.14a–c	38.80ab	23.19ab	0.56c	3.70bc	0.03cd	1.81cd
5	20.99b–e	17.03b–d	3.96a–c	36.00ab	21.11a–c	0.79bc	3.95b	0.03cd	2.20a–d
6	20.94c–e	17.09a–d	3.85a–c	37.51ab	23.80ab	0.59c	3.98b	0.04b–d	1.48c–e
7	21.31a–c	16.76b–d	4.54ab	34.60a–c	20.12bc	0.65bc	3.92b	0.01d	1.47c–e
8	21.36a–c	16.99b–d	4.36ab	37.81ab	22.25ab	0.59c	4.00b	0.03cd	1.29de
9	22.47a	18.78a	3.69bc	35.15a–c	22.49ab	0.90bc	3.96b	0.04b–d	2.19a–d
10	21.12b–d	17.27a–d	3.85a–c	35.87ab	24.09ab	0.54c	4.92a	0.03cd	1.62cd
11	20.79c–e	17.78a–c	3.01c	37.10ab	22.04ab	0.56c	3.46bc	0.02d	2.80ab
12	20.47c–e	17.33a–d	3.14c	34.97a–c	23.01ab	1.22ab	3.89b	0.04b–d	1.96b–d
13	22.19ab	18.38ab	3.82a–c	37.09ab	21.45ab	1.52a	3.80b	0.03cd	2.30a–c
14	21.46a–c	17.78a–c	3.70bc	38.24ab	23.52ab	0.55c	3.89b	0.05bc	1.81cd
15	21.29a–c	17.89a–c	3.41bc	38.18ab	21.19a–c	0.91bc	3.72bc	0.07b	3.01a
16	21.11b–d	17.41a–d	3.70bc	40.05a	21.97ab	0.46c	3.43bc	0.04b–d	1.96b–d
CP	21.20b–d	16.23cd	4.97a	38.90ab	25.59a	0.55c	4.31ab	0.02d	1.96b–d

* Similar letter after means in each treatment indicates insignificant difference for the dependent variable at $\alpha = 0.05$, LSD. CP denotes center point. AIL stands for Acid-Insoluble Lignin; ASL stands for Acid-Soluble Lignin.

3.2. Effects of LMAA Pretreatment on Glucan Digestibility

Figure 2 shows the overall results of enzymatic digestibility for the 17 treatments mentioned in previous experimental design; moreover, the results of avicel (used as a reaction blank for the substrate) and untreated corn stover are indicated in Figure 2 as well. All the enzymatic digestibility results have been organized from the highest digestibility to the lowest in Figure 3. The Lineweaver-Burke linear regressions used to determine enzymatic digestibility kinetic constants are demonstrated in Table 5. As observed in Figure 2, the combinations of the four factors resulted in various digestibility. More clearly, in Figure 3, the highest glucose digestibility (57.23%) compared with the lowest one (29.02%) showed that LMAA pretreated corn stover was 1.97 times higher.

According to the research of Yoo et al. [24], the optimal pretreatment temperature was 80 °C and the pretreatment time was 84 h. In our study, among the 17 treatments, treatment CP, which contained 50 w.b.% moisture content with 0.9–2.0 mm particle size, achieved the highest glucose digestibility with the conditions of 96 h pretreatment time and 75 °C pretreatment temperature. The results are similar to those of Yoo et al. [24], thus indicating that the consistency remained in small and large scale reactors.

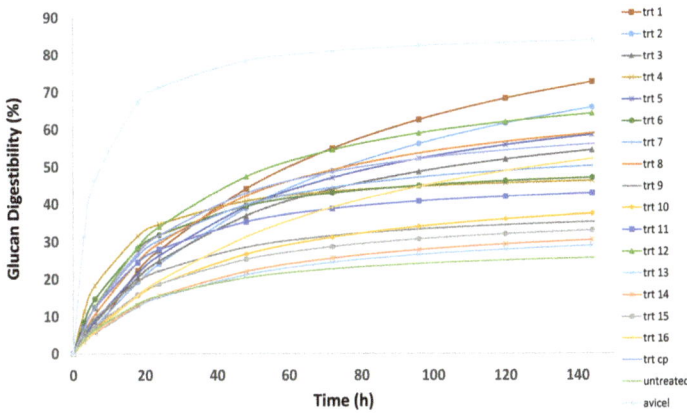

Figure 2. Glucan digestibility results for all treatments. Trt denotes treatment; CP denotes center point.

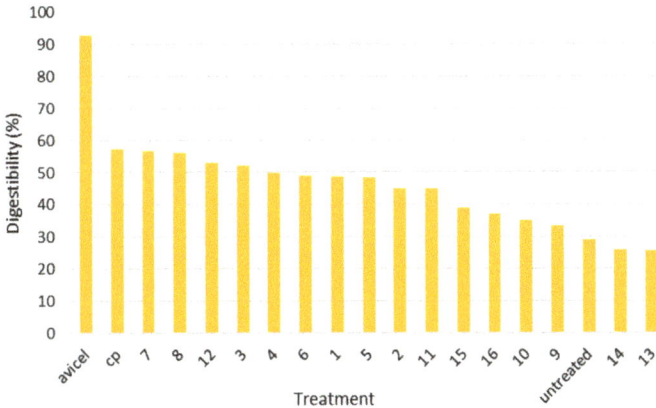

Figure 3. Enzymatic digestibility results of glucan (from highest to lowest) for all the treatments. CP denotes center point.

Table 5. Lineweaver-Burke linear regressions used to determine enzymatic digestibility kinetic constants. Dynamic changes in enzymatic digestibility over time are provided in Figure 2. *

Treatment	Equation
1	$y = 0.6417 \times x + 0.0093$ ($R^2 = 0.955$)
2	$y = 0.7603 \times x + 0.0099$ ($R^2 = 0.983$)
3	$y = 0.6274 \times x + 0.0140$ ($R^2 = 0.996$)
4	$y = 0.2067 \times x + 0.0202$ ($R^2 = 0.982$)
5	$y = 0.6021 \times x + 0.0129$ ($R^2 = 0.985$)
6	$y = 0.2939 \times x + 0.0192$ ($R^2 = 0.990$)
7	$y = 0.3751 \times x + 0.0173$ ($R^2 = 0.965$)
8	$y = 0.4849 \times x + 0.0136$ ($R^2 = 0.984$)
9	$y = 0.4805 \times x + 0.0250$ ($R^2 = 0.992$)
10	$y = 0.7780 \times x + 0.0213$ ($R^2 = 0.986$)
11	$y = 0.3611 \times x + 0.0208$ ($R^2 = 0.987$)
12	$y = 0.3995 \times x + 0.0128$ ($R^2 = 0.943$)
13	$y = 0.9155 \times x + 0.0282$ ($R^2 = 0.981$)
14	$y = 0.9004 \times x + 0.0266$ ($R^2 = 0.968$)
15	$y = 0.6683 \times x + 0.0256$ ($R^2 = 0.945$)
16	$y = 0.9196 \times x + 0.0128$ ($R^2 = 0.983$)
CP	$y = 0.3939 \times x + 0.0151$ ($R^2 = 0.957$)
Untreated	$y = 0.7294 \times x + 0.0339$ ($R^2 = 0.913$)
Avicel	$y = 0.0609 \times x + 0.0115$ ($R^2 = 0.986$)

* y stands for the reverse of digestibility; x stands for the reverse of time.

In this study, four independent variables were tested: moisture content, particle size, pretreatment temperature, and pretreatment time. Among the four variables, pretreatment temperature was regarded as the most critical due to the smallest *p*-value (0.0013). Table 6 shows the difference in average glucan digestibility between high and low levels of pretreatment temperature when other factors were kept constant, in particular, the other main effects. From Table 6, it is clear that higher pretreatment temperature led to decreased glucan digestibility in this study.

Table 6. Main effects on glucan digestibility results (at $t = 144$ h).

Factor	Levels	Digestibility (%)
Temperature (°C)	20	47.76 (16.11)
	75	56.07 (−)
	130	51.02 (9.56)
Time (h)	24	53.14 (13.83)
	96	56.07 (−)
	168	45.65 (11.55)
Moisture Content (%)	20	57.51 (8.47)
	50	56.07 (−)
	80	41.28 (11.60)
Size	S	47.02 (14.80)
	M	56.07 (−)
	L	51.77 (11.17)

As for pretreatment time, the difference between the longest time and the shortest one when other factors were kept constant was also significant as shown in Table 6. The average glucose digestibility at 168 h (47.76%) was relatively lower than the average for 24 h of pretreatment time (51.02%). This could be explained by the longer pretreatment times causing the collapse of the LCC structure of corn stover. It is observed from Figure 2 that from 6 to 18 h, there was an average of 92.7% increase in glucan digestibility, which was the maximum increase rate during all the enzymatic digestibility tests.

In terms of moisture content, higher moisture content resulted in lower glucan digestibility. The reason for this may be that the reduction of retaining ammonia with higher moisture content could result in lower delignification within its structure. As for the effect of particle size, there were some differences between the smallest size and the largest size of corn stover, as observed in Table 6. Larger corn stover particles tend to be more digestible than smaller ones.

4. Conclusions

In this study, the effect of the LMAA pretreatment method with four independent factors was investigated. As a result, LMAA pretreatment showed the potential to achieve higher glucose yield due to higher glucan digestibility. When corn stover (50 w.b. % moisture content) was pretreated at 75 °C for 96 h, the maximum enzymatic digestibility for glucan was obtained. What's more, because there was no washing step involved during the study, the LMAA pretreatment method has the potential to eliminate water consumption compared to other ammonia-based pretreatment methods.

Acknowledgments: The authors would like extend gratitude to the North Central Regional Sun Grant Center and Iowa State University for providing facilities, equipment, and financial support for this study.

Author Contributions: Minliang Yang and Weitao Zhang performed the experiment; Minliang Yang wrote the paper; Kurt A. Rosentrater designed the experiments and edited the paper.

Conflicts of Interest: The authors declare no conflict of interest.

References

1. Mosier, N.; Wyman, C.; Dale, B.; Elander, R.; Lee, Y.Y.; Holtzapple, M.; Ladisch, M. Features of promising technologies for pretreatment of lignocellulosic biomass. *Bioresour. Technol.* **2005**, *96*, 673–686. [CrossRef] [PubMed]
2. Sims, R.E.H.; Mabee, W.; Saddler, J.N.; Taylor, M. An overview of second generation biofuel technologies. *Bioresour. Technol.* **2010**, *101*, 1570–1580. [CrossRef] [PubMed]
3. Cheng, J.J.; Timilsina, G.R. Status and barries of advanced biofuel technologies: A review. *Renew. Energy* **2011**, *36*, 3541–3549. [CrossRef]
4. Naik, S.N.; Goud, V.V.; Rout, P.K.; Dalai, A.K. Production of first and second generation biofuels: A comprehensive review. *Renew. Sustain. Energy Rev.* **2010**, *14*, 578–597. [CrossRef]
5. Kim, K.H.; Hong, J. Supercritical CO_2 pretreatment of lignocellulose enhances enzymatic cellulose hydrolysis. *Bioresour. Technol.* **2001**, *77*, 139–144. [CrossRef]
6. Alinia, R.; Zabihi, S.; Esmaeilzadeh, F.; Kalajahi, J.F. Pretreatment of wheat straw by supercritial CO_2 and its enzymatic hydrolysis for sugar production. *Biosyst. Eng.* **2010**, *107*, 61–66. [CrossRef]
7. Srinivasan, N.; Ju, L.-K. Pretreatment of guayule biomass using supercritical carbon dioxide-based method. *Bioresour. Technol.* **2010**, *101*, 9785–9791. [CrossRef] [PubMed]
8. Narayanaswamy, N.; Faik, A.; Goetz, D.J.; Gu, T. Supercritical carbon dioxide pretreatment of corn stover and switchgrass for lignocellulosic ethanol production. *Bioresour. Technol.* **2011**, *102*, 6995–7000. [CrossRef] [PubMed]
9. Yourchisin, D.M.; Walsum, G.P.V. Comparison of Microbial Inhibition and Enzymatic Hydrolysis Rates of Liquid and Solid Fractions Produced From Pretreatment of Biomass with Carbonic Acid and Liquid Hot Water. Available online: http://link.springer.com/chapter/10.1007/978-1-59259-837-3_87 (accessed on 16 February 2017).
10. Wan, C.; Li, Y. Effect of hot water extraction and liquid hot water pretreatment on the fungal degradation of biomass feedstocks. *Bioresour. Technol.* **2011**, *102*, 9788–9793. [CrossRef] [PubMed]
11. Screenath, H.K.; Koegel, R.G.; Moldes, A.B.; Jeffreies, T.W.; Straub, R.J. Enzymatic saccharification of alfalfa fibre after liquid hot water pretreatment. *Process Biochem.* **1999**, *35*, 33–41.
12. Zhang, B.; Shahbazi, A.; Wang, L. Hot-water pretreatment of cattails for extraction of cellulose. *J. Ind. Microbiol. Biotechnol.* **2011**, *38*, 819–824. [CrossRef] [PubMed]
13. Alizadeh, H.; Teymouri, F.; Gilbert, T.I.; Dale, B.E. Pretreatment of Switchgrass by Ammonia Fiber Explosion (AFEX). Available online: http://link.springer.com/chapter/10.1007/978-1-59259-991-2_94 (accessed on 16 February 2017).

14. Hanchar, R.J.; Teymouri, F.; Nielson, C.D.; McCalla, D.; Stowers, M.D. Separation of Glucose and Pentose Sugars by Selective Enzymes Hydrolysis of AFEX-Treated Corn Fiber. Available online: http://link.springer.com/chapter/10.1007/978-1-60327-181-3_28 (accessed on 16 February 2017).

15. Lee, J.M.; Jameel, H.; Venditti, R.A. A comparison of the autohydrolysis and ammonia fiber explosion (AFEX) pretreatments on the subsequent enzymatic hydrolysis of coastal Bermuda grass. *Bioresour. Technol.* **2010**, *101*, 5449–5458. [CrossRef] [PubMed]

16. Mathew, A.; Parameshwaran, B.; Sukumaran, R.; Pandey, A. An evaluation of dilute acid and ammonia fiber explosion pretreatment for cellulosic ethanol production. *Bioresour. Technol.* **2016**, *199*, 13–20. [CrossRef] [PubMed]

17. Bals, B.; Dale, B.; Balan, V. Enzymatic hydrolysis of Distiller's Dry Grain and Solubles (DDGS) using ammonia fiber expansion pretreatment. *Energy Fuels* **2006**, *20*, 2732–2736. [CrossRef]

18. Gao, D.; Chundawat, S.P.S.; Uppugundla, N.; Balan, V.; Dale, B.E. Binding characteristics of trichoderma reesei cellulases on untreated, ammonia fiber expansion (AFEX), and dilute-acid pretreatment lignocellulosic biomass. *Biotechnol. Bioeng.* **2011**, *108*, 1788–1800. [CrossRef] [PubMed]

19. Garlock, R.J.; Chundauat, S.P.S.; Balan, V.; Dale, B.E. Optimizing harvest of corn stover fractions based on overall sugar yields following ammonia fiber expansion pretreatment and enzymatic hydrolysis. *Biotechnol. Biofuels* **2009**, *2*, 1–14. [CrossRef] [PubMed]

20. Hoover, A.N.; Tumuluru, J.S.; Teymouri, F.; Moore, J.; Gresham, G. Effect of pelleting process variables on physical properties and sugar yields of ammonia fiber expansion pretreated corn stover. *Bioresour. Technol.* **2014**, *164*, 128–135. [CrossRef] [PubMed]

21. Jurado, E.; Skiadas, I.V.; Gavala, H.N. Enhanced methane productivity from manure fibers by aqueous ammonia soaking pretreatment. *Appl. Energy* **2013**, *109*, 104–111. [CrossRef]

22. Gupta, R.; Lee, Y.Y. Pretreatment of hybrid poplar by aqueous ammonia. *Biotechnol. Prog.* **2009**, *25*, 357–364. [CrossRef] [PubMed]

23. Kim, T.H.; Lee, Y.Y. Pretreatment of corn stover by soaking in aqueous ammonia at moderate temperatures. *Appl. Biochem. Biotechnol.* **2007**, *136–140*, 81–92.

24. Yoo, C.G.; Nghiem, N.P.; Hicks, K.B.; Kim, T.H. Pretreatment of corn stover using low-moisture anhydrous ammonia (LMAA) process. *Bioresour. Technol.* **2011**, *102*, 10028–10034. [CrossRef] [PubMed]

25. Yang, M.; Rosentrater, K.A. Comparison of sealing and open conditions for long term storage of corn stover using low-moisture anhydrous ammonia pretreatment method. *Ind. Crops Prod.* **2016**, *91*, 377–381. [CrossRef]

26. Cayetano, R.D.A.; Kim, T.H. Effects of low moisture anhydrous ammonia (LMAA) pretreatment at controlled ammoniation temperatures on enzymatic hydrolysis of corn stover. *Appl. Biochem. Biotechnol.* **2016**. [CrossRef] [PubMed]

27. NREL. Determination of Structural Carbohydrates and Lignin in Biomass. Laboratory Analytical Procedure (LAP). NREL/TP-510-42618; 2011. Available online: http://www.nrel.gov/biomass/analytical_procedures.html (accessed on 16 February 2017).

28. NREL. Enzymatic Saccharification of Lignocellulosic Biomass. Laboratory Analytical Procedure (LAP). NREL/TP-510-42629; 2008. Available online: http://www.nrel.gov/biomass/analytical_procedures.html (accessed on 16 February 2017).

![fermentation logo] *fermentation*

MDPI

Article

A Sequential Steam Explosion and Reactive Extrusion Pretreatment for Lignocellulosic Biomass Conversion within a Fermentation-Based Biorefinery Perspective

José Miguel Oliva, María José Negro, Paloma Manzanares, Ignacio Ballesteros, Miguel Ángel Chamorro, Felicia Sáez, Mercedes Ballesteros and Antonio D. Moreno *

Centro de Investigaciones Energéticas, Medioambientales y Tecnológicas (CIEMAT), Department of Energy, Biofuels Unit, Avda. Complutense 40, 28040 Madrid, Spain; josemiguel.oliva@ciemat.es (J.M.O.); mariajose.negro@ciemat.es (M.J.N.); p.manzanares@ciemat.es (P.M.); ignacio.ballesteros@ciemat.es (I.B.); chamopiano@gmail.com (M.A.C.); felicia.saez@ciemat.es (F.S.); m.ballesteros@ciemat.es (M.B.)
* Correspondence: david.moreno@ciemat.es; Tel.: +34-91-346-6054

Academic Editor: Thaddeus Ezeji
Received: 16 March 2017; Accepted: 14 April 2017; Published: 20 April 2017

Abstract: The present work evaluates a two-step pretreatment process based on steam explosion and extrusion technologies for the optimal fractionation of lignocellulosic biomass. Two-step pretreatment of barley straw resulted in overall glucan, hemicellulose and lignin recovery yields of 84%, 91% and 87%, respectively. Precipitation of the collected lignin-rich liquid fraction yielded a solid residue with high lignin content, offering possibilities for subsequent applications. Moreover, hydrolysability tests showed almost complete saccharification of the pretreated solid residue, which when combined with the low concentration of the generated inhibitory compounds, is representative of a good pretreatment approach. *Scheffersomyces stipitis* was capable of fermenting all of the glucose and xylose from the non-diluted hemicellulose fraction, resulting in an ethanol concentration of 17.5 g/L with 0.34 g/g yields. Similarly, *Saccharomyces cerevisiae* produced about 4% (v/v) ethanol concentration with 0.40 g/g yields, during simultaneous saccharification and fermentation (SSF) of the two-step pretreated solid residue at 10% (w/w) consistency. These results increased the overall conversion yields from a one-step steam explosion pretreatment by 1.4-fold, showing the effectiveness of including an extrusion step to enhance overall biomass fractionation and carbohydrates conversion via microbial fermentation processes.

Keywords: lignocellulosic biomass; steam explosion; extrusion; *Scheffersomyces stipitis*; *Saccharomyces cerevisiae*; simultaneous saccharification and fermentation

1. Introduction

Uncertainties about future energy supplies and the current effects of global warming promoted by massive greenhouse gas emissions make it imperative to develop and implement competitive technologies for establishing a sustainable bio-based economy.

Lignocellulosic biomass is the major renewable organic matter in nature. It is composed of cellulose, hemicellulose and lignin polymers, bonded through non-covalent and covalent cross-linkages to form a complex and recalcitrant structure. Similar to current petroleum-based refineries, future biorefineries will efficiently convert the different components of lignocellulosic biomass into fuels, materials, high value-added chemicals, and other energy forms [1].

Biochemical conversion of lignocellulose includes pretreatment, enzymatic hydrolysis and fermentation steps. Pretreatment is needed to alter the structural characteristics of lignocellulose and increase the accessibility of cellulose and hemicellulose polymers to the hydrolytic enzymes, which are responsible for breaking down these polysaccharides into fermentable sugars. From a biorefinery point of view, pretreatment processes must guarantee optimal and efficient biomass fractionation in order to maximize the potential value obtained from each component (cellulose, hemicellulose and lignin). Over the last four decades, different chemical, physical, physicochemical and biological methods have been developed for the pretreatment of lignocellulose [2,3]. Among pretreatment processes, hydrothermal-based technologies such as steam explosion or liquid hot water (with or without the addition of catalysts) have proven to be effective in deconstructing biomass structure. In the case of steam explosion, biomass accessibility is enhanced mainly by opening lignocellulosic fibers, solubilizing hemicellulosic sugars, and promoting partial solubilization and redistribution of lignin polymers [4]. This hydrothermal pretreatment is usually performed at elevated temperatures and pressures, with varying residence times. In general, temperatures ranging from 200 to 230 °C with short residence times (2–10 min) results in high cellulose saccharification yields (>70%; however, saccharification yields are highly dependent on biomass feedstock), but also in extensive hemicellulose degradation [4]. This side effect lowers the amount of sugars available for fermentation, and releases several biomass-derived products (aliphatic acids, furan derivatives and phenolic compounds), which inhibit hydrolytic enzymes and fermenting microorganisms [5,6]. In addition to hemicellulose degradation, the residual lignin present in the resulting pretreated solid material promotes the unspecific adsorption of hydrolytic enzymes, decreasing saccharification yields [7].

Besides hydrothermal methods, extrusion has been considered as another cost-effective pretreatment technology [2]. Extrusion represents a promising pretreatment method for industrial applications, since it has a highly versatile configuration process for the use of lignocellulosic feedstocks. This physical pretreatment provides effective mixing, rapid heat transfer, and high shear stress, which increases biomass accessibility by (1) promoting defibrillation and shortening of fibers; (2) increasing the surface area available to hydrolytic enzymes; and (3) reducing the crystallinity index and the degree of polymerization of cellulose [8,9]. Furthermore, chemical and or biological catalysts can be integrated in the process to boost saccharification processes. For instance, the addition of alkali during extrusion pretreatment has been shown to promote lignin solubilization and provoke a water-swollen effect, which leads to higher sugar yields in the subsequent saccharification step [8,9].

The combination of both hydrothermal and extrusion technologies can contribute to the balancing of biomass accessibility and biomass degradation, by using milder pretreatment conditions, while offering efficient biomass fractionation. In this context, the present work sequentially combines a mild acid-catalyzed steam explosion with an alkali-based extrusion process for optimal fractionation of lignocellulosic biomass. Using barley straw as a lignocellulosic source, the two-step pretreatment was designed to obtain (1) a liquid fraction containing mainly hemicellulosic sugars; (2) a lignin-rich liquid fraction; and (3) a solid fraction with a high cellulose content. To explore the full potential of the two-step pretreatment process in terms of subsequent applications, collected fractions were studied by analytical techniques and/or fermentation processes. First, the chemical compositions of collected fractions were analyzed to determine recovery yields. Second, the precipitated solid residue (PSR) from collected lignin-rich liquid fraction was analyzed by attenuated total reflectance-Fourier transform infrared spectroscopy (ATR-FTIR) to evaluate lignin purity. Finally, the corresponding water-insoluble solid fractions obtained from steam explosion (WIS) and extrusion (LE-WIS) were subjected to saccharification and fermentation processes, to evaluate their hydrolysability and fermentability in the context of bioethanol production.

2. Materials and Methods

2.1. Raw Material and Pretreatment Process

Barley straw, supplied by CEDER-CIEMAT (Soria, Spain), was used as lignocellulosic feedstock. It had the following composition in terms of percentage dry weight (DW): cellulose, 31.1 ± 0.8; hemicelluloses, 27.2 ± 0.4 (xylan, 22.3 ± 0.2; arabinan, 3.6 ± 0.1; galactan, 1.3 ± 0.1); Klason lignin,

18.8 ± 0.2; ashes, 3.9 ± 0.1; extractives, 10.5 ± 0.6; others components (including acid soluble lignin, acetyl groups, etc.), ~6%.

In order to collect hemicellulosic sugars, raw material was first pretreated by acid-catalyzed steam explosion. Prior to steam explosion, barley straw was milled in a laboratory cutting mill (Cutting Mill Type SM2000; Retsch GmbH, Haan, Germany) to obtain a chip size between 2 and 10 mm. Milled material was then impregnated with H_2SO_4 at an acid/biomass ratio of 10 mg/g, and pretreated in a 10 L steam explosion reactor (CIEMAT, Madrid, Spain) at mild conditions: 180 °C (~9 bar), 3.5 min. This condition was selected on the basis of preliminary studies showing a good balance between cellulose accessibility and hemicelluloses solubilization (data not shown). The recovered slurry was vacuum filtered to obtain a WIS fraction rich in cellulose and lignin, and a liquid fraction rich in hemicellulosic sugars and biomass-derived inhibitors. One portion of the WIS residue was stored for comparison purposes in the hydrolysability and fermentability tests.

Since the lignin polymer remains in the recovered solid fraction after steam explosion pretreatment, the corresponding WIS was subsequently subjected to an alkali-based extrusion process for lignin solubilization. Reactive extrusion was performed in a twin-screw extruder (Clextral Processing Platform Evolum® 25 A110, Clextral, Firminy, France) at 100 °C, 1 min of residence time (rotor speed: 150 rpm), with a biomass feeding rate of 2.5 kg/h, and at a final NaOH/biomass ratio of 80 mg/g (2 L/h of 10% (*w/v*) NaOH). Extrusion conditions and screw configuration were adapted from Duque et al. [10]. Similar to steam-pretreated slurry, extruded slurry was vacuum filtered to obtain a lignin-rich liquid fraction and a lignin-extracted solid residue (LE-WIS), which contained mainly cellulose and the remaining lignin polymers. The resulting lignin-rich liquid fraction was subsequently supplemented with H_2SO_4 (1N) to reach a final pH of 2, to produce a PSR fraction. The PSR was collected by centrifugation at 5000 g in a fixed-angle rotor for 10 min, washed once with distilled water, and lyophilized with a LyoQuest lyophilizer (Telstar, Terrassa, Spain).

Compositional analysis of raw material and collected fractions was determined as described in Section 2.6.1. Before usage, all collected liquid and solid fractions were stored at 4 °C.

2.2. Microorganisms and Growth Conditions

Scheffersomyces stipitis CBS 6054 (Westerdijk Fungal Biodiversity Institute, Utrecht, The Netherlands) and *Saccharomyces cerevisiae* Ethanol Red (Fermentis, Marcq-en-Baroeul, France) were used as fermenting microorganisms in the present study. Active cultures for inoculation were obtained in 100-mL flasks containing 50 mL of growth medium: 30 g/L sugar (*S. cerevisiae* was grown on glucose, while xylose was used for growing *S. stipitis*), 5 g/L yeast extract, 2 g/L NH_4Cl, 1 g/L KH_2PO_4, and 0.3 g/L $MgSO_4·7H_2O$. Flasks were incubated in an orbital shaker at 150 rpm and under controlled temperatures (35 °C for *S. cerevisiae* and 30 °C for *S. stipitis*) for 16 h (reagents for culture medium were purchased from Merck; Darmstadt, Germany). After incubation, cells were harvested by centrifugation at 5000 g in a fixed-angle rotor for 5 min, washed once with distilled water and diluted accordingly to obtain an inoculum concentration of 1 g/L cell dry weight (CDW).

2.3. Enzymes

Saccharification processes were carried out by using the commercial cocktails Celluclast + Novozyme 188 or Cellic CTec2 (Novozymes, Bagsvard, Denmark). Both Celluclast and Cellic CTec2 are mainly cellulase preparations. Due to its low β-glucosidase activity, Celluclast requires supplementation with Novozyme 188 (β-glucosidase) for the hydrolysis of cellobiose into glucose monomers. In contrast to Celluclast, Cellic CTec2 incorporates β-glucosidase activity, and does not therefore require supplementation with additional cocktails. Moreover, Cellic CTec2 also contains endoxylanase activity, which aids in hydrolyzing hemicellulosic sugars.

Overall cellulase activity, measured as filter paper units (FPU), was determined using filter paper (Whatman No. 1 filter paper strips), while β-glucosidase and xylanase activities were determined using cellobiose and birchwood xylan (filter paper, cellobiose and birchwood xylan were purchased from Sigma-Aldrich Quimica SL; Madrid, Spain), respectively [11,12]. One unit of enzyme activity was defined as the amount of enzyme that transformed 1 μmol of substrate per minute.

2.4. Fermentation of the Hemicellulosic-Rich Liquid Fraction

The non-diluted liquid fraction obtained after filtration of steam-pretreated slurry was subjected to fermentation with *S. stipitis*, to evaluate its inhibitory capacity during revalorization of hemicellulosic sugars. Before inoculation, an enzymatic hydrolysis with Cellic CTec2 was carried out to hydrolyze both glucan and xylan oligomers. Enzymatic saccharification was performed in 100 mL shake flasks containing 50 mL of the corresponding liquid fraction. After adjusting the pH to 5, the liquid was supplemented with 2% (*v/v*) Cellic CTec2, and then incubated at 50 °C and 150 rpm for 24 h. Once oligomers were hydrolyzed, the pH was adjusted to 6, and nutrients (5 g/L yeast extract, 2 g/L NH$_4$Cl, 1 g/L KH$_2$PO$_4$, and 0.3 g/L MgSO$_4$·7H$_2$O) and 1 g/L CDW of *S. stipitis* were added. Fermentation assays were performed at 30 °C and 150 rpm for 72 h. Samples were withdrawn periodically during fermentation for analytical purposes. Assays were performed in triplicate, and the corresponding average and standard deviation values were calculated to present the results.

2.5. Hydrolysability and Fermentability Studies of the Pretreated Solid Fractions

2.5.1. Enzymatic Hydrolysis

WIS and LE-WIS fractions were subjected to enzymatic hydrolysis to evaluate the pretreatment process in terms of hydrolysability potential. In this case, 2.5 g of the corresponding solid residue were first diluted in 100 mL shake flasks to a final substrate concentration of 5% (*w/v*). Saccharification was performed at pH 5, 50 °C and 150 rpm for 72 h, with an enzyme loading of 15 FPU/g DW substrate of Celluclast and 15 IU/g DW substrate of Novozyme 188. Assays were performed in triplicate, and the corresponding average and standard deviation values were calculated to present the results.

2.5.2. Simultaneous Saccharification and Fermentation

In addition to hydrolysability tests, collected WIS and LE-WIS residues were also subjected to SSF processes with *S. cerevisiae* to evaluate the fermentability potential of these residues. SSF processes were performed at 35 °C and pH 5 for 72 h in an orbital shaker (150 rpm). For this method, 5 g of the corresponding solid residue was first diluted to a final substrate concentration of 10% (*w/w*) and supplemented with 15 FPU/g DW substrate of Celluclast and 15 IU/g DW substrate of Novozyme 188, and 1 g/L CDW of *S. cerevisiae* Ethanol Red. Samples were withdrawn periodically during SSF for analytical purposes. Assays were performed in triplicate, and the corresponding average and standard deviation values were calculated to present the results.

2.6. Analytical Methods

2.6.1. Compositional Analysis of Biomass

Chemical composition of raw and pretreated material was determined using the Laboratory Analytical Procedures (LAP) for biomass analysis, provided by the National Renewable Energies Laboratory (NREL, Golden, CO, USA) [13]. Sugars and degradation compounds contained in the liquid fraction were also measured. For analysis of the oligomeric forms in the liquid fraction, a mild acid hydrolysis (4% (*w/w*) H$_2$SO$_4$, 120 °C and 30 min) was required to determine the concentration of all monomeric sugars. Monomeric sugars and degradation compounds were analyzed as described in Section 2.6.3.

2.6.2. ATR-FTIR Analysis of Solid Residues

Raw material, WIS, LE-WIS and PSR were analyzed by ATR–FTIR to determine chemical changes during pretreatment process. Dried biomass was analyzed in a FTIR spectrometer (Thermo Scientific

Nicolet 6700 spectrometer; Thermo Fisher Scientific Inc., Waltham, MA, USA), using an attenuated total reflectance (ATR) accessory and a deuterated triglycine sulfate detector. Spectra were collected at room temperature in the 4000–600 cm^{-1} range with a 1.928 cm^{-1} resolution and with an average of 64 scans.

2.6.3. Identification and Quantification of Metabolites

Ethanol was analyzed by gas chromatography (GC), while high-performance liquid chromatography (HPLC) was used to analyze sugars and biomass degradation compounds. In the case of ethanol, a 7890A GC System (Agilent, Waldbronn, Germany) equipped with an Agilent 7683B series injector, a flame ionization detector and a Carbowax 20 M column was used. The column oven was kept constant at 85 °C, while injector and detector temperatures were maintained at 175 °C. The carrier gas, helium, was set at a flow rate of 30 mL/min.

Sugars were analyzed by HPLC (Waters, Mildford, MA, USA) using a CarboSep CHO-682 carbohydrate analysis column (Transgenomic, San Jose, CA, USA). The operating temperature was 80 °C and the flow rate of the mobile phase (ultrapure water) was 0.5 mL/min. The identification of sugars was performed with a refractive index detector (Waters, Mildford, MA, USA).

Syringaldehyde, vanillin, ferulic acid, *p*-coumaric acid, furfural and 5-hydroxymethylfurfural (5-HMF) were analyzed and quantified by HPLC (Agilent, Waldbronn, Germany). The system was equipped with a Coregel 87H3 column (Transgenomic, San Jose, CA, USA). The operating temperature was 65 °C, and the mobile phase was 89% 5 mM H_2SO_4 and 11% acetonitrile, with a flow rate of 0.7 mL/min. All these compounds were identified by a 1050 photodiode-array detector (Agilent, Waldbronn, Germany). Finally, formic acid and acetic acid were also quantified by HPLC (Waters, Mildford, MA, USA). The system was equipped with a Bio-Rad Aminex HPX-87H column (Bio-Rad Labs, Hercules, CA, USA) and a 2414 refractive index detector (Waters, Mildford, MA, USA) for the separation and identification of acids, respectively. The operating temperature was 65 °C, and the flow rate of the mobile phase (5 mM H_2SO_4) was 0.6 mL/min.

3. Results and Discussion

3.1. Pretreatment of Barley Straw

Many pretreatment technologies have already been studied and developed to overcome the recalcitrant structure of lignocellulosic biomass. However, improvements are still necessary to maximize sugar recovery and establish a competitive lignocellulosic-based biorefinery process [2,3]. A two-step pretreatment process was designed for improving lignocellulosic biomass fractionation and facilitating its conversion into value-added compounds via fermentation processes (Figure 1).

Pretreatment consisted of a mild acid-catalyzed steam explosion, and an alkali-based extrusion process. First, steam explosion of acid impregnated barley straw resulted in a slurry with a total solid content of 20.4% (*w/w*) (12.7% and 7.7% insoluble and soluble solids, respectively). Steam explosion increased the cellulose and lignin content in the WIS fraction from 31.1% (*w/w*) and 18.8% (*w/w*), to 55.1% (*w/w*) and 32.1% (*w/w*), respectively (Table 1). This result is explained by an extensive hemicellulose solubilization, indicated by the low hemicellulose content in the pretreated WIS fraction (less than 10% (*w/w*)), and the high content of xylan and xylose in the recovered liquid fraction (Table 1). Biomass degradation compounds including acetic acid, furfural, 5-HMF and certain phenolic compounds (such as vanillin, syringaldehyde, *p*-coumaric acid and ferulic acid) were also identified in the liquid fraction of steam-exploded barley straw. Acetic acid is released by hydrolysis of the acetyl groups present in hemicelluloses. Formic acid derives from furfural and 5-HMF degradation, which results from the degradation of pentoses (mainly xylose) and hexoses respectively. Finally, phenols are released during partial solubilization and degradation of the lignin polymer [14,15].

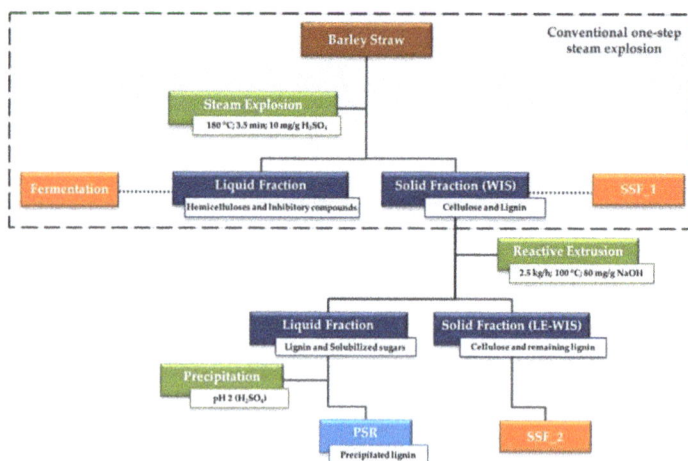

Figure 1. Process scheme depicting the two-step pretreatment process followed in the present study. SSF_1, simultaneous saccharification and fermentation (SSF) of the solid fraction obtained after steam explosion (WIS); SSF_2, SSF of the solid fraction obtained by the two-step pretreatment process (LE_WIS).

Table 1. Composition of steam-exploded barley straw.

WIS Fraction				
Component			% (w/w)	
Cellulose			55.1 ± 0.3	
Hemicellulose			8.8 ± 0.2	
Lignin			32.1 ± 1.9	
Ashes			2.5 ± 0.3	
Others			~1.5	
Liquid Fraction				
Sugar	Monomeric Form % (w/w) [a]	Oligomeric Form % (w/w) [a]	Inhibitor	% (w/w) [a]
Glucan	0.7 ± 0.1 (1.7)	2.8 ± 0.2 (7.6)	Acetic ac.	0.23 ± 0.04 (0.6)
Xylan	7.2 ± 0.4 (18.0)	13.9 ± 1.2 (31.9)	Formic ac.	n.d.
Arabinan	2.5 ± 0.3 (6.2)	1.1 ± 0.2 (2.9)	Furfural	0.17 ± 0.03 (0.4)
Galactan	0.7 ± 0.2 (1.8)	0.7 ± 0.1 (1.7)	5-HMF	0.04 ± 0.01 (0.1)
			Vanillin	<0.01 (12×10^{-3})
			Syringaldehyde	<0.01 (7×10^{-3})
			p-courmaric ac.	0.01 ± 0.00 (15×10^{-3})
			Ferulic ac.	0.01 ± 0.00 (21×10^{-3})

5-HMF, hydroxymethylfurfural; n.d., not determined; WIS, water insoluble solids; [a] Values expressed in g/L are listed in brackets.

Temperatures above 200 °C are needed during steam explosion pretreatment for enhancing biomass accessibility. Under these severe conditions, extensive biomass degradation—mainly hemicellulosic sugars—is also promoted, resulting in higher concentrations of inhibitory compounds. The use of lower pretreatment temperatures increases the recovery of hemicelluloses, and decreases the amount of lignocellulose-derived inhibitors in pretreated streams. However, at lower temperatures, longer pretreatment times (20–60 min) are needed to obtain similar saccharification yields in the subsequent enzymatic hydrolysis step, which increases pretreatment costs [4]. In order to reduce pretreatment time and lower the concentration of inhibitory compounds, an acid catalyst can be added to boost hemicellulose solubilization at temperatures below 200 °C. Thus, the pretreatment condition used in the present work for steam explosion was fairly sufficient for the solubilization of a major fraction of hemicellulosic sugars, reducing the hemicellulose content from 27.2% to 8.8% (w/w)

(Table 1). Furthermore, it is important to highlight the very low concentration of lignocellulose-derived inhibitors in the resulting liquid fraction, which can be indicative of a good pretreatment balance.

In the second stage of the pretreatment process, the recovered WIS fraction was subjected to an alkali-based extrusion process. The obtained extruded slurry contained 25% (w/w) insoluble solids out of 30% (w/w) total solids. After the extrusion process, the cellulose content of collected LE-WIS increased to 64.2% (w/w) (Table 2). Such an increase was promoted by lignin solubilization, even though similar lignin content was measured for both WIS and LE-WIS residues (Tables 1 and 2). The effectiveness of alkali-catalyzed extrusion processes to solubilize lignin has been previously observed. Duque et al. [10] reported a minimum NaOH/biomass ratio of 2.5–5% (w/w) to promote lignin solubilization in barley straw. Furthermore, these authors showed the highest lignin solubilization when using similar NaOH/biomass ratio and temperatures (7.5% (w/w) and 100 °C) to those used in the present study.

Another advantage of reactive extrusion with alkali is the possibility of lignin revalorization. Lignin represents an economic raw material for a wide range of applications. Although it has not yet been converted into high-value products at large scales, lignin has been utilized for the production of fertilizers, bioplastics or carbon fibers, among others products [16]. In this context, solubilized lignin was recovered by precipitation from the corresponding lignin-rich liquid fraction, resulting in a PSR fraction with about 85% and 3.5% (w/w) of lignin and sugar content, respectively (Table 2).

Table 2. Composition of extruded barley straw.

LE-WIS Fraction	
Component	% (w/w)
Cellulose	64.2 ± 2.0
Hemicellulose	6.8 ± 0.1
Lignin	29.3 ± 0.6
Ashes	2.1 ± 0.0
PSR Fraction	
Component	% (w/w)
Glucan	0.9 ± 0.1
Xylan	2.5 ± 0.2
Lignin	85.1 ± 1.5
Ashes	8.6 ± 0.6

LE-WIS, lignin-extracted water insoluble solids; PSR, precipitated solid residue.

In addition to determining the chemical composition of each collected fraction, the global mass balance for each component was estimated by comparing both raw and pretreated biomass yields. As listed in Table 3, high overall recovery yields were observed for glucan (84% (w/w)), hemicellulose (91% (w/w)) and lignin (87%, (w/w)), when considering all collected fractions. As well as the low concentration of biomass degradation compounds, these high recovery yields are representative of the well-balance pretreatment strategy, which offers high potential for the revalorization of lignocellulose.

Table 3. Mass balance during the two-step pretreatment process.

Component	Steam Explosion		Extrusion	
	Solid [a]	Liquid	Solid [b]	Liquid
Glucan	90	9	75	n.d.
Hemicellulose	17	82	9	n.d.
Lignin	87	n.d.	55	32 [c]

Values expressed as g/100 g DW of initial biomass; n.d., not determined; [a] Solid refers to WIS fraction; [b] Solid refers to LE-WIS fraction; [c] Value considering precipitated lignin in PSR fraction.

3.2. Characterization of Solid Residues by Attenuated Total Reflectance-Fourier Transform Infrared (ATR-FTIR) Spectroscopy

Chemical changes promoted during pretreatment process were analyzed by ATR-FTIR on each solid residue. Figure 2 shows the absorbance in the mid-infrared region (2000–800 cm^{-1}) for all collected solid fractions. In the case of the non-pretreated barley straw (Figure 2a), typical peaks related to lignocellulosic biomass were observed [17]. The carbohydrate region (1370–890 cm^{-1}), including peaks characteristic of C–H deformation (900 cm^{-1}), C–O stretching (1105–1050 cm^{-1}), C–O–C vibration (1159 cm^{-1}) and C–H stretching (1375 cm^{-1}), showed the highest absorbance values. In addition, a lignin region (1595–1261 cm^{-1})—including signal for aromatic rings vibration (1595, 1510, 1421, 1329 and 1261 cm^{-1}) and C–H symmetric deformation (1498 cm^{-1})—and a peak related to ester groups in hemicelluloses (1731 cm^{-1}) could be also identified. This peak pattern of barley straw was modified during the two-step pretreatment process. First, the WIS fraction obtained after steam explosion pretreatment showed a significant reduction in the carbohydrate region, and at band 1731 cm^{-1} (Figure 2a). This reduction was supported by the extensive hemicellulose solubilization induced during the first stage of the pretreatment process (Table 1). In the case of extrusion pretreatment, an increase in the peak intensity of the carbohydrate region was noted when comparing WIS and LE-WIS fractions (Figure 2a). The higher absorbance in the carbohydrate region of LE-WIS can be explained by lignin solubilization, which increased the glucan/lignin ratio (Table 2).

A completely different absorbance profile was obtained with the collected PSR fraction (Figure 2b). This spectrum presented clearly defined peaks in the lignin region, which shows evidence of the high lignin content of this residue (Table 2). This result, combined with the high lignin content measured for the PSR fraction, offers possibilities for the subsequent revalorization of this residue from a biorefinery point of view. Nevertheless, further studies are needed to evaluate the actual potential of PSR utilization, since lignin polymers are usually altered during steam explosion pretreatment (e.g., cleavage of the β–O–4 ether bonds and other acid labile linkages) [4].

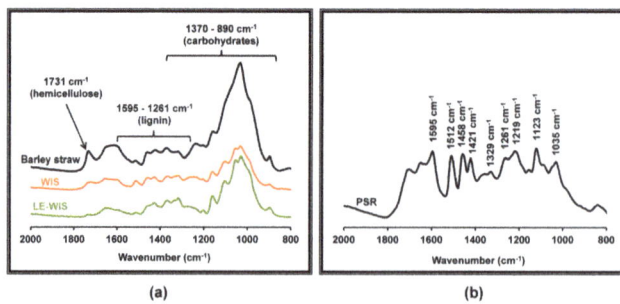

Figure 2. Infrared absorption spectra (cm^{-1}) of non-pretreated barley straw and pretreated collected fractions. (**a**) Solid fractions collected during the two-step pretreatment process: (black) non-pretreated barley straw, (orange) WIS fraction obtained after steam explosion, (green) LE-WIS fraction obtained after extrusion pretreatment; (**b**) PSR obtained by precipitation of the liquid fraction collected after extrusion pretreatment.

3.3. Saccharification of Pretreated Solid Residues

Saccharification is a key step during lignocellulosic biomass conversion as it highly influences overall production yields [18]. In this context, an efficient saccharification step is essential to obtain higher concentrations of fermentable sugars. After steam explosion pretreatment, 75% of potential sugars were enzymatically hydrolyzed from the collected WIS fraction (Figure 3a). This sugar yield was increased to about 100% by introducing the extrusion process, showing the effectiveness of this second stage for improving biomass accessibility.

Both steam explosion and extrusion are considered effective pretreatment technologies for enhancing biomass accessibility to the hydrolytic enzymes [2,3,8,9]. Moreover, these methods are highly versatile with regards to biomass feedstock and process configuration (such as the use of chemical catalysts). By combining steam explosion and extrusion processes, steam explosion can be performed at lower temperatures, decreasing the amount of released biomass degradation compounds without compromising biomass recovery and accessibility (Figure 3, Tables 1 and 3). Extrusion has been previously combined with other pretreatment technologies, with the aim of reaching high sugar yields and use milder process conditions (such as using lower temperatures and pressures, reducing the amounts of chemicals or solvents required during the process, decreasing enzyme loadings, etc.) [8,19,20]. For instance, Chen et al. [20] obtained an enzymatic hydrolysis yield of 80% (with about 84% xylan recovery), when subjecting rice straw to a combined extrusion and dilute acid pretreatment process. Similarly, Lee et al. [19] combined extrusion with hot-compressed water to pretreat Douglas fir, obtaining five-fold higher sugar yields.

In addition of increasing the concentration of fermentable sugars, higher saccharification yields also benefit the potential utilization of the remaining lignin polymer. Thus, the 55% (w/w) of the lignin that was left in the LE-WIS fraction could be recovered after an enzymatic hydrolysis step, increasing the overall lignin recovery yield from 32% (w/w) (in the PSR) to 87% (w/w) (Table 3).

When considering the initial sugar content, however, similar overall saccharification yields were observed for both WIS and LE-WIS fractions (Figure 3b). This result can be explained by the fact that some glucan and hemicellulose is co-solubilized with lignin during extrusion pretreatment, as indicated by the lower glucan and hemicellulose recovery yields for the LE-WIS fraction [10] (Table 3).

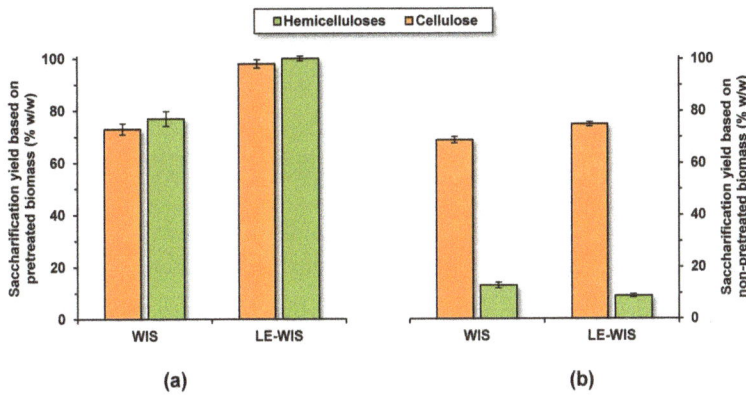

Figure 3. Saccharification yields obtained by enzymatic hydrolysis (72 h) of the WIS and LE-WIS fractions at 5% (w/v) substrate loadings. (**a**) cellulose and hemicellulose yields based on the composition of each pretreated fraction; (**b**) overall saccharification yields based on the initial composition of non-pretreated barley straw.

3.4. Conversion of Lignocellulosic Sugar by Microbial Fermentation Processes

From a biorefinery perspective, several biofuels and biochemicals (ethanol, methane, lactic acid, lipids, etc.) can be obtained via microbial fermentation of lignocellulosic sugars [21]. Among biofuels, lignocellulosic bioethanol is considered to be a promising alternative for the partial replacement of fossil fuels in the short to medium prospect. In this context, the two-step pretreatment process was evaluated in terms of ethanol production from pretreated sugar fractions: the hemicellulose-rich liquid fraction and the solid WIS and LE-WIS fractions. Results related to these assays are discussed in the following subsections.

3.4.1. Fermentation of the Hemicellulose-Rich Liquid Fraction

The presence of biomass degradation compounds in pretreated biomass is one of the main limitations for the fermentation of lignocellulosic sugars. These compounds have a negative impact on cell growth by inhibiting specific intracellular enzymes, causing an energy imbalance, and/or affecting the integrity of cell membranes [6,14,22,23]. After steam explosion pretreatment, the collected liquid fraction contained, in addition to solubilized hemicellulosic sugars, those compounds released from biomass degradation (Table 1). With the aim of evaluating the inhibitory capacity of this stream, the hemicellulose-rich liquid fraction was subjected to fermentation with *S. stipitis*. This yeast was chosen as a fermentative microorganism since it is capable of assimilating and converting xylose, the major component of this fermentation medium (Table 1). Most of the non-*Saccharomyces* yeast strains, including *S. sitipitis*, are known to be more sensitive to the inhibitory compounds released from biomass [24]. This means that lower concentrations of lignocellulose-derived compounds are needed to inhibit these fermentative microorganisms. To overcome microbial inhibition, different physical, chemical and biological detoxification processes have been developed to lower the concentration of degradation compounds [6,23,25]. Typical detoxification methods include filtration and washing, vacuum evaporation, and the use of resins and/or chemical/biological catalysts [26,27]. These processes, however, should be avoided since they usually require higher quantities of freshwater, the use of extra equipment, produce a loss of soluble sugars, and increase wastewater and overall process costs [25].

Biomass degradation promoted by steam explosion pretreatment can be reduced by using milder pretreatment conditions. As discussed above, the liquid fraction resulted from the first pretreatment stage (steam explosion) showed low concentrations of lignocellulose-derived compounds (Table 1). Nevertheless, the synergistic interaction between degradation compounds might cause the inhibition of the fermenting microorganisms, even at low concentrations [28,29], depending mainly on the inhibitory mixture and the inoculum size. In this case, the non-diluted liquid fraction caused no inhibition on *S. stipitis*, confirming the low inhibitory potential of this collected fraction. During the fermentation process, a maximum ethanol concentration of 17.5 g/L and a maximum ethanol volumetric productivity of 0.46 g/L·h were obtained, showing glucose and xylose depletion within 24 h and 72 h, respectively (Figure 4, Table 4). The observed ethanol concentration corresponds to a final ethanol yield of 0.34 g/g, which represents to about 70% of the theoretical ethanol that can be produced from the initial concentration of glucose and xylose.

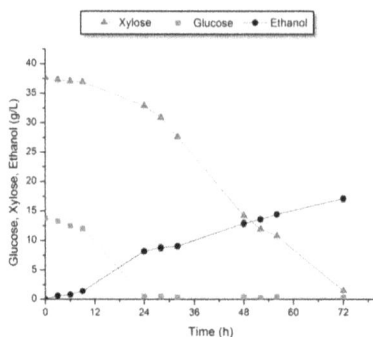

Figure 4. Fermentation of the non-diluted hemicellulose-rich liquid fraction (equivalent to about 13% WIS (*w*/*w*)). Time course of glucose and xylose consumption and ethanol production by the yeast *S. stipitis* CBS 6054. Prior to inoculation, the liquid fraction was enzymatically hydrolyzed with Cellic CTec2 at 50 °C for 24 h. Mean values and standard deviations were calculated from replicates to present the results. Note: glucose concentration is higher than that reported in Table 1 due to the presence of glucose in Cellic CTec2 preparation.

From a biorefinery point of view, the resulting ethanol concentration was below the minimum required for scaling up the process [30]. In this context, the low inhibitory capacity of the obtained hemicellulose-rich liquid fraction may offer possibilities for alternative microbial-based processes, such as the production of xylitol, lactic acid or microbial oils [31–33].

3.4.2. SSF of Pretreated Solid Fractions

Taking into account the good hydrolysability of LE-WIS (Figure 3a), this fraction was subjected to SSF processes to evaluate the fermentability potential of this pretreated material. The WIS fraction collected after steam explosion pretreatment was also subjected to SSF for comparison purposes. Due to its superior fermentation capacity of hexose sugars, the yeast *S. cerevisiae* was chosen as the fermentative microorganism for SSF processes. When using the WIS fraction as substrate, a maximum ethanol concentration of 19.6 g/L was obtained after 72 h of SSF process (Figure 5a, Table 4). This value was increased up to 31.7 g/L when using the LE-WIS fraction instead (Figure 5b, Table 4). With a 16% higher glucan content (Tables 1 and 2), higher ethanol concentrations during SSF of LE-WIS were expected. Nevertheless, the obtained ethanol concentrations respectively correspond to 0.29 g/g and 0.40 g/g overall yields, which were equivalent to 57% and 78% of the theoretical ethanol yield (Table 4). Both higher ethanol concentrations and yields were consequently observed for the LE-WIS fraction, being representative of the better hydrolysability of the two-step pretreated solid fraction. The differences in the glucan content, however, had an effect on the corresponding increase in ethanol concentration and yield. Thus, ethanol concentration increased by 60%, while overall ethanol yields increased by 1.4-fold.

In addition to ethanol concentration and yields, slightly higher maximum ethanol volumetric productivities were also observed during SSF of LE-WIS (0.96 g/L·h, compared to 0.83 g/L·h for WIS). In SSF processes, ethanol volumetric productivities are highly influenced by hydrolysis rates. Therefore, these small differences could be justified by the differences in the hydrolysability capacity of pretreated fractions.

The better fermentation parameters observed for LE-WIS fraction could be explained by the better hydrolysability of LE-WIS, as indicated by the higher glucose concentration within the first 12 h of SSF processes, and the higher overall yields (Figure 5, Table 4). However, although hydrolysability tests showed 75% and 98% saccharification yields for the WIS and LE-WIS fraction, respectively, only 57% and 78% ethanol yields were obtained –even though glucose concentration remained below 0.5 g/L after 72 h of SSF (Figure 5). This result hints at enzymatic hydrolysis as the main impeding factor for reaching higher conversion yields. Differences between saccharification yields during hydrolysability tests and SSF could be explained by the increase in substrate concentration (from 5% (*w/v*) to 10% (*w/w*)) and the lower temperature (35 °C instead of 50 °C) used during SSF processes. The increase in substrate loadings influences enzymatic hydrolysis by promoting (1) end-product inhibition of hydrolytic enzymes; (2) unproductive adsorption of proteins to the remaining lignin polymer; (3) protein deactivation or denaturalization and (4) the decline in the binding capacity of enzymes to cellulose [34,35]. For instance, Moreno et al. [36] reported a 35% decrease on the overall ethanol yields after increasing the substrate concentration from 10% to 20% (*w/w*) during SSF processes. Another factor that highly influences saccharification yields is SSF temperature. Enzymatic hydrolysis has an optimal temperature around 50 °C, while most fermenting yeasts work at 30–37 °C. In this context, the use of thermotolerant strains that can ferment at temperatures above 40 °C, may contribute to obtain increased overall conversion yields [37,38].

Energy balance is another important aspect for evaluating the economic feasibility of the process [39]. In this context, the present work provides the basic scenario to set optimal conditions for the future success of the process. Also, it is remarkable to mention that a final ethanol concentration of 4% (*v/v*) was obtained with the present two-stage pretreatment strategy. Notwithstanding, with the aim of increasing final ethanol concentration and overall yields, different experiments at higher substrate concentrations and using novel enzyme cocktails are now being performed.

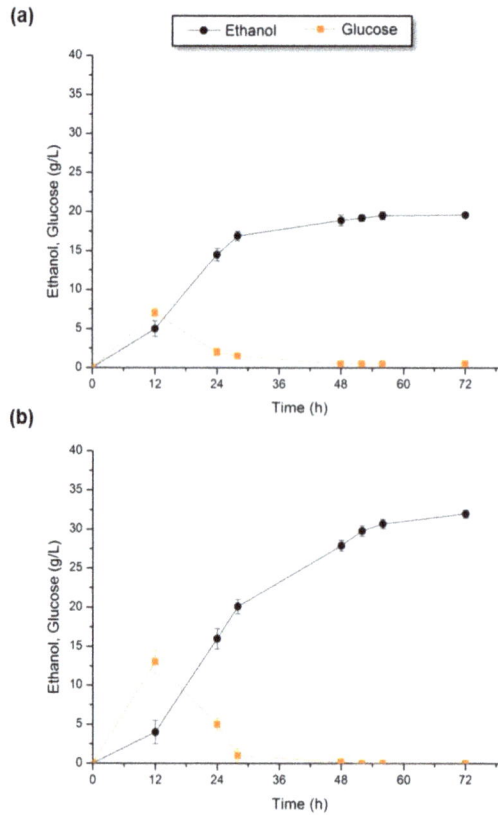

Figure 5. Simultaneous saccharification and fermentation (SSF) of (**a**) WIS and (**b**) LE-WIS at 10% (*w/w*) substrate loading. Time course of glucose and xylose consumption and ethanol production by the yeast *S. cerevisiae* Ethanol Red. Mean values and standard deviations were calculated from replicates to present the results.

Table 4. Summary of the fermentation parameters obtained for collected sugar fractions.

Substrate (*w/w*)	Yeast	EtOH$_{max}$ (g/L)	Y$_{E/S}$ (g/g)	Y$_{E/ET}$ (%)	Q$_{Emax}$ (g/L·h)
Liquid fraction [a]	*S. stipitis*	17.5 ± 0.2	0.34 ± 0.01 [b]	66.7	0.46 ± 0.01
10% WIS	*S. cerevisiae*	19.6 ± 0.1	0.29 ± 0.00 [c]	56.9	0.83 ± 0.04
10% LE-WIS	*S. cerevisiae*	31.7 ± 0.3	0.40 ± 0.01 [c]	78.4	0.96 ± 0.09

[a] The liquid fraction used was equivalent to about 13% (*w/w*) WIS. EtOH$_{max}$, maximum ethanol concentration reached at 72 h; Y$_{E/G}$, ethanol yield based on [b] initial glucose and xylose concentration or [c] potential available glucose (considering the glucan content of substrate); Y$_{E/ET}$, percentage of the theoretical ethanol, assuming maximum ethanol yields of 0.51 g/g for both glucose and xylose; Q$_{Emax}$, maximum volumetric ethanol productivity, estimated within 12–24 h. Ethanol yield was calculated with the assumption that the liquid volume of the SSF system is constant [40].

4. Conclusions

By combining an acid-catalyzed steam explosion and an alkali-based extrusion process, lignocellulosic biomass (barley straw) can be fractionated with high overall recovery yields, producing (1) a solid residue with high lignin content, (2) a non-inhibitory liquid fraction containing hemicellulosic sugars and (3) a solid residue with high glucan content. From a sugar platform perspective, the majority of uses for sugar are via microbial fermentation. The present two-step pretreatment process has demonstrated not only the possibility for maximizing lignin and sugar recovery, but also for enhancing the hydrolysability and fermentability of collected residues. Thus, this pretreatment favors the revalorization of each lignocellulosic component when considering a fermentation-based biorefinery.

Acknowledgments: Authors thank the Regional Government of Madrid (Spain) for funding the present work via Project S2013/MAE-2882. Antonio D. Moreno acknowledges the Spanish Ministry of Economy and Competitiveness and the specific "Juan de la Cierva" Subprogramme for contract FJCI-2014-22385. Novozymes is also gratefully acknowledge for providing enzymatic cocktails.

Author Contributions: All authors have participated in the design of the study. José Miguel Oliva, María José Negro and Mercedes Ballesteros conceived and designed the experiments; Ignacio Ballesteros and Paloma Manzanares performed pretreatment of barley straw; José Miguel Oliva and Miguel Ángel Chamorro performed fermentation experiments and analyzed the data; Felicia Sáez contributed with biomass analysis; Antonio D. Moreno wrote the paper.

Conflicts of Interest: The authors declare no conflict of interest.

Abbreviations

The following abbreviations are used in this manuscript:

5-HMF	5-hydroxymethylfurfural
ATR-FTIR	Attenuated Total Reflectance-Fourier Transform Infrared spectroscopy
CDW	Cell Dry Weight
DW	Dry Weight
$EtOH_{max}$	Maximum Ethanol concentration
FPU	Filter Paper Units
GC	Gas Chromatography
HPLC	High Performance Liquid Chromatography
LE-WIS	Lignin-Extracted Water Insoluble Solid fraction
NREL-LAP	National Renewable Energies Laboratory-Laboratory Analytical Procedures
PSR	Precipitated Solid Residue
Q_E	Ethanol Volumetric Productivity
WIS	Water Insoluble Solid fraction
$Y_{E/ET}$	Ethanol Yield based on the maximum theoretical ethanol
$Y_{E/S}$	Ethanol Yield based on potential sugars

References

1. Olsson, L.; Saddler, J. Biorefineries, using lignocellulosic feedstocks, will have a key role in the future bioeconomy. *Biofuels Bioprod. Biorefin.* **2013**, *7*, 475–477. [CrossRef]
2. Alvira, P.; Tomás-Pejó, E.; Ballesteros, M.; Negro, M.J. Pretreatment technologies for an efficient bioethanol production process based on enzymatic hydrolysis: A review. *Bioresour. Technol.* **2010**, *101*, 4851–4861. [CrossRef] [PubMed]
3. Mussatto, S.I.; Dragone, G.M. Biomass pretreatment, biorefineries, and potential products for a bioeconomy development. In *Biomass Fractionation Technologies for a Lignocellulosic Feedstock Based Biorefinery*; Mussatto, S.I., Ed.; Elsevier: Amsterdam, The Netherlands, 2016; pp. 1–22.
4. Duque, A.; Manzanares, P.; Ballesteros, I.; Ballesteros, M. Steam explosion as lignocellulosic biomass pretreatment. In *Biomass Fractionation Technologies for a Lignocellulosic Feedstock Based Biorefinery*; Mussatto, S.I., Ed.; Elsevier: Amsterdam, The Netherlands, 2016; pp. 349–368.
5. Ximenes, E.; Kim, Y.; Mosier, N.; Dien, B.; Ladisch, M. Inhibition of cellulases by phenols. *Enzyme Microb. Technol.* **2010**, *46*, 170–176. [CrossRef]
6. Jönsson, L.J.; Martín, C. Pretreatment of lignocellulose: Formation of inhibitory by-products and strategies for minimizing their effects. *Bioresour. Technol.* **2016**, *199*, 103–112. [CrossRef] [PubMed]
7. Chandra, R.P.; Bura, R.; Mabee, W.E.; Berlin, A.; Pan, X.; Saddler, J.N. Substrate pretreatment: The key to effective enzymatic hydrolysis of lignocellulosics? *Adv. Biochem. Eng. Biotechnol.* **2007**, *108*, 67–93. [PubMed]
8. Zheng, J.; Rehmann, L. Extrusion pretreatment of lignocellulosic biomass: A review. *Int. J. Mol. Sci.* **2014**, *15*, 18967–18984. [CrossRef] [PubMed]

9. Vandenbossche, V.; Brault, J.; Vilarem, G.; Hernández-Meléndez, O.; Vivaldo-Lima, E.; Hernández-Luna, M.; Barzana, E.; Duque, A.; Manzanares, P.; Ballesteros, M.; et al. A new lignocellulosic biomass deconstruction process combining thermo-mechano chemical action and bio-catalytic enzymatic hydrolysis in a twin-screw extruder. *Ind. Crops Prod.* **2014**, *55*, 258–266. [CrossRef]

10. Duque, A.; Manzanares, P.; Ballesteros, I.; Negro, M.J.; Oliva, J.M.; Sáez, F.; Ballesteros, M. Optimization of integrated alkaline–Extrusion pretreatment of barley straw for sugar production by enzymatic hydrolysis. *Process Biochem.* **2013**, *48*, 775–781. [CrossRef]

11. Ghose, T.K. Measurement of cellulase activities. *Pure Appl. Chem.* **1987**, *59*, 257–268. [CrossRef]

12. Bailey, M.J.; Biely, P.; Poutanen, K. Interlaboratory testing of methods for assay of xylanase activity. *J. Biotechnol.* **1992**, *23*, 257–270. [CrossRef]

13. NREL. Chemical Analysis and Testing Laboratory Analytical Procedures. National Renewable Energy Laboratory: Golden, CO, USA, 2008. Available online: https://www.nrel.gov/bioenergy/biomass-compositional-analysis.html (accessed on 1 March 2017).

14. Palmqvist, E.; Hahn-Hägerdal, B. Fermentation of lignocellulosic hydrolysates. II: Inhibitors and mechanisms of inhibition. *Bioresour. Technol.* **2000**, *74*, 25–33. [CrossRef]

15. Oliva, J.M.; Sáez, F.; Ballesteros, I.; González, A.; Negro, M.J.; Manzanares, P.; Ballesteros, M. Effect of lignocellulosic degradation compounds from steam explosion pretreatment on ethanol fermentation by thermotolerant yeast *Kluyveromyces marxianus*. *Appl. Biochem. Biotechnol.* **2003**, *105–108*, 141–153. [CrossRef]

16. Norgren, M.; Edlund, H. Lignin: Recent advances and emerging applications. *Curr. Opin. Colloid Interface Sci.* **2014**, *19*, 409–416. [CrossRef]

17. Sun, X.F.; Xu, F.; Sun, R.C.; Fowler, P.; Baird, M.S. Characteristics of degraded cellulose obtained from steam-exploded wheat straw. *Carbohydr. Res.* **2005**, *340*, 97–106. [CrossRef] [PubMed]

18. Gurram, R.N.; Al-Shannag, M.; Lecher, N.J.; Duncan, S.M.; Singsaas, E.L.; Alkasrawi, M. Bioconversion of paper mill sludge to bioethanol in the presence of accelerants or hydrogen peroxide pretreatment. *Bioresour. Technol.* **2015**, *192*, 529–539. [CrossRef] [PubMed]

19. Lee, S.H.; Inoue, S.; Teramoto, Y.; Endo, T. Enzymatic saccharification of woody biomass micro/nanofibrillated by continuous extrusion process II: Effect of hot-compressed water treatment. *Bioresour. Technol.* **2010**, *101*, 9645–9649. [CrossRef] [PubMed]

20. Chen, W.H.; Xu, Y.Y.; Hwang, W.S.; Wang, J.B. Pretreatment of rice straw using an extrusion/extraction process at bench-scale for producing cellulosic ethanol. *Bioresour. Technol.* **2011**, *102*, 10451–10458. [CrossRef] [PubMed]

21. E4tech; RE-CORD; WUR. From the Sugar Platform to Biofuels and Biochemicals. Final report for the European Commission Directorate-General Energy, contract No. ENER/C2/423-2012/SI2.673791: European Union, 2015. Available online: http://ibcarb.com/wp-content/uploads/EC-Sugar-Platform-final-report.png (accessed on 15 January 2017).

22. Palmqvist, E.; Hahn-Hägerdal, B. Fermentation of lignocellulosic hydrolysates. I: Inhibition and detoxification. *Bioresour. Technol.* **2000**, *74*, 17–24. [CrossRef]

23. Taherzadeh, M.J.; Karimi, K. Fermentation inhibitors in ethanol processes and different strategies to reduce their effects. In *Biofuels Alternative Feedstocks and Conversion Processes*; Pandey, A., Larroche, C., Ricke, S.C., Dussap, C.-G., Gnansounou, E., Eds.; Academic Press: Amsterdam, The Netherlands, 2011; pp. 287–311.

24. Agbogbo, F.K.; Coward-Kelly, G. Cellulosic ethanol production using the naturally occurring xylose-fermenting yeast, *Pichia stipitis*. *Biotechnol. Lett.* **2008**, *30*, 1515–1524. [CrossRef] [PubMed]

25. Moreno, A.D.; Ibarra, D.; Alvira, P.; Tomás-Pejó, E.; Ballesteros, M. A review of biological delignification and detoxification methods for lignocellulosic bioethanol production. *Crit. Rev. Biotechnol.* **2015**, *35*, 342–354. [CrossRef] [PubMed]

26. Larsson, S.; Reimann, A.; Nilvebrant, N.-O.; Jönsson, L.J. Comparison of different methods for the detoxification of lignocellulose hydrolyzates of spruce. *Appl. Biochem. Biotechnol.* **1999**, *77*, 91–103. [CrossRef]

27. Moreno, A.D.; Ibarra, D.; Fernández, J.L.; Ballesteros, M. Different laccase detoxification strategies for ethanol production from lignocellulosic biomass by the thermotolerant yeast *Kluyveromyces marxianus* CECT 10875. *Bioresour. Technol.* **2012**, *106*, 101–109. [CrossRef] [PubMed]

28. Oliva, J.M.; Ballesteros, I.; Negro, M.J.; Manzanares, P.; Cabañas, A.; Ballesteros, M. Effect of binary combinations of selected toxic compounds on growth and fermentation of *Kluyveromyces marxianus*. *Biotechnol. Prog.* **2004**, *20*, 715–720. [CrossRef] [PubMed]

29. Alvira, P.; Moreno, A.D.; Ibarra, D.; Sáez, F.; Ballesteros, M. Improving the fermentation performance of *Saccharomyces cerevisiae* by laccase during ethanol production from steam-exploded wheat straw at high-substrate loadings. *Biotechnol. Prog.* **2013**, *29*, 74–82. [CrossRef] [PubMed]

30. Zacchi, G.; Axelsson, A. Economic evaluation of preconcentration in production of ethanol from dilute sugar solutions. *Biotechnol. Bioeng.* **1989**, *34*, 223–233. [CrossRef] [PubMed]

31. Mussatto, S.I.; Roberto, I.C. Xylitol production from high xylose concentration: Evaluation of the fermentation in bioreactor under different stirring rates. *J. Appl. Microbiol.* **2003**, *95*, 331–337. [CrossRef] [PubMed]

32. Ilmén, M.; Koivuranta, K.; Ruohonen, L.; Suominen, P.; Penttilä, M. Efficient production of L-lactic acid from xylose by *Pichia stipitis*. *Appl. Environ. Microbiol.* **2007**, *73*, 117–123. [CrossRef] [PubMed]

33. Díaz-Fernández, D.; Lozano-Martínez, P.; Buey, R.M.; Revuelta, J.L.; Jiménez, A. Utilization of xylose by engineered strains of *Ashbya gossypii* for the production of microbial oils. *Biotechnol. Biofuels* **2017**, *10*, 3. [CrossRef] [PubMed]

34. Kristensen, J.B.; Felby, C.; Jørgensen, H. Yield-determining factors in high-solids enzymatic hydrolysis of lignocellulose. *Biotechnol. Biofuels* **2009**, *2*, 11. [CrossRef] [PubMed]

35. Wang, W.; Kang, L.; Wei, H.; Arora, R.; Lee, Y.Y. Study on the decreased sugar yield in enzymatic hydrolysis of cellulosic substrate at high solid loading. *Appl. Biochem. Biotechnol.* **2011**, *164*, 1139–1149. [CrossRef] [PubMed]

36. Moreno, A.D.; Tomás-Pejó, E.; Ibarra, D.; Ballesteros, M.; Olsson, L. In Situ laccase treatment enhances the fermentability of steam-exploded wheat straw in SSCF processes at high dry matter consistencies. *Bioresour Technol.* **2013**, *143*, 337–343. [CrossRef] [PubMed]

37. Abdel-Banat, B.M.; Hoshida, H.; Ano, A.; Nonklang, S.; Akada, R. High-temperature fermentation: How can processes for ethanol production at high temperatures become superior to the traditional process using mesophilic yeast? *Appl. Microbiol. Biotechnol.* **2010**, *85*, 861–867. [CrossRef] [PubMed]

38. Moreno, A.D.; Ibarra, D.; Ballesteros, I.; Fernández, J.L.; Ballesteros, M. Ethanol from laccase-detoxified lignocellulose by the thermotolerant yeast *Kluyveromyces marxianus*-Effects of steam pretreatment conditions, process configurations and substrate loadings. *Biochem. Eng. J.* **2013**, *79*, 94–103. [CrossRef]

39. Gurram, R.N.; Al-Shannag, M.; Knapp, S.; Das, T.; Singsaas, E.; Alkasrawi, M. Technical possibilities of bioethanol production from coffee pulp: A renewable feedstock. *Clean Technol. Environ.* **2016**, *18*, 269–278. [CrossRef]

40. Zhang, J.; Bao, J. A modified method for calculating practical ethanol yield at high lignocellulosic solids content and high ethanol titer. *Bioresour. Technol.* **2012**, *116*, 74–79. [CrossRef] [PubMed]

fermentation

MDPI

Review

Laccases as a Potential Tool for the Efficient Conversion of Lignocellulosic Biomass: A Review

Úrsula Fillat [1], David Ibarra [1,*], María E. Eugenio [1], Antonio D. Moreno [2], Elia Tomás-Pejó [3] and Raquel Martín-Sampedro [1]

[1]   Instituto Nacional de Investigación y Tecnología Agraria y Alimentaria–Centro de Investigación Forestal (INIA–CIFOR), Forestry Products Department, Ctra. de La Coruña Km 7.5, Madrid 28040, Spain; fillat.ursula@inia.es (Ú.F.); mariaeugenia@inia.es (M.E.E.); martin.raquel@inia.es (R.M.-S.)

[2]   Centro de Investigaciones Energéticas, Medioambientales y Tecnológicas (CIEMAT), Department of Energy, Biofuels Unit, Avda. Complutense 40, Madrid 28040, Spain; david.moreno@ciemat.es

[3]   Instituto Madrileño de Estudios Avanzados (IMDEA)–Energía, Biotechnological Processes for Energy Production Unit, Móstoles 28935, Spain; elia.tomas@imdea.org

*   Correspondence: ibarra.david@inia.es; Tel.: +34-91-347-3948

Academic Editor: Thaddeus Ezeji

Received: 26 March 2017; Accepted: 26 April 2017; Published: 2 May 2017

Abstract: The continuous increase in the world energy and chemicals demand requires the development of sustainable alternatives to non-renewable sources of energy. Biomass facilities and biorefineries represent interesting options to gradually replace the present industry based on fossil fuels. Lignocellulose is the most promising feedstock to be used in biorefineries. From a sugar platform perspective, a wide range of fuels and chemicals can be obtained via microbial fermentation processes, being ethanol the most significant lignocellulose-derived fuel. Before fermentation, lignocellulose must be pretreated to overcome its inherent recalcitrant structure and obtain the fermentable sugars. Usually, harsh conditions are required for pretreatment of lignocellulose, producing biomass degradation and releasing different compounds that are inhibitors of the hydrolytic enzymes and fermenting microorganisms. Moreover, the lignin polymer that remains in pretreated materials also affects biomass conversion by limiting the enzymatic hydrolysis. The use of laccases has been considered as a very powerful tool for delignification and detoxification of pretreated lignocellulosic materials, boosting subsequent saccharification and fermentation processes. This review compiles the latest studies about the application of laccases as useful and environmentally friendly delignification and detoxification technology, highlighting the main challenges and possible ways to make possible the integration of these enzymes in future lignocellulose-based industries.

Keywords: lignocellulosic biorefinery; delignification; detoxification; ethanol; fermentation; inhibitory compounds; laccase; lignin; pretreatment; saccharification

1. Introduction

Renewable fuels are considered promising alternatives to mitigate global warming and reduce our dependence on fossil fuels. In the particular case of transportation, ethanol is one of the few alternatives for the diversification of this sector in the short term, since it can be easily integrated into current fuel distribution systems [1]. Traditionally, certain food-related products including sugar crops and starch-based feedstocks have been used to produce ethanol. Alternatively, lignocellulosic biomass is an abundant and low-cost raw material that has no directly influence on food production [2]. Among them, forestry and agricultural residues (e.g., pine harvest forest, wheat straw, olive tree pruning, etc.), dedicated crops (e.g., elephant grass, forage sorghum, poplar, etc.), and municipal solid wastes are considered potential materials for ethanol production. Lignocellulosic biomass, in addition, is expected to provide a wide range of different renewable products such as food and feed additives, chemicals and materials. This lignocellulose-based industry—also known as biorefinery—is likely to become increasingly important in the future society as a complement and/or alternative to the current petroleum-based industry.

Biochemical conversion of lignocellulose represents the most favorable route among all developed technologies [3]. It includes three major steps: pretreatment, enzymatic hydrolysis, and fermentation. Pretreatment increases the accessibility of lignocellulose to hydrolytic enzymes by removing or modifying lignin and hemicellulose polymers, and by altering cellulose structure. Enzymatic hydrolysis or saccharification breaks down carbohydrates into fermentable sugars by the combined action of different enzyme activities. Finally, microorganisms convert sugars into alcohols, organic acids, alkenes, lipids or other chemicals through fermentation processes.

Focusing on pretreatment processes, several physical and/or chemical technologies have been developed and optimized for improving the conversion of a high number of lignocellulosic feedstocks [4]. During pretreatment, high pressures and temperatures and/or the addition of chemicals and solvents are in general required. These harsh pretreatment conditions lead to biomass degradation and generation of different enzymatic (mainly phenolic compounds) and microbial inhibitors (weak acids, furan derivatives and phenols), which limits the subsequent saccharification and fermentation steps [5]. Another factor that limits enzymatic hydrolysis is the residual lignin that remains in pretreated materials. Lignin hampers the accessibility of carbohydrates to hydrolytic enzymes by acting as a physical barrier; but also, it promotes the non-specific adsorption of hydrolytic enzymes to the lignin polymer, lowering the number of enzymes available for hydrolyzing carbohydrates and therefore decreasing saccharification yields [6].

To overcome the effects of lignocellulose-derived inhibitors and lignin, different detoxification and delignification processes have been evaluated [7,8]. Among them, the utilization of laccase enzymes has been widely investigated, showing to be effective in removing and/or modifying the lignin polymer, and in reducing the phenolic content of pretreated lignocellulosic materials [9,10]. The present work focuses on review the use of laccases as delignification and detoxification agents for the efficient conversion of lignocellulosic biomass into value-added products, with special accent in the lignocellulosic ethanol production.

2. Lignocellulosic Biomass Conversion: The Sugar Platform

The implementation of a sugar platform offers the possibility to obtain a high number of fuel and chemical products (alcohols, organic acids, alkenes, lipids and other chemicals) via fermentation processes [11]. With a high carbohydrate content, lignocellulosic biomass represents a promising sugar source for such an aim. Lignocellulosic sugars can be obtained either by acidolysis or via enzymatic hydrolysis, being the latter a preferred choice since it is more selective, it requires less energy (lower temperatures are needed), and it releases no harmful by-products [3]. However, the recalcitrant structure of lignocellulose hinders the accessibility of carbohydrates to hydrolytic enzymes and prevents the release of fermentable sugars. In this context, a pretreatment process is therefore needed to alter the structure of lignocellulose and thus facilitate an efficient enzymatic hydrolysis of carbohydrates [12].

The effectiveness of pretreatment processes for improving enzymatic hydrolysis of lignocellulosic biomass has been attributed to (1) hemicellulose removal; (2) lignin removal and redistribution [13]; (3) a reduction in the degree of polymerization and crystallinity of cellulose [14]; and/or (4) an increment in the porosity of pretreated materials [15]. Over the years, many different pretreatment methods have been investigated on a wide variety of feedstocks, being classified into physical, chemical, physicochemical, and biological pretreatments [4,16]. It is important to highlight that there is no best pretreatment technology and that the choice of the pretreatment method depends very much on the type and composition of the feedstock to be processed [17]. Among pretreatment technologies, chemical and physicochemical pretreatments are the most effective and promising processes for industrial applications [1]. Chemical methods, especially alkali- and acid-based pretreatments, are low cost processes and have shown to effectively remove hemicellulose and lignin from lignocellulosic feedstocks. Physicochemical pretreatments (e.g., steam explosion, liquid hot water, ammonia fiber explosion/expansion, wet oxidation, etc.), on the other hand, are also low cost technologies but with a lower environmental impact compared to chemical technologies [4]. These methods are capable of solubilizing hemicellulose, disrupting the structure of lignocellulose and increasing the accessible surface area of pretreated substrates. Other pretreatment technologies including milling, organosolv, and ionic liquids (ILs) can also significantly improve the digestibility of lignocellulosic materials [1]. Nevertheless, their high operational costs represent an important limitation for their commercial applications.

After pretreatment, enzymatic hydrolysis is responsible for breaking down lignocellulose-contained carbohydrates. It is in overall a crucial step that highly influences final process yields. Due to the complex structure and the heterogeneous composition of lignocellulose, a high number of enzymatic activities including cellulases, hemicellulases, and ligninases are needed for its complete hydrolysis [18]. Cellulases (endoglucanases, cellobiohydrolases, and β-glucosidases) hydrolyze cellulose into glucose monomers, while hemicellulases (e.g., xylanases, β-xilosidases, α-L-arabinofuranosidases, esterases, etc.) and ligninases (e.g., laccases, peroxidases, reductases, oxidases generating H_2O_2, etc.) depolymerize hemicellulose and lignin, respectively. Major limitations of the enzymatic hydrolysis are the costs for enzyme production and the necessity of providing the appropriate enzyme mixtures. Although significant advances have been achieved to overcome these limitations, the enzymatic mixtures and the enzyme production process still need to be optimized. This optimization involves the use of low-cost substrates and/or the inclusion of novel enzymatic activities, such as the non-hydrolytic proteins swollenins and expansins, and the polysaccharide monooxygenases (LPMOs) [19,20]. In addition, recent studies also aim at increasing the catalytic efficiency of hydrolytic enzymes, by screening and/or engineering of enzyme-producing microorganisms, while other studies aim at cost reduction by enzyme recycling [18].

The corresponding sugars obtained after enzymatic hydrolysis can be potentially converted to a large number of products via microbial fermentation processes. Among them, the sugar-to-ethanol conversion process has been the most widely studied. Three main process configurations have been described for ethanol production, including separate hydrolysis and fermentation (SHF), simultaneous saccharification and (co)fermentation (SSF/SSCF) and consolidating bioprocessing (CBP) [21]. SSF/SSCF processes integrate the enzymatic hydrolysis and the fermentation stages in a single step, which has shown to be beneficial for improving conversion efficiencies. During these processes, the introduction of a presaccharification step (PSSF/PSSCF) to liquefy the media prior yeast addition is especially suitable when working at high substrate loadings [18]. Several yeast, bacterial or fungal strains have been used for fermentation of lignocellulosic-based streams. Among them, the yeast *Saccharomyces cerevisiae* is the most commonly employed microorganism, especially in the alcohol industry. *S. cerevisiae* can utilize all kind of hexoses to produce ethanol, reaching conversion yields close to the theoretical. However, its inability to metabolize pentoses has led to the exploration and development of novel fermenting microorganisms with the capacity to convert all kind of sugars to ethanol [22]. Besides the capacity of utilizing a wide range of sugars, it is important that the fermenting microorganism also shows high tolerance to inhibitory compounds, temperatures, ethanol and/or mechanical and osmotic stress.

3. Inhibitors and Lignin in Pretreated Materials

Pretreatment of lignocellulose often involves side reactions resulting in the release of certain biomass-derived by-products that are inhibitors of downstream biochemical processes [5]. They mainly include furan derivatives, aliphatic acids, and phenolic and other aromatic compounds (Figure 1). Extractives (mainly terpenes, fats, waxes, and phenolics) and inorganic compounds may also promote inhibition of enzymes and microorganisms in the subsequent steps [1]. The nature and concentration of all these inhibitory products is strongly dependent on the feedstock as well as the pretreatment process [23].

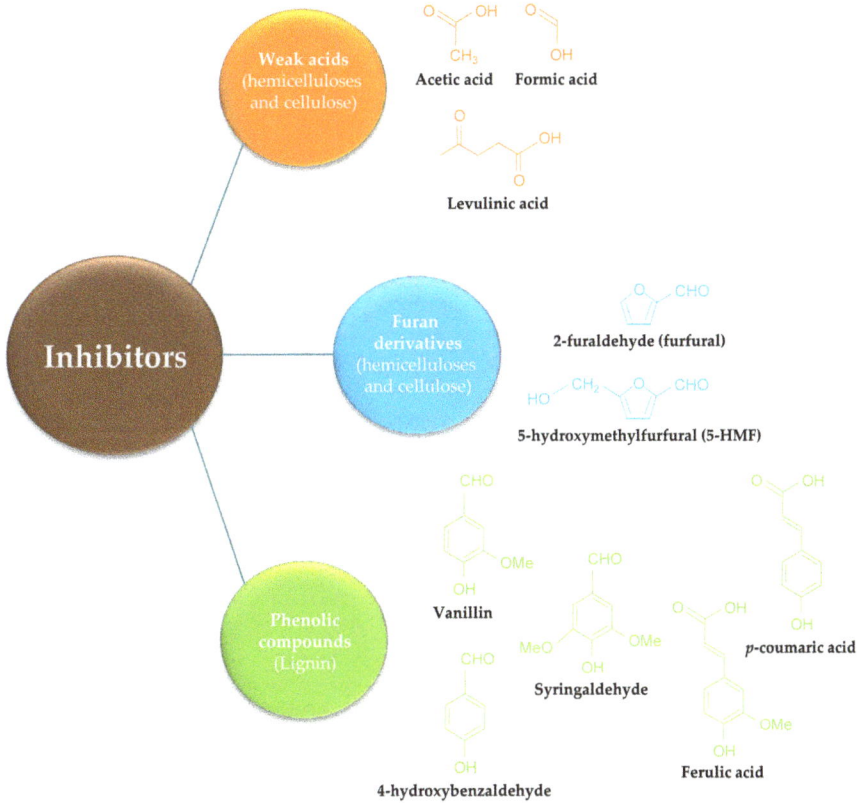

Figure 1. Common inhibitory compounds present in lignocellulosic pretreated materials, indicating main sources of its formation.

During pretreatment processes, the pentoses resulting from hemicellulose can undergo dehydration with formation of furfural, while hexoses can be dehydrated to 5-hydroxymethylfurfural (5-HMF). In addition, furan derivatives can be further degraded to form levulinic acid and formic acid, depending on the severity of the pretreatment process. From hemicelluloses, acetic acid can be also generated from the acetyl groups, while a large number of phenolic compounds, such as 4-hydroxybenzoic acid, 4-hydroxybenzaldehyde, vanillin, dihydro-coniferyl alcohol, coniferyl aldehyde, syringaldehyde, syringic acid, *p*-coumaric acid, ferulic acid, and Hibber's cetones, can be produced from lignin [5,24].

The inhibitory effects caused by degradation compounds can be observed in both hydrolytic enzymes and fermentative microorganisms [5,24–26]. Furan derivatives are one of the most important microbial inhibitors during fermentation. They affect cell viability and growth rates, extend the lag phase at the initial stage of the fermentation process, and lower ethanol yields and productivities. These effects derived from the inhibition of several intercellular enzymes (such as alcohol dehydrogenase and pyruvate dehydrogenase) and from the damage promoted to cell membranes and/or to genetic materials [5,24]. Carboxylic acids also affect biomass growth and ethanol production by mainly promoting the intracellular accumulation of H$^+$ ions. Among the main biomass-derived carboxylic acids, formic acid has a greater inhibitory effect than levulinic acid, which in turn has shown to have a greater impact than acetic acid [5,24]. The undissociated form of carboxylic acids can diffuse through cell membranes and once inside the cell they are dissociated due to an increase in the pH (the pH increases from about 5 to 7). As a consequence, H^+ ions are accumulated, lowering the intracellular pH and causing an imbalance in the ATP/ADP ratio by the increase in the activity of ATP/H^+ pumps. At last, phenolic compounds have shown to affect microbial growth and reduce ethanol production rates, but not ethanol yield. Usually, this group of lignocellulosic-derived compounds causes loss of membrane integrity and affects specific intracellular enzymatic activities [5,24]. Regarding to hydrolytic enzymes, phenols are the main degradation compounds that inhibit and deactivate them, reducing both rates and yields during the saccharification step [25,26]. Thus, vanillin and syringaldehyde have shown to inhibit cellulases—and in particular β-glucosidases—, while ferulic acid and *p*-coumaric acid are capable of deactivate them. Nonetheless, cellobiose, glucose, and sugars from hemicellulose have been also shown to inhibit hydrolytic enzymes [18].

In addition to the inhibitory compounds, the residual lignin present in pretreated materials represents an important limiting factor during enzymatic hydrolysis of carbohydrates. Lignin constitutes a physical barrier that may unspecifically adsorb hydrolytic enzymes, decreasing the enzyme concentration during the saccharification process [6]. Lignin polymer is built up of *p*-hydroxyphenyl (H) (derived from *p*-coumaryl alcohol), guaiacyl (G) (derived from coniferyl alcohol), and syringyl (S) (derived from sinapyl alcohol) phenylpropanoid units and their acylated forms [27]. The G:S:H unit proportion varies depending on biomass feedstock. Softwood lignin is mainly composed of G units with small proportions of H units, whereas lignin in hardwood contains mainly S and G units. Lignin from non-woody plants, such as agricultural residues, also contains H units together with G and S units [27]. As can be observed in Figure 2, lignin units are linked through a variety of inter-unit linkages including C–C and ether bonds [28]. Among them, the most abundant inter-unit linkages are β-O-4′ (aryl ether), β-5′ (phenylcoumaran), and β-β′ (resinol) bonds. Other structural links such as β-1′ (spirodienone), 5-5′-O-4 (dibenzodioxocin), 5-5′ and 4-O-5′ bonds have been also described. In addition to the interaction between lignin units, lignin-carbohydrate complexes (LCC) are also formed in plant cell walls [28]. The main types of LCC linkages in lignocellulosic materials are phenyl glycoside, ether, or ester bonds.

It has been suggested that the chemical and physical structure of lignin plays an important role during enzymatic hydrolysis. Lignin structure is, in turn, highly dependent on biomass feedstock and/or on pretreatment conditions [29]. For instance, steam-explosion pretreatment produces great reductions in β-O-4′ linkages, resulting in partial lignin solubilization and the release of free phenolic groups [30,31]. Moreover, lignin repolymerization can also take place [32], increasing the number of aromatics substitutions at the C_6. Depending on pretreatment temperature and time, an increase of phenolic hydroxyl groups and a decrease in aliphatic hydroxyl groups can also be observed [33].

Different mechanisms including hydrophobic, electrostatic and hydrogen bonding interactions have been proposed to explain the inhibition of hydrolytic enzymes by lignin [15]. However, the actual mechanism by which hydrolytic enzymes interact with lignin and become inhibited has yet to be fully elucidated. One of the most common accepted explanations is related to an increase in lignin phenolic groups and hydrophobicity (resulted by a lower amount of carboxylic groups and aliphatic hydroxyl groups), which promotes enzyme adsorption to the lignin polymer [15]. This hypothesis is supported by Sewalt et al. [34], who reversed the inhibitory mechanism of organosolv-pretreated lignin by hydroxypropylation of the phenolic groups. Moreover, the addition of surfactants and certain polymers (e.g., tween, bovine serum albumin, polyethylene glycol, gelatin, etc.) has shown to reduce the unspecific adsorption of hydrolytic enzymes to lignin as they can bind to the adsorption sites, improving saccharification yields [34,35].

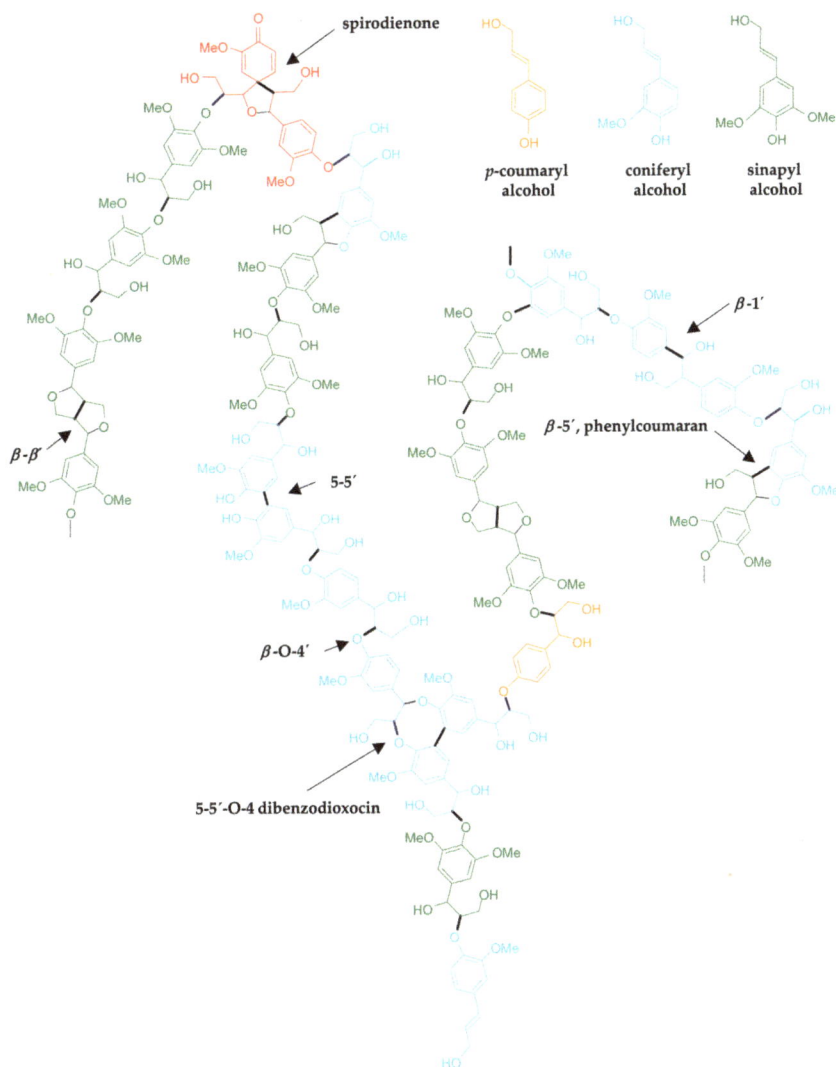

Figure 2. Schematic representation of lignin structure showing the main interunit linkages originated from *p*-coumaryl, coniferyl, and sinapyl alcohols.

The presence of inhibitors and residual lignin makes detoxification and delignification processes powerful tools for improving saccharification and fermentation of pretreated lignocellulosic biomass.

3.1. Detoxification of Pretreated Materials

A detoxification step prior to enzymatic hydrolysis and/or fermentation of pretreated materials may reduce the concentration of inhibitory compounds, enhancing saccharification and conversion yields. Filtration and washing processes have been widely used for this purpose. However, these methods involve additional and expensive steps, waste of water and loss of soluble sugars [36]. As alternative to filtration and washing, several detoxification technologies have been developed to overcome the effects of inhibitory compounds of pretreated materials [7–9].

Vacuum evaporation is capable of reducing volatile compounds such as furfural, acetic acid and vanillin [37]. Solvents (e.g., ethyl acetate) and active charcoal and/or ion-exchange resins reduce the concentration of inhibitors by extraction or adsorption, respectively [38–40]. Chemical transformation of inhibitors is also possible by addition of reducing agents (dithionite and sulfite) [41] and chemical catalysts, being overliming (treatment with $Ca(OH)_2$) the most efficient chemical detoxification method for removing phenols and furan derivatives [37].

Biological detoxification involves the use of microorganisms and/or their enzymes to decrease the inhibitory effects of degradation compounds. In comparison to physico-chemical detoxification processes, biological detoxification methods are advantageous as they have lower energy requirements, they take place at milder reaction conditions, they need no chemical addition and they have fewer side-reactions [9,10]. Among different microorganisms, fungi such as *Trichoderma reesei* have the ability to remove different inhibitory compounds. Larsson et al. [37] evaluated this fungus to detoxify a diluted-acid hydrolysate from spruce, observing an important removal of furans and a small proportion of phenols. Furthermore, *T. reesei* can produce hydrolytic enzymes while detoxification takes place. In this sense, Palmqvist et al. [42] used *T. reesei* to remove phenolic compounds, furan derivatives and aliphatic acids from acid-catalyzed steam-pretreated willow, simultaneously obtaining 0.2–0.6 IU/mL of cellulase activity. Besides fungi, several bacteria and yeasts have been also used for detoxification purposes [10]. For instance, the thermophilic bacterium *Ureibacillus thermophaercus* was employed to remove furfural and 5-HMF and phenolic compounds from a waste house wood hydrolysate [43], increasing markedly the ethanol production rate by *S. cerevisiae* in a subsequent fermentation stage. The yeast *S. cerevisiae* has also the natural ability to assimilate some of these inhibitory compounds –mainly furfural, 5-HMF and aromatic aldehydes such as vanillin, syringaldehyde or 4-hydroxybenzaldehyde– and convert them into less inhibitory forms [44,45]. Furthermore, this innate capacity can be improved by subjecting *S. cerevisiae* to evolutionary engineering in the presence of inhibitory compounds, boosting its fermentation performance in lignocellulosic pretreated materials [46]. Strategies such as genetic modification also offer the possibility to introduce a particular characteristic that is not present naturally in a certain microorganism. The yeast tolerance towards inhibitors has been improved by homologous or heterologous overexpression of certain genes. Larsson et al. [47] improved the tolerance of *S. cerevisiae* to phenylacrylic acids by overexpression of *Pad1p* gene (encoding a phenylacrylic acid decarboxylase). This genetically modified strain was capable of metabolizing different cinnamic acids from a spruce hydrolysate, showing higher growth rates and ethanol productivities. Similarly, Petersson et al. [48] overexpressed the gene *ADH6p* (which encodes an NADPH-dependent alcohol dehydrogenase enzyme with ability to reduce furfural and 5-HMF) on *S. cerevisiae*, increasing microbial conversion rates of 5-HMF in both aerobic and anaerobic cultures. Besides evolutionary or genetic engineering modifications, strategies such as cell retention, flocculation, and encapsulation of the fermenting microorganism have been also assessed to increase the intrinsic tolerance or the inherent detoxification capacity of some strains [9].

3.2. Delignification of Pretreated Materials

Together with detoxification processes, delignification is considered an important step for improving enzymatic saccharification of lignocellulosic biomass. Some traditional pretreatments methods such as alkaline, organosolv, and oxidative processes have been developed to target lignin removal. Biological delignification has also shown to be efficient in reducing the lignin content of lignocellulosic feedstocks. In contrast to physico/chemical delignification processes, biological methods are promising alternatives due to the lower environmental impact and the resulting higher product yield in the subsequent saccharification and fermentation steps. Biodelignification involves lignin removal/modification, the increase in the number of pores and the available surface area, and the reduction in the non-productive binding of hydrolytic enzymes. Wood-decaying fungi are the sole organisms in nature capable of degrading the lignin polymer, making the carbohydrates of lignocellulose accessible to cellulolytic enzymes [49]. Microbial lignin attack is an extracellular and oxidative process that involves different oxidoreductase enzymes: ligninolytic peroxidases (lignin peroxidase (LiP), manganese peroxidase (MnP), versatile peroxidase (VP), and dye-decolorizing peroxidase (DyP)), laccases, oxidases for the production of extracellular H_2O_2 (glyoxal oxidase, pyranose-2 oxidase, and aryl-alcohol oxidase), and dehydrogenases (aryl-alcohol dehydrogenase, and quinone reductase). Along with oxidoreductases, certain low molecular weight compounds play an important role, acting as mediators in some reactions [49]. Among peroxidases, LiP and MnP were first discovered in *Phanerochaete chrysosporium* and are capable of degrading non-phenolic (about 70–90%) and phenolic lignin units [49–51]. Regarding VP, it was first described in *Pleorotus* sp. [52,53], and combines properties from both LiP and MnP enzymes. DyP has been recently discovered during fungal pretreatment of wheat straw with *Irpex lacteus* [54], showing the ability to degrade non-phenolic lignin compounds. Finally, laccases can only address direct oxidation of phenolic compounds due to their lower redox potential [49]. However, in the presence of redox mediators, laccases can also degrade non-phenolic lignin units, as it is discussed in the following section.

Different wood-decaying fungi have been widely explored for biological delignification, being "white-rot" basidiomycetes (e.g., *P. chrysosporium*, *Trametes versicolor*, *Ceriporiopsis subvermispora*, *I. lacteus*, *Pleurotus ostreatus*, *Cyathus stercoreus*, etc.) the most efficient microorganisms for this purpose [9,10]. *T. versicolor* was grown on steam-exploded wheat straw for 40 days, resulting in 55.4% lignin degradation compared with the 20% obtained after steam-explosion treatment alone [55]. Salvachúa et al. [56] combined mild alkaline extraction with microbial delignification to reduce the lignin content of wheat straw. When using *C. subvermispora* and *I. lacteus*, 30% and 34% lower lignin content was measured, respectively, after 21 days of incubation. The lower lignin content increased the cellulose available for subsequent processing and conversion to around 66–69%, allowing to obtain 69% ethanol yields during the fermentation process. Microbial delignification was also studied with *P. ostreatus* on H_2O_2-pretreated rice hull [57]. This pretreatment combination increased the delignification range about two times, leading to 49.6% of glucose yield in the subsequent saccharification step. Although only "white-rot" basidiomycetes can degrade lignin extensively, certain ascomycetes can also colonize lignocellulosic biomass, showing to be beneficial for the subsequent saccharification step. Martín-Sampedro et al. [58] reported for the first time the ability of new endophytic fungi to enhance saccharification of autohydrolysis-pretreated eucalypt wood. Two of the evaluated fungi, *Ulocladium* sp. and *Hormonema* sp., produced a slight delignification in comparison to autohydrolysis pretreament alone, showing 8.5 and 8.0 times higher saccharification yields. Eventually, certain bacterial strains such as *Bacillus macerans*, *Cellulomonas cartae*, and *Zymomonas mobilis* are also capable of delignifying lignocellulosic feedstocks [59], yielding lignin degradation up to 50%.

In spite of the ability of ligninolytic microorganisms for delignification, treatment time as well as white-rot pattern must be taken into consideration for an efficient microbial delignification. Incubation time can vary from days to weeks, which depends on the strain used. An increment of lignin removal from 17% to 47% was reported when the residence time of wheat straw treatment with *Panus tigrinus* was increased from 7 days to 3 weeks (from 15% to 34% using *Coriolopsis rigida*) [56]. In terms of pattern lignocellulose deconstruction by microorganisms, selective delignification (sequential decay) should be favored against simultaneous cellulose and lignin degradation (simultaneous rot) to avoid carbohydrate consumption during microbial treatment [49]. These patterns vary among species and strains. Then, some fungi, such as *P. tigrinus* and *Phlebia radiata*, degraded lignin and sugars simultaneously in wheat straw; whereas *Pleurotus eryngii* was able to remove lignin selectively and faster than the carbohydrate components [56].

4. Outline of Laccase Enzymes

The use of ligninolytic enzymes, especially laccases, is an attractive method and an alternative to the use of microorganisms for detoxification and delignification of pretreated materials (Figure 3). These enzymes are substrate specific and offer the possibility to increase conversion rates and yields during saccharification and fermentation processes, reducing detoxification and delignification times from weeks to hours and avoiding carbohydrate consumption [9]. Laccases enzyme was first isolated from sap of the Japanese lacquer tree *Rhus vernicifera* [60]. Afterwards, laccases have been widely described in higher plants, fungi, insects, and bacteria [61], being their production a characteristic distinctive of "white-rot" basidiomycetes [49], and some ascomycetes [62]. In plants, laccases are involved in the biosynthesis of lignin by inducing radical polymerization of the phenylpropanoid units. In contrast, in wood-decaying fungi laccases play a key role in lignin degradation [27].

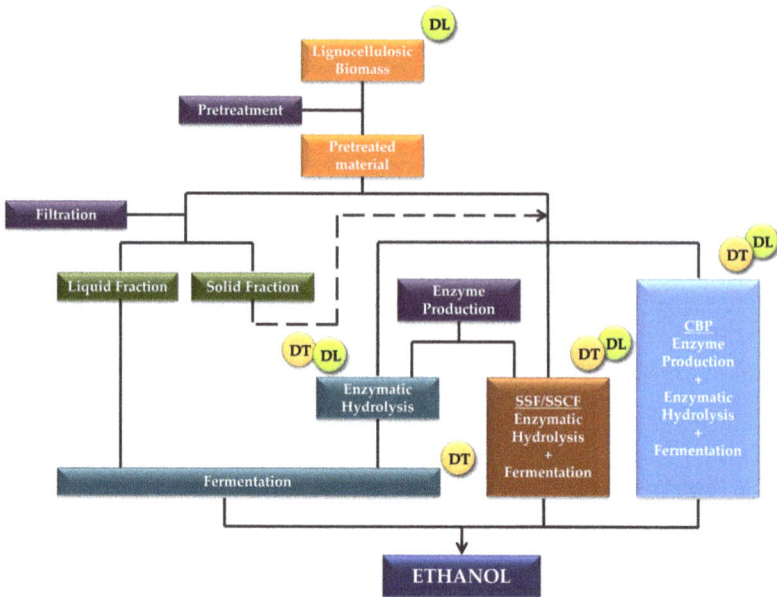

Figure 3. Schematic representation of lignocellulosic ethanol production showing (1) the different process configurations, and (2) the points where laccase delignification (DL) and laccase detoxification (DT) can be applied. The scheme can also be extended to the generation of several fermentation-based products including different alcohols, lipids, alkenes and other chemicals. SSF, simultaneous saccharification and fermentation; SSCF, simultaneous saccharification and co-fermentation; CBP, consolidated bioprocessing. Dashed line arrow represents the flow of the solid fraction after a water washing step.

Laccases (benzenediol:oxygen oxidoreductases, EC 1.10.3.2) are multicopper-containing oxidases with phenoloxidase activity, which catalyze the oxidation of substituted phenols, anilines and aromatic thiols, at the expense of molecular oxygen [63]. The catalytic site of laccases involves four copper ions. Type-T1 copper (blue copper) is implicated in the oxidation of the reducing substrate, acting as the primary electron acceptor. Type-T2 copper together with two type-T3 coppers form a tri-nuclear copper cluster where the transferred electrons reduce the molecular oxygen to water. Electrochemical potential of type-T1 copper is one of the most significant features of laccases and might vary from 0.4 to 0.8 V [49]. Plant and bacterial laccases have comparatively low redox potential, whereas the highest values are generally reported for fungal laccases [64]. This redox potential allows the direct oxidation of some substrates by laccases, including the phenolic part of lignin (less than 20% of lignin polymer). However, potential substrates too large to enter the laccase catalytic site or with redox potential about 1.3 V cannot be oxidized directly by laccases.

Laccase-Mediator Systems (LMS)

The inability of laccases for the oxidation of complex lignocellulosic substrates or with high redox potential, such as non-phenolic lignin, can be overcome by using redox mediators in the so-called laccase-mediator systems (LMS). Certain low molecular compounds forming stable radicals that act as redox mediators, expand the catalytic activity of laccases towards more recalcitrant compounds which are not oxidized by laccase alone [65,66]. ABTS (2,2′-azino-bis (3 ethylbenzothiazoline-6-sulfonic acid)) was the first chemical molecule described as laccase mediator for oxidation of non-phenolic lignin model compounds [66], following the electro transfer (ET) route for the oxidation of the target substrate [67]. Since then, new chemical mediators have been proposed for this purpose. Among them, the N–OH mediators such as 1-hydroxybenzotriazole (HBT), N-hydroxyphthalimide (HPI), violuric acid (VLA) or N-hydroxyacetanilide (NHA) have been described as the most efficient chemical mediators for the oxidation of recalcitrant compounds [68,69], performing the radical hydrogen atom transfer (HAT) route as oxidation mechanism [67]. These N–OH compounds have been successfully applied for delignification and bleaching of paper pulps, being the laccase-HBT system particularly effective in woody and non-woody pulp bleaching and delignification [70,71]. Moreover, decolorization of industrial dyes or detoxification of pollutants are another fields where the applicability potential of laccase-mediator systems has been comprehensively demonstrated [67,72].

Nevertheless, the high cost of chemical mediators and the generation of possible toxic species hamper the use of laccase-mediator systems at industrial scale. Consequently, the search of cheaper and environmental-friendly natural mediators has increased in the last years [67]. In this context, lignin-derived phenolic compounds obtained from lignocellulose biodegradation or as by-product or residue during the own industrial process of biomass conversion (e.g., from the black liquors of paper pulp industry) have been identified as potential natural mediators. A set of such compounds, including acetosyringone, syringaldehyde, vanillin, and *p*-hydroxycinnamic acids have been successfully applied in dye decolorization, delignification and bleaching of paper pulps, and removal of lipophilic extractives [73–75]. Similar to HBT, the HAT route is the mechanism by which the phenoxy radicals from these natural mediators oxidize the target substrate [67].

5. Application of Laccases for Detoxification of Pretreated Materials

5.1. Detoxification Mechanism

Laccases have been largely used to diminish the toxicity of different pretreated substrates (Table 1). These enzymes catalyze the selective oxidation of phenolic compounds generating unstable phenoxy radicals without affecting furan derivatives and aliphatic acids [37]. These phenoxy radicals further interact with each other and lead to the polymerization into aromatic compounds with lower inhibitory capacity [76]. It is important to highlight that not all phenolic compounds are susceptible to oxidation by laccase enzymes. Kolb et al. [77] described different catalytic activities for *T. versicolor* laccase when acting on phenolic compounds released from liquid hot water pretreatment of wheat straw. Thus, complete removal of syringaldehyde, *p*-coumaric acid and ferulic acid was achieved within 1-hour treatment, while vanillin was only removed after 24-h treatment, and 4-hydroxybenzaldehyde did not vary its concentration within 1-week reaction time in the presence of laccase. These reaction mechanisms are determined by the structure of the different phenolic compounds [67]. Laccase activity toward phenols is improved by the presence of electron-donating substituents in the ring and these substituents decrease the electrochemical potential of the corresponding phenols. Then, an additional methoxy group (the structural difference between vanillin and syringaldehyde) increases the affinity of the phenolic compounds toward laccase. Furthermore, the presence of ethylene groups in para-substituted phenols, such as *p*-coumaric and ferulic acids, also increases the activity of laccase [73,78].

Table 1. Application of laccase enzymes for detoxification of different pretreated materials.

Pretreated Material	Laccase Treatment	Effects Observed	Benefits Produced	Reference
Steam-exploded rice straw	*Coltricia perennis*	Removal of phenolic compounds by 76%	Increased saccharification yield by 48%	[79]
Steam-exploded wheat straw	*Pycnoporus cinnabarinus* or *Trametes villosa*	Removal of phenols identified (vanillin, syringaldehyde, ferulic acid and *p*-coumaric acid) by 93–95% with both laccases	Improved the fermentation performance of *Kluyveromyces marxianus* CECT 10875, shortening its lag phase and enhancing the ethanol yields	[80]
SO_2 steam-pretreated willow	*Trametes versicolor*	Removal of phenolic compounds (93–95%), revealing an oxidative polymerization mechanism by SEC analysis	Higher yeast growth, glucose consumption rate, ethanol productivity and ethanol yield using *Saccharomyces cerevisiae*	[81]
Dilute acid steam-pretreated spruce	*T. versicolor*	Removal of phenolic compounds by 93–95%	Ethanol yield produced by *S. cerevisiae* comparable with that obtained after detoxification with anion exchange chromatography at pH 10	[37]
Steam-exploded wheat straw	Commercial bacterial laccase MetZyme®	Phenol reduction of 18% (laccase alone) and 21% (simultaneous laccase and presaccharification)	Improved the fermentation performance of *K. marxianus* CECT 10875 during SSF and PSSF processes, shortening the adaptation phases and the overall fermentation times	[82]
Water and acid-impregnated steam-exploded wheat straw	*T. versicolor* or *Coriolopsis rigida*	Removal of phenolic compounds by 93–95% with both laccases	Reduction of the toxic effects on *S. cerevisiae*, resulting in higher yeast growth and improved ethanol production	[76]
Steam-exploded wheat straw	*P. cinnabarinus*	Phenol reduction around 67% (laccase alone) and 73% (simultaneous laccase and presaccharification)	Laccase detoxification allowed to obtain ethanol concentrations and yields with *K. marxianus* CECT 10875 comparable to those obtained with *S. cerevisiae*	[83]
Steam-exploded wheat straw	*P. cinnabarinus*	Removal of phenolic compounds by 95%	Improvement of cell growth and ethanol production of *S. cerevisiae* during SSF process	[84]
Steam-exploded sugarcane bagasse	*T. versicolor*	Approximately 80% of the phenolic compounds removal	Improvements in ethanol yield and ethanol volumetric using a xylose-utilizing *S. cerevisiae*	[85]
Steam-exploded sugarcane bagasse	*Ganoderma lucidum* 77002	84% of the phenolic compounds in prehydrolysate	Ethanol yield was improved when *S. cerevisiae* was used on detoxified prehydrolysate	[86]
Alkali-extracted sugarcane bagasse	*Aspergillus oryzae*	Not observed	Laccase improved the fermentation efficiency by 6.8% for one-pot SSF and 5.7% for SSF	[87]
Acid hydrolyzed from sugarcane bagasse	*Cyathus stercoreus*	Reduction of 77.5% of total phenols	Improvements in the performance of *Candida shehatae* NCIM 3501	[88]
Steam-exploded wheat straw	*P. cinnabarinus*	Phenol reduction around 44% (laccase alone) and 95% (simultaneous laccase and presaccharification) at 12% (w/v) of substrate loading	Laccase detoxification triggered the fermentation by *K. marxianus* of steam-exploded material at 12% (w/v), resulting in an ethanol concentration of 16.7 g/L during SSF process	[89]
Steam-exploded wheat straw	*P. cinnabarinus*	Reduction of total phenolic compounds by 50–80%	Laccase detoxification allowed the fermentation of pretreated material at 20% (w/v) of substrate loading using the evolved xylose-consuming yeast *S. cerevisiae* F12, producing more than 22 g/L during SSCF process	[90]
Steam-exploded wheat straw	*P. cinnabarinus*	Approximately 73–81% of the phenolic compounds removal	Laccase detoxification improved cell viability of the evolved xylose-recombinant *S. cerevisiae* KI6-12, and increased the ethanol production up to 32 g/L when fed-batch SSCF process was used at 16% (w/v) of substrate loading	[91]
Steam-exploded wheat straw	*P. cinnabarinus*	Phenols removal by 53% during simultaneous laccase and presaccharification at 25% (w/v) of substrate loading	Ethanol production of 58.6 g/L at 48 h with detoxified material at 25% (w/v) of substrate loading during PSSF process with *S. cerevisiae*	[92]

Table 1. *Cont.*

Pretreated Material	Laccase Treatment	Effects Observed	Benefits Produced	Reference
Dilute-acid spruce hydrolysate	*T. versicolor* expressed in a recombinant *S. cerevisiae* strain	Reduction of low-molecular of phenolic compounds	Laccase-producing transformant was able to ferment at a faster rate than the control transformant	[93]
Organosolv pretreated wheat straw	*T. versicolor* immobilized on both active epoxide and amino carriers	Higher phenols removal (82%) efficiency with laccase immobilized on active amino carrier	Better performance of *Pichia stipitis* during fermentation and reusability of immobilized laccase	[94]
Steam-exploded wheat straw	*T. villosa* or a bacterial laccase from *Streptomyces ipomoeae*	Phenol content reduction of 29% and 90% with bacterial and fungal laccases, respectively	Improvement performance of *S. cerevisiae* during SSF and PSSF process	[95]

SEC, Size exclusion chromatography; SSF, simultaneous saccharification and fermentation process; PSSF, presaccharification and simultaneous saccharification and fermentation process; SSCF, simultaneous saccharification and co-fermentation process. Generally, laccases source is fungal, except in those cases where it is indicated.

Fermentation **2017**, *3*, 17

Incomplete phenols removal has been widely described with different high redox fungal laccases. Kalyani et al. [79] achieved a phenol removal of 76% when steam-exploded whole slurry from rice straw was treated with *Coltricia perennis* laccase. Moreno et al. [80] reported higher phenol reductions (93–95%) when *Pycnoporus cinnabarinus* and *Trametes villosa* laccases were used to detoxify steam-exploded wheat straw. Similar ranges were observed by Jönsson et al. [81] with acid steam-pretreated willow and *T. versicolor* laccase, and by Jurado et al. [76] with both water and acid-impregnated steam-exploded wheat straw and *T. versicolor* and *C. rigida* laccases. Together with the structure of phenols, the redox potential of laccases also determines the grade of action toward them. Then, low redox potential laccases, a particular property of bacterial laccases [96], show minor reactivities on phenols [97]. In this sense, Moreno et al. [82] described a lower phenol reduction of 21% when a commercial bacterial laccase (MetZyme®, Kaarina, Finland) was used to reduce the toxicity of a whole slurry from steam-exploded wheat straw. Finally, other factors, such as the viscosity of the medium in which the laccase detoxification is implemented also affects the laccase efficiency. Higher viscosity when higher solids content is used difficult the blending of laccase with the pretreated material, consequently reducing the laccase efficiency [90].

Laccase detoxification is usually performed either by using a partially purified laccase [88], or with a totally purified enzyme [83]. Nevertheless, culture enriched in laccase activity has been also successfully proved [79]. The treatments can be carried out at a wide range of optimal pH and temperature depending of laccases source. Then, the treatment of steam-exploded wheat straw with a fungal laccase from *T. villosa* at optimal pH 4 removed 90% of phenols, while a reduction in the phenol content of 29% was achieved with a bacterial laccase from *S. ipomoea* at optimal pH 8 [95]. Regarding to temperature, Moreno et al. [80] reported phenols reduction around of 94% when steam-exploded wheat straw was treated with laccases from *P. cinnabarinus* and *T. villosa* at their optimal temperatures of 50 and 30 °C, respectively. The treatment time and the enzyme loading at which the detoxification is carried out are also two important factors. Moreno et al. [92] obtained similar phenols reduction, 65% and 53%, in steam-exploded wheat straw using *P. cinnabarinus* laccase after 3 h and 12 h of treatment, respectively. In terms of enzyme loading, laccase can be added at low or high loadings, depending on process optimization and material type. Then, only 1.5 U/mL of a laccase from *C. perennis* was enough to remove 77.5% of total phenols from acid steam-exploded rice straw [79]; whereas a higher enzyme loading (100 times more) of *C. stercoreus* laccase was necessary to remove the same phenols range from sugarcane bagasse hydrolysate [88].

5.2. Detoxification and Fermentation

S. cerevisiae, the most commonly employed microorganism for ethanol production, has been also largely used to evaluate the effects generated by laccase detoxification. Jönsson et al. [81] and Larsson et al. [37] reported higher yeast growth together with higher glucose consumption rate, ethanol productivity, and ethanol yield when liquid fractions from acid steam-exploded wood were detoxified by *T. versicolor* laccase. Similarly, Moreno et al. [83,84] used *P. cinnabarinus* laccase to detoxify steam-exploded wheat straw, observing higher cell viability and shorter lag phase during SSF and PSSF processes. Jurado et al. [76] also described a greater influence on ethanol concentration and yeast growth when both enzymatic hydrolyzed from water and acid-impregnated steam-exploded wheat straw were treated with *T. versicolor* and *C. rigida* laccases. On the other hand, Martín et al. [85] explored the use of *T. versicolor* laccase to detoxify a steam-exploded sugarcane bagasse hydrolysate, resulting in improved ethanol yield and ethanol volumetric productivity by using a recombinant xylose-utilizing *S. cerevisiae* strain. Steam-exploded sugar cane bagasse prehydrolysate was also detoxified by Fang et al. [86] with *Ganoderma lucidum* laccase, resulting in improved yeast growth and ethanol yield. Finally, one-pot SSF process with alkali-extracted sugar cane bagasse was carried out with *Aspergillus oryzae* laccase, improving the fermentation efficiency by 6.8% [87].

In addition to *S. cerevisiae*, similar effects derived from laccase detoxification have been also reported in other fermenting yeasts. Chandel et al. [88] observed an improvement in the performance of *Candida shehatae* during the fermentation of an acid hydrolysate from sugarcane bagasse treated with *C. stercoreus* laccase. Moreno et al. [83] described similar ethanol concentrations and yields comparable to those obtained by *S. cerevisiae* when steam-exploded wheat straw was detoxified by *P. cinnabarinus* laccase and fermented with the thermotolerant yeast *Kluyveromyces marxianus* CECT 10875. This thermotolerant yeast was also used by Moreno et al. [82] during both SSF and PSSF processes of steam-exploded wheat straw detoxified with the bacterial laccase MetZyme®. In this case, a shorter adaptation phase and an increase in cell viability could be observed in laccase-treated samples. This result is of special relevance, since the use of thermotolerant yeasts lead to a better integration of both saccharification and fermentation processes. Saccharification has an optimum temperature around of 50 °C, whereas most fermenting yeasts have an optimum temperature ranging from 30 to 37 °C [98]. The use of thermotolerant microorganisms such as *K. marxianus*, with capacity of growing and fermenting at temperature above 40 °C, represents therefore an advantage to obtain higher saccharification and fermentation yields [99]. In addition, the use of thermotolerant strains has shown to reduce overall process costs due to the reduction cooling costs.

Another strategy to reach higher ethanol concentrations and make the process more economically viable is to operate saccharification and fermentation processes at high-substrate consistencies. This approach offers possibilities to reduce freshwater consumption and downstream processing, and minimize energy consumption during subsequent distillation—due to the higher ethanol concentrations after fermentation—and evaporation stages [100]. Nevertheless, increasing the substrate consistency presents some disadvantages such as accumulation of glucose and cellobiose (that inhibits hydrolytic enzymes), mixing and mass transfer limitations, and larger concentration of inhibitors in the fermentation medium [101]. In this context, laccase detoxification enables the fermentation of inhibitory hydrolysates at higher substrate consistencies, improving final ethanol concentrations and yields. Moreno et al. [89] used laccase from *P. cinnabarinus* to detoxify steam-exploded wheat straw at 12% (*w/v*) substrate loadings, triggering its fermentation by *K. marxianus* CECT 10875 during SSF processes and yielding an ethanol concentration of 16.7 g/L. These authors also described the fermentability of steam-exploded wheat straw at 20% (*w/v*) substrate loadings. At this consistency, the evolved xylose-consuming yeast *S. cerevisiae* F12 was unable to growth. However, this inhibition was overcome by *P. cinnabarinus* laccase, allowing *S. cerevisiae* F12 to produce more than 22 g/L of ethanol during a SSCF process [90]. The evolved xylose-recombinant *S. cerevisiae* KE6-12 was also explored to produce ethanol from steam-exploded wheat straw at 16% (*w/v*) of substrate loading. In this case, *P. cinnabarinus* laccase reduced the toxicity of this media improving cell viability and increasing the ethanol production up to 32 g/L during a fed-batch SSCF process [91]. Finally, a water insoluble solids (WIS) fraction from steam-exploded wheat straw was used at 25% (*w/v*) of substrate loading for ethanol production. This material, detoxified by *P. cinnabarinus* laccase, was then subjected to PSSF processes with *S. cerevisiae*, obtaining an ethanol production of 58.6 g/L [92].

5.3. Detoxification and Saccharification

Laccase detoxification processes have been also evaluated in terms of enzymatic hydrolysis, showing contradictory effects. Kalyani et al. [79] observed an enhancement in the saccharification yield by 48% of acid-pretreated rice straw due to a phenols reduction by *C. perennis* laccase. Contrary, Tabka et al. [102], Jurado et al. [76] and Moreno et al. [80,89] described lower glucose concentration after enzymatic hydrolysis of steam-exploded wheat straw treated with *P. cinnabarinus*, *T. villosa* and *C. rigida* laccases. This negative phenomenon was attributed to the formation of laccase-derived compounds from phenols that inhibit cellulolytic enzymes. In this sense, Oliva-Taravilla et al. [103] showed a strong inhibition due to oligomeric products derived from the oxidative polymerization of vanillin and syringaldehyde by *Myceliophthora thermophila* laccase. The presence of these resulting oligomers caused a decrement on enzymatic hydrolysis yield of a model cellulosic substrate (Sigmacell) of 46.6% and 32.6%, respectively. Moreover, a decrease in more than 50% of cellulase and β-glucosidase activities was observed in presence of laccase and vanillin.

Negative effects on xylose production has been also reported by phenolic oligomers formed from vanillin, syringaldehyde and ferulic acid, as was observed by Oliva-Taravilla et al. [104] with a WIS fraction from steam-exploded wheat straw treated with *M. thermophila* laccase in the presence of the mentioned phenols. Finally, an increase in the competition of cellulose binding sites between hydrolytic enzymes and laccases has been also suggested as a reason for the reduction in glucose recovery [105].

5.4. Other Comments

Although significant advances have been demonstrated about the use of laccases for detoxification, the high enzyme production cost is one of the most important limitations for its application at industrial scale. An alternative approach to adding directly laccase to pretreated materials could be the genetic engineering of fermenting yeast for laccase production. This would allow detoxification and ethanolic fermentation processes simultaneously, thus reducing the cost and time associated with laccase production and detoxification step, respectively. In this matter, Larsson et al. [93] designed a recombinant *S. cerevisiae* strain carrying the laccase gene from the white-rot fungus *T. versicolor*. This strain had the ability to decrease the content of low-molecular phenolic compounds and ferment a dilute-acid spruce hydrolysate, showing higher ethanol productivity compared to control. On the other hand, laccase recycling by enzyme immobilization or co-immobilization could also represent a cost effective approach. Ludwig et al. [94] immobilized a laccase from *T. versicolor* on both active epoxide and amino carriers (Sepabeads® EC-EP and EC-EA, respectively) for detoxification of a wheat straw organosolv fraction. With the immobilized laccase phenolic compounds could be efficiently removed (higher with EC-EA), observing a better performance of *Pichia stipitis* during the fermentation of the detoxified fraction. Additionally, reusability of the immobilized laccase was demonstrated.

6. Application of Laccases for Delignification of Pretreated Materials

The modification or partial removal of lignin by laccases has been shown to be effective for improving enzymatic hydrolysis of different lignocellulosic materials. Different strategies have been assayed with this purpose, either using laccases alone or in combination with mediators (LMS). Consequently, lignin oxidation is produced leading to the formation of aromatic lignin radicals that give rise to a variety of reactions, such as ether and C–C bonds degradation, and aromatic ring cleavage, and finally resulting in lignin degradation [49].

6.1. Delignification by Laccase Alone

Although the direct action of laccases on lignin is, in principle, restricted to phenolic units—which only represent a small percentage of the total polymer—, different studies have showed the ability of laccase alone for delignifying different pretreated materials, improving the subsequent enzymatic hydrolysis (Table 2). Kuila et al. [106,107] explored the use of a laccase from *Pleurotus* sp. to treat milled materials from Indian Thorny bamboo (*Bambusa bambos*) and Spanish flag (*Lantana camara*). A range of delignification between 84–89% was obtained for both materials, observing an increment of the saccharification performance because of the better accessibility of hydrolytic enzymes. The same laccase was used by Mukhopadhyay et al. [108] to treat a milled material from *Ricinus communis*, reporting a delignification yield of about 86%, which increased the yields on reducing sugars by 2.68-fold. Similar lignin removal (81.6%) was achieved by Rajak and Banerjee [109] using a laccase produced by *Lentinus squarrosulus* MR13 to delignify karn grass (*Saccharum spontaneum*), resulting in a sugar production increase by 7.03 fold. On the other hand, lower lignin loss (18%) was obtained when milled material from wheat straw was treated with *P. cinnabarinus* laccase followed by an alkaline peroxide extraction [110]. Then, 24–25% increase in glucose and xylose release was produced. In the same way, Rico et al. [111] compared laccases from *M. thermophila* and *P. cinnabarinus* to treat milled eucalypt wood followed by an alkaline peroxide extraction in a multistage sequence (four cycles of enzyme-alkaline extraction). Whereas the treatment with *M. thermophila* decreased the lignin content of about 20%, *P. cinnabarinus* laccase did not affect the lignin content. Concerning glucose release, the treatment with *M. thermophila* and *P. cinnabarinus* laccases produced an increase of glucose liberation of 9% and 4%, respectively. Finally, Singh et al. [112] has recently described the use of a small bacterial laccase from *Amycolatopsis* sp. to delignify steam-pretreated poplar, obtaining a 6-fold increase in terms of the release of acid insoluble lignin. Then, glucose production from laccase-treated sample was increased by 8%.

Table 2. Application of laccase alone for delignification of different pretreated materials.

Pretreated Material	Laccase Treatment	Effects Observed	Benefits Produced	Reference
Milled material from Thorny bamboo and Spanish flag	*Pleurotus* sp.	Range of delignification between 84–89%, revealing the lignin removal by FTIR, XRD, and SEM analysis	Better accessibility of hydrolytic enzymes	[106,107]
Milled material from *Ricinus communis*	*Pleurotus* sp.	86% of lignin loss, resulting in a degradation of the surface tissues (SEM analysis)	Reducing sugar yields increased 2.68-fold	[108]
Milled material from karm grass	*Lentinus squarrosulus* MR13	Lignin removal of 81.6%. Porosity analysis evidenced the specific action of laccase on lignin	Increase of sugar production of 7.03 fold	[109]
Milled material from wheat straw	*P. cinnabarinus* laccase followed by alkaline peroxide extraction	18% decrease in lignin after sequential treatment	24-25% increase in glucose and xylose production	[110]
Milled wood from *Eucalyptus globulus*	Four cycles of *Myceliophthora thermophila* laccase -alkaline extraction	Up to 20% of lignin loss after four cycles treatment	Increase of glucose production by 9%	[111]
Steam-pretreated poplar	Bacterial laccase from *Amycolatopsis* sp.	Increment of acid insoluble lignin release by 6 fold, observing a reduction of molar mass lignin (approx. 50%) by SEC analysis	8% increment of glucose production	[112]
Alkali-extracted corn straw	*Trametes hirsuta*	Increment of porosity and surface area in laccase-treated samples	2-fold increment in sugar production	[113]
Alkali-extracted straw from *Brassica campestris*	*Ganoderma lucidum*	Higher number and density of holes with greater width and depth after laccase treatment	Saccharification yield increased 1.7-fold	[114]
Steam-exploded wheat straw	*Sclerotium* sp.	Loosening of lignin-carbohydrate complex	16.8% increase in cellulose hydrolysis	[115]
Acid steam-pretreated spruce	*T. hirsuta*	Reduction of lignin hydrophobicity and enrichment of carboxylic groups revealed by ESCA (electron spectroscopy) for chemical analysis	13% increase in sugar yield	[116]
Acid steam-pretreated spruce	*Cerrena unicolor* and *T. hirsuta* laccases	Reduced binding of hydrolytic enzymes by lignin modification	Improvement of hydrolysis yield by 12%	[117,118]
Steam-exploded sugarcane bagasse	*G. lucidum*	Delignification	75% increase in glucose production	[119]
Corncob residue	*Trametes* sp. AH28-2 heterologously expressed in *Trichoderma reesei*	Not investigated	Up to 71.6% increase in reducing sugar yields	[120]
Milled wheat straw	Bacterial laccase from *Thermobifida fusca* incorporated into a designer cellulosome including two cellulases and xylanase	Not investigated	Reducing sugar yields increased 2.0-fold	[121]
Milled sugarcane bagasse	Bacterial laccase from *T. fusca*	SEM analysis of laccase-treated sample shows smaller shatters	2-fold increment in sugar production	[122]
Steam-exploded wheat straw	Alkaline extraction followed by a commercial bacterial laccase MetZyme®	Slight delignification (2%) after alkaline extraction-laccase sequence	Increment of glucose and xylose production by 21% and 30%, respectively	[82]
Steam-exploded wheat straw	Alkaline extraction followed by *Trametes villosa* laccase or bacterial laccase from *Streptomyces ipomoea*	Slight delignification (4%) after alkaline extraction-laccase sequence. No delignification observed by *T. villosa*	Increment of glucose and xylose production by 16% and 6%, respectively. No positive effects observed by *T. villosa*	[95]

FTIR, Fourier transform infrared spectroscopy; XRD, X-ray diffraction; SEM, Scanning electron microscopy; SEC, Size exclusion chromatography; Generally, laccases source is fungal, except in those cases where it is indicated.

In addition to lignin removal, the improvement of enzymatic hydrolysis due to lignin and/or microfiber structure modification by laccase has been also reported. Properties such as porosity, surface area, and hydrophobicity can be altered, resulting in the reduction of unproductive binding of hydrolases. Li et al. [113] observed an increment in the porosity and surface area of alkali-extracted corn straw after a treatment with *Trametes hirsuta* laccase, doubling the sugar production. The same effect was observed on alkali-extracted straw from *Brassica campestris* [114]. Then, the treatment of this material with a laccase from *Ganoderma lucidum* increased saccharification yields 1.7-fold. Regarding steam-exploded materials, laccase treatment has shown contradictory results. Qiu and Chen [115] explored the use of a laccase from *Sclerotium* sp. to treat steam-exploded wheat straw. Fourier transform infrared spectroscopy (FTIR) and scanning electron microscopy (SEM) analysis indicated that laccase oxidized lignin, which contributed to loose the compact wrap of lignin-carbohydrate complexes and consequently enhancing the cellulose hydrolysis. Palonen and Viikari [116] also reported lignin modification of acid steam-pretreated spruce (*Picea abies*) by treatment with *T. hirsuta* laccase. This modification consisted in a reduction of lignin hydrophobicity together with an enrichment of carboxylic groups, which reduced the unproductive binding of hydrolytic enzymes to lignin. Consequently, an enhancement of saccharification yield by 13% was observed during the subsequent enzymatic hydrolysis. Similar results were attained by Moilanen et al. [117,118] when acid steam-pretreated spruce was treated with *C. unicolor* and *T. hirsuta* laccases. However, using acid steam-pretreated giant read (*Arundo donax*), *C. unicolor* laccase reduced the hydrolysis yield by 17% [117]. In this case, the lower sugar production was explained by an increase of the unproductive adsorption of hydrolytic enzymes onto the lignocellulosic fibers and a major strengthening of lignin-carbohydrate complexes. Moreno et al. [84] also reported a reduction of glucose recovery by almost 6–7% after 72 h of enzymatic hydrolysis of steam-exploded wheat straw treated with *P. cinnabarinus* laccase. These authors observed a slight lignin content increment after laccase treatment due to a grafting phenomenon. Grafting takes place when the lignin-derived phenols resulting from steam explosion pretreatment are oxidized by laccase to phenoxy radicals, which can undergo polymerization by radical coupling or being grafted onto steam-exploded material (via radical coupling to lignin residues) [123]. This lignin content increment by grafting phenomenon might prevent the accessibility of hydrolytic enzymes to cellulose, either by reducing the number and/or the size pores or hindering the processivity of cellulases. Moreover, the grafting process could also lead to an increase of the lignin surface area, thereby limiting the accessibility of hydrolytic enzymes to cellulose, and consequently reducing sugar recovery yields. Oliva et al. [105] also suggested the grafting effect to support the lower sugar recovery obtained after treatment of steam-exploded wheat straw with *P. cinnabarinus* laccase. For the first time, these authors observed by FTIR spectroscopy the incorporation of *p*-hydroxycinnamic acids into the fibers of laccase-treated samples.

6.2. Delignification by Laccase-Mediator System (LMS)

Compared to laccase alone, laccase in the form of LMS can oxidize both phenolic and non-phenolic component of lignin moieties, producing an extensive cleavage of covalent bonds in lignin. Different pretreated materials have been subjected to the LMS action for delignification in order to improve the enzymatic hydrolysis, being chemical mediators mainly used (Table 3). Milled material from oil palm empty fruit bunch (OPEFB) was treated with an enzymatic crude extract from *Pycnoporus sanguineus* and HBT and ABTS as mediators [124]. This process leads to lignin removal of 8% and 8.7% when using HBT and ABTS as mediators, respectively. As a consequence, the LMS treatment resulted in a fermentable sugars production of 30 g/L, in comparison to the crude ligninolytic extract without mediator, which showed a maximum concentration of fermentable sugars of 19.1 g/L. Higher delignification range (up to 97%) was reported in liquid hot water pretreated wheat straw and corn stover when using *P. sanguineus* laccase and violuric acid (VIO) as mediator [125]. Al-Zuhair et al. [126] treated milled materials from palm trees fronds and seaweed with a laccase from *T. versicolor* and using HBT as mediator, achieving 9% and 24% of lignin removal, respectively. Consequently, the subsequent enzymatic hydrolysis was improved from 0.04% to 3.1%. Furthermore, when combining laccase-HBT system with the ionic liquid [C_2 mim] [OAc] (1-ethyl-3-methylimidazolium acetate), saccharification yields increased up to 13%. Moniruzzaman and Ono [127] also combined LMS treatment with ionic liquids. These authors reported 50% delignification yields when wood chips from hinoki cypress (*Chamaecyparis obtusa*) pretreated with [C_2 mim] [OAc] (1-ethyl-3-methylimidazolium acetate) were treated with the commercial laccase Y120 (*Trametes* sp.) and HBT as mediator. The same laccase-mediator system was also applied on OPEFB biomass pretreated with the hydrophilic ionic liquid [EMIM] [DEP] (1-ethyl-3-methylimidazolium diethyl phosphate) [128], resulting in a delignification range of 35%. On the other hand, a sequential combination of ultrasonication, liquid hot water and a commercial LMS (PrimaGreen® EcoFade LT100 composed principally by a laccase from modified strains of *C. unicolor* and the mediator 3,5-dimethoxy-4-hydroxybenzonitrile) was performed on cotton gin trash [129]. This process led up to 15% lignin removal, increasing glucose and ethanol yields by 23% and 31%, respectively. A new sequential pretreatment combines an alkaline ultrasonication with liquid hot water and the commercial LMS PrimaGreen® EcoFade LT100, was again evaluated [130]. When applied to cotton gin trash, the delignification range was increased to 27%, resulting in increments of 41% and 64% of glucose and ethanol yields, respectively. Ultrasound pretreatment was also applied on elephant grass (*Pennisetum purpureum*) by Nagula and Pandit [131]. The pretreated material was then subjected to a LMS treatment consisting of *T. hirsuta* crude laccase supernatant and ABTS as mediator, resulting in a delignification range of 69%. In another study, Gutiérrez et al. [132] evaluated the ability of *T. villosa* laccase, together with HBT as mediator and a subsequent alkaline extraction, to remove lignin from milled eucalypt wood and elephant grass. 48% and 32% of the eucalypt and elephant grass lignin were removed, respectively. Consequently, the glucose yield was increased by 61% and 12% from both lignocellulosic materials, respectively, as compared to those without LMS treatment. Additionally, lignin structural changes were observed by two-dimensional nuclear magnetic resonance (2D NMR), as a result of the laccase-HBT system. A significant decrease of aromatic lignin units (with preferential degradation of guaiacyl over syringyl units) and aliphatic (mainly β-O-4'-linked) side-chains of lignin after LMS treatment was showed, leading to residual lignin with mainly oxidized syringyl units. These authors also described similar lignin structural changes when four cycles of a sequential treatment of LMS (including *P. cinnabarinus* laccase-HBT) followed by an alkaline peroxide extraction were applied on milled eucalypt wood [111]. Rencoret et al. [110] also reported lignin structural variations in milled wheat straw treated with the same laccase-mediator system. Moreover, a substantial lignin removal (37%) was produced by *P. cinnabarinus* laccase in the presence of HBT, which was increased up to 48% when a subsequent alkaline peroxide extraction was applied. This LMS treatment increased glucose yields by 60% after enzymatic hydrolysis.

Table 3. Application of laccase-mediator systems for delignification of different pretreated materials.

Pretreated Material	LMS Treatment	Effects Observed	Benefits Produced	Reference
Oil palm empty fruit bunch milled	*Pycnoporus sanguineus* laccase with HBT and ABTS as mediators	Klason lignin reduction of 8% and 8.7% for HBT and ABTS, respectively	Increment of sugar yield by 16–17% compared to laccase alone	[124]
Wheat straw and corn stover pretreated with liquid hot water	*P. sanguineus* H275 laccase with VIO as mediator	Up to 97% lignin loss	19.98% increase in sugar production	[125]
Milled material from palm trees and seaweed	*Trametes versicolor* laccase with HBT as mediator	Lignin removal of 9% and 24% for palm trees and seaweed, respectively	Better enzymatic hydrolysis with a ionic liquid [C_2 mim] [OAc] (1-ethyl-3-methylimidazolium acetate) treatment prior to laccase-HBT	[126]
Wood chips swollen with ionic liquid [C_2 mim] [OAc] (1-ethyl-3-methylimidazolium acetate)	*Trametes* sp. Y120 laccase with HBT as mediator	50% delignification, revealing structural lignin changes by SEM and FTIR analysis	Pretreated material with cellulose more accessible	[127]
Oil palm empty fruit bunch pre-treated with ionic liquid [EMIM] [DEP] (1-ethyl-3-methylimidazolium diethyl phosphate)	*Trametes* sp. Y120 laccase with HBT as mediator	35% decrease in lignin	Cellulose rich-material	[128]
Cotton gin trash pretreated with a sequential combination of ultrasonication and liquid hot water	*Cerrena unicolor* laccase with 3,5-dimethoxy-4-hydroxybenzonitrile as mediator	Up to 15% lignin loss	Up to 23% and 31% increase in glucose and ethanol yields, respectively	[129]
Cotton gin trash pretreated with a sequential combination of alkaline ultrasonication and liquid hot water	*C. unicolor* laccase with 3,5-dimethoxy-4-hydroxybenzonitrile as mediator	27% reduction in lignin, observing lignin aromatic change structure by FTIR	41% and 64% increase in glucose and ethanol yields, respectively	[130]
Elephant grass pretreated with ultrasound	*Trametes hirsuta* laccase with ABTS as mediator	Delignification range of 69%	Better accessibility of cellulose	[131]
Milled materials from eucalypt wood and elephant grass	*Trametes villosa* laccase with HBT as mediator and a subsequent alkaline extraction	Up to 48% and 32% lignin removal for eucalypt and elephant grass, respectively	Increase in glucose yield (61% and 12% for eucalypt and elephant grass, respectively) and ethanol production (over 4 g/L in eucalypt and 2 g/L in elephant)	[132]
Eucalypt wood milled	Four cycles of *Myceliophthora thermophila* laccase with methyl syringate as mediator and a subsequent alkaline peroxide extraction	50% delignification, observing by Py/GC-MS and 2D NMR analysis a significant reduction of both aromatic and aliphatic lignin with high presence of oxidized syringyl units	Increases (approximately 40%) in glucose and xylose yields after enzymatic hydrolysis	[133]
Eucalypt wood milled	Comparing four cycles of *Pycnoporus cinnabarinus* laccase with HBT as mediator (or *M. thermophila* laccase with methyl syringate as mediator) and a subsequent alkaline peroxide extraction	50% decrease in lignin with both LMS after four cycles, Slight delignification observed after the first cycle with *P. cinnabarinus* laccase and HBT, but not after *M. thermophila* laccase and methyl syringate	Increased glucose yield (30%) with both LMS after four cycles. Saccharification increment of 10% after the first cycle with *P. cinnabarinus* laccase and HBT, but not after *M. thermophila* laccase and methyl syringate	[111]

Table 3. *Cont.*

Pretreated Material	LMS Treatment	Effects Observed	Benefits Produced	Reference
Acid steam-pretreated spruce	*T. hirsuta* laccase with acetosyringone as mediator	Reduction of unproductive hydrolases adsorption due to an increment of syringyl/guaiacyl ratio	Downstream cellulose hydrolysis was improved 36%	[118]
Acid steam-pretreated spruce	*T. hirsuta* laccase with ABTS, HBT, and TEMPO as mediators	Lignin modification resulting in a decrease of unproductive cellulases adsorption, except with HBT. TEMPO also oxidized cellulose	Increment of enzymatic hydrolysis by 54% and 49% with ABTS and TEMPO, respectively. No positive effects with HBT	[118]
Milled material from date palm waste	*T. versicolor* laccase with HBT as mediator	Reduced binding of hydrolytic enzymes by lignin modification	Improvement of sugar production 8 times	[134]
Ensiled corn stover	*T. versicolor* laccase with HBT as mediator	Lignin side chain oxidation	Downstream cellulose hydrolysis was improved 7%	[135]
Acid steam-exploded wheat straw	*T. versicolor* laccase with HBT as mediator followed by alkaline peroxide extraction	Lignin oxidation revealed by Py/GC–MS TMAH	Increment of glucose release by up to 2.3 g/L	[136]
Acid steam-pretreated spruce	*T. hirsuta* laccase with NHA as mediator	Lignin modification showing both modified hydrophobicity and surface charge	Enzymatic hydrolysis yield increased 1.61-fold compared to laccase alone	[116]
Steam-exploded eucalypt wood	*M. thermophila* laccase and HBT as mediator	Lignin oxidation let to an increment of both secondary OH groups and degree condensation	Slightly increase of sugar production	[137,138]

HBT, 1-hydroxybenzotriazole; VIO, violuric acid; ABTS, 2,2′-azino-bis (3 ethylbenzothiazoline-6-sulfonic acid; TEMPO, (2,2,6,6-Tetramethylpiperidin-1-yl)oxyl; FTIR, Fourier transform infrared spectroscopy; SEM, Scanning electron microscopy; Py/GC–MS, Pyrolysis/gas chromatography-mass spectrometry; TMAH, tetramethylammonium hydroxide; Generally, laccases source is fungal, except in those cases where it is indicated.

The use of lignin-derived soluble phenols, such as vanillin, acetosyringone, *p*-hydroxycinnamic acids, etc., as natural mediators for ethanol production would offer environmental and economic advantages compared to chemical mediators. Although their use could be compromised by laccase-mediated coupling reactions, several studies have also shown their potential for improving delignification and cellulose hydrolysis. Rico et al. [111] evaluated the use of methyl syringate as natural mediator in the presence of *M. thermophila* laccase to delignify milled eucalypt wood. Four cycles of LMS-alkaline peroxide extraction were performed, resulting in a lignin content reduction of about 50%, and an increase in glucose yields of 30%. These results were comparable to those obtained with *P. cinnabarinus* laccase-HBT as LMS. Moilanen et al. [118], in contrast, observed an increase in the lignin content of acid-steam pretreated spruce when using acetosyringone mediator together with *T. hirsuta* laccase. In spite of this effect, laccase-acetosyringone treatment improved the hydrolysis yield by 36%. This result was explained by an increment of the syringyl/guaiacyl ratio promoted by the enzymatic treatment, which let to reduce the unproductive adsorption of cellulases.

In addition to the reduction in the lignin content, LMS has been also reported for improving enzymatic hydrolysis by lignin modification. By using ABTS and TEMPO as mediators of *T. hirsuta* laccase, Moilanen et al. [118] increased the hydrolysis yields of acid steam-exploded spruce by 54% and 49%, respectively. These improvements were explained to be based on the reduction of the unspecific adsorption of hydrolases on enzyme-treated lignin. Similar results were obtained by Al-Zuhair et al. [134], which showed an increment of sugar production from 5.6% to 45.6% after treatment of a milled material from date palm lignocellulosic waste with *T. versicolor* laccase and the mediator HBT. Using the same LMS and the pyrolysis/gas chromatography-mass spectrometry (Py/GC–MS) with tetramethylammonium hydroxide (TMAH) thermochemolysis analysis, Chen et al. [135] described a significant lignin modification (lignin side chain oxidation) after treatment of ensiled corn stover. This resulted in an increment of the subsequent hydrolysis yield of 7%. The *T. versicolor* laccase-HBT system was also used by Heap et al. [136] for improving the saccharification yield of acid steam-exploded wheat straw. In a first assay, LMS impaired the enzymatic hydrolysis of acid steam-exploded material. However, when a subsequent alkaline peroxide extraction was carried out after LMS, the released glucose concentration increased by up to 2.3 g/L (35%) compared to untreated control. Py/GC–MS with TMAH analysis also revealed lignin oxidation via C_α–C_β sidechain cleavage at the C_α position. In another study, the use of *N*-hydroxy-*N*-phenylacetamide (NHA) as mediator of *T. hirsuta* laccase increased the saccharification yield of acid steam-exploded spruce from 13% to 21% compared to the treatment with laccase alone [116]. Nevertheless, a filtration and washing step had to be performed between laccase-mediator treatment and enzymatic hydrolysis due to inhibitory effect of oxidized NHA on cellulases. In this sense, Moreno et al. [84] also observed a direct inhibition on hydrolytic enzymes activities of different oxidized radicals generated by *P. cinnabarinus* laccase from HBT, VIO, and ABTS mediators. A decrease of about 34% was observed for overall cellulase activity in the presence of the different chemical mediators. However, enzymatic deactivation was even more remarkable in the case of β-glucosidase activity, showing a reduction of about 50%. Martín-Sampedro et al. [137] also observed lignin changes after treatment with LMS (*M. thermophila* laccase and the mediator HBT) of steam-exploded eucalyptus wood chips. By using 2D NMR and ^{13}C NMR, these authors reported an increase in the amount of secondary OH groups and in the degree of lignin condensation. In a subsequent enzymatic hydrolysis, this LMS-treated material showed an increase in the glucose yield from 24.7% to 27.1% [138].

6.3. Other Comments

As previously discussed for laccase detoxification, the genetic engineering of microorganisms for the simultaneous production of laccase and hydrolytic enzymes would allow better processes integration for delignification and saccharification of lignocellulose biomass, and thus reducing the cost and time associated with laccase production and delignification step, respectively. Zhang et al. [120] observed higher saccharification yields during the hydrolysis of corn residue by the heterologous expression of *Trametes* sp. AH28-2 laccase in *T. reesei*. With a similar concept, Davidi et al. [121] have recently incorporated laccase activity into a cellulase- and xylanase-containing cellulosome. For that, authors designed a dockerin-fused variant of a recently characterized laccase from the aerobic bacterium *Thermobifida fusca* [122]. The resulting cellulosome complex yielded a 2-fold increase in the amount of reducing sugars released from wheat straw compared with the same system lacking laccase activity.

7. Laccases for Detoxification and Delignification in a Lignocellulose-based Biorefinery

On the basis of the current review, laccase enzymes have been largely evaluated as specific, effective and environmental friendly tools for detoxification and delignification of lignocellulosic feedstocks. After laccase treatment, higher saccharification and fermentation yields are usually observed, which offer high potential to reduce overall process costs. For instance, by modifying or partially removing lignin, the unspecific adsorption of hydrolases is reduced and lower enzymes loadings are therefore required for the enzymatic hydrolysis of lignocellulose. This fact represents an important breakthrough, since the costs of hydrolytic enzymes is one of the major economical bottlenecks in the conversion of lignocellulosic biomass. The cost of laccase and/or of mediators should also take into account. Another relevant advantage is the possibility of having a better water economy. After detoxification with laccase, pretreated material contains lower inhibitory compounds, avoiding the necessity of including a filtration and washing step and therefore saving freshwater and reducing the amount of wastewater. Also, by having less inhibitory pretreated materials, conversion processes can be performed at higher substrate loadings, giving the possibility of reaching higher product concentrations with shorter fermentation times.

With the aim of implementing laccases in the current conversion processes, in situ laccase treatment with saccharification and/or fermentation offers some advantages as they do not require extra equipment and thus generates benefits in terms of lower capital and operating costs. Simultaneous delignification and detoxification with laccase is another interesting strategy to consider. However, little is known about the existence of laccases with capacity for simultaneous delignification and detoxification. Furthermore, it should be noted that lower sugar yields are usually observed during saccharification of detoxified feedstocks [76,82,89,102]. Searching for novel laccases with ability to delignify and detoxify simultaneously or designing new ones with the required properties need to be further explored. In this sense, Moreno et al. [82] has recently evaluated the commercial bacterial laccase MetZyme® for enhancing saccharification and ethanol fermentation of steam-exploded wheat straw. When the pretreated material was subjected to laccase action, a modest increase of about 5% in the sugar recovery yield was observed. In contrast, when performing an alkaline extraction prior to laccase treatment, the glucose and xylose recovery increased by 15% and 23%, respectively, compared to alkaline treatment alone. A modest phenols removal could be also observed during treatment of steam-exploded wheat straw with Metzyme® laccase. The lower phenolic content allowed to improve the fermentation performance of the thermotolerant yeast *K. marxianus* CECT 10875 during SSF processes, shortening its adaptation phase and reducing fermentation times. Similarly, De La Torre et al. [95] compared the use of both bacterial *Streptomyces ipomoeae* and fungal *T. villosa* laccases for delignification and detoxification of steam-exploded wheat straw. When using the bacterial laccase, no significant effects were observed on delignification or saccharification of laccase-treated biomass. However, the use of fungal laccase resulted in higher lignin content and lower sugar recoveries. By combining an alkali extraction with *S. ipomoeae* laccase, a 4% reduction in the lignin content was observed compared to alkaline treatment alone, increasing the glucose and xylose concentrations in the resulting hydrolysate by 16% and 6%, respectively. These positive effects were however not observed when using *T. villosa* laccase. In addition to delignification, the capacity of these bacterial and fungal laccases for detoxification of pretreated material was also evaluated. A reduction in the phenol content of 29% and 90% were achieved with the bacterial and fungal laccases, respectively. This reduction resulted in an improved fermentation performance of *S. cerevisiae* during SSF processes.

Fermentation **2017**, *3*, 17

Cost-effectiveness in future biorefineries goes through valorization of all components of lignocellulosic biomass. In this context, biorefineries have to deal with producing not only high-volume and low-cost fuels but also low-volume and high-value compounds, minimizing downstream wastes. With this purpose, in addition to carbohydrate fermentation processes such as ethanol and/or organic acids production, alternative value-added products and chemicals can be also obtained from lignin. Laccases can also contribute to such an aim, assisting in certain processes during the manufacture of new value-added products. For instance, laccases have been typically applied in the pulp and paper industry (1) for pulp bleaching, removing the residual lignin responsible of pulp color [71], (2) for controlling pitch deposits that reduce pulp quality [75], or (3) for detoxification of bleaching effluents rich in phenolic compounds [139]. Laccases have been also evaluated for the synthesis of new materials and products from lignocellulosic feedstocks. Laccases can limit and/or avoid the use of toxic synthetic adhesives (such as formaldehyde-based resins) during production of fiberboards and other materials, by catalyzing the cross-linking reactions of phenolic residues in lignin based-materials [140]. Tailoring of lignocellulosic materials by laccase-assisted biografting of phenols and other compounds is another emerging area. Also, laccase-assisted functionalization of wood and non-wood fibers to modify different properties has been achieved, obtaining new physico-mechanical, optical and antimicrobial properties [141–143]. Finally, laccases are also a promising approach to decompose the lignin polymer into several phenolic and aromatic compounds that are currently produced from fossil fuels [144].

8. Conclusions

In the current biorefinery concept, laccases constitute a powerful biotechnological tool for the complete utilization of lignocellulosic biomass to new added-value products and fuels, with lower energy demand, better economy and less environmental impact. Laccases act selectively to remove lignin-derived phenolic compounds released from biomass pretreatment, diminishing the impact of these inhibitors on the subsequent saccharification and ethanol fermentation stages. Then, a reduction of phenols by laccase-aided polymerization promotes microbial growth, glucose consumption and increase notably the ethanol production. Laccases and laccase-mediator systems can also be effective in oxidative modification and/or partially depolymerization of lignin, increasing the final hydrolysis yields of different pretreated materials. Nevertheless, the costs for enzyme production and the use of expensive synthetic mediators are current challenges to overcome for the successful implementation of laccases in these lignocellulose-based industries. Screening of microorganism cultures and genomes for novel laccases or engineering of existing ones by direct evolution and related approaches are solutions to consider. Moreover, the search of new, cheap and environmentally friendly mediators can also push these biocatalysts toward their application on an industrial level.

Acknowledgments: Authors wish to thank Spanish Ministry of Economy and Competitiveness (MINECO) for funding this study via Projects CTQ2013-47158-R and LIGNOYEAST (ENE2014-315 54912-R). Antonio D. Moreno acknowledges the MINECO and the specific "Juan de la Cierva" Subprogramme for contract FJCI-2014-22385.

Author Contributions: Úrsula Fillat, David Ibarra, María E. Eugenio, Antonio D. Moreno, Elia Tomás-Pejó and Raquel Martín-Sampedro conceived the study, participated in their design and commented on the manuscript. David Ibarra wrote the manuscript. All the authors read and approved the final manuscript.

Conflicts of Interest: The authors declare no conflict of interest.

References

1. Moreno, A.D.; Alvira, P.; Ibarra, D.; Tomás-Pejó, E. Production of ethanol from lignocellulosic biomass. In *Production of Chemicals from Sustainable Resources, Biofuels and Biorefineries 7*; Fang, Z., Smith, R.L., Jr., Richard, L., Xinhua, Q., Eds.; Springer: Berlin/Heidelberg, Germany, 2017.
2. Balan, V.; Chiaramonti, D.; Kumar, S. Review of US and EU initiatives toward development, demonstration, and commercialization of lignocellulosic biofuels. *Biofuels Bioprod. Biorefin.* **2013**, *7*, 732–759. [CrossRef]

3. Ballesteros, M. Enzymatic hydrolysis of lignocellulosic biomass. In *Bioalcohol Production. Biochemical Conversion of Lignocellulosic Biomass*; Waldron, K., Ed.; Woodhead Publishing: Cambridge, UK, 2010; pp. 159–177.

4. Alvira, P.; Tomás-Pejó, E.; Ballesteros, M.; Negro, M.J. Pretreatment technologies for an efficient bioethanol production process based on enzymatic hydrolysis: A review. *Bioresour. Technol.* **2010**, *101*, 4851–4861. [CrossRef] [PubMed]

5. Palmqvist, E.; Hahn-Hägerdal, B. Fermentation of lignocellulosic hydrolysates. II: Inhibitors and mechanism of inhibition. *Bioresour. Technol.* **2000**, *74*, 25–33. [CrossRef]

6. Berlin, A.; Balakshin, M.; Gilkes, N.; Kadla, J.; Maximenko, V.; Kubo, S. Inhibition of cellulase, xylanase and β-glucosidase activities by sofwood lignin preparations. *J. Biotechnol.* **2006**, *125*, 198–209. [CrossRef] [PubMed]

7. Chandel, A.K.; Silvério da Silva, S.; Singh, O.V. Detoxification of lignocellulosic hydrolysates for improved bioethanol production. In *Biofuel-Production-Recent Development Prospect*; Aurelio, M., Bernardes, S., Eds.; InTechOpen: Rijeka, Croatia, 2011; pp. 225–246.

8. Palmqvist, E.; Hahn-Hägerdal, B. Fermentation of lignocellulosic hydrolysates. I: Inhibition and detoxification. *Bioresour. Technol.* **2000**, *74*, 17–24. [CrossRef]

9. Moreno, A.D.; Ibarra, D.; Alvira, P.; Tomás-Pejó, E.; Ballesteros, M. A review of biological delignification and detoxification methods for lignocellulosic bioethanol production. *Crit. Rev. Biotechnol.* **2015**, *35*, 342–354. [CrossRef] [PubMed]

10. Parawira, W.; Tekere, M. Biotechnological strategies to overcome inhibitors in lignocellulose hydrolysates for ethanol production: Review. *Crit. Rev. Biotechnol.* **2011**, *31*, 20–31. [CrossRef] [PubMed]

11. E4tech; RE-CORD; WUR. From the Sugar Platform to Biofuels and Biochemicals. Final Report for the European Commission Directorate-General Energy, Contract No. ENER/C2/423-2012/SI2.673791: European Union. 2015. Available online: http://ibcarb.com/wp-content/uploads/EC-Sugar-Platform-final-report.png (accessed on 19 March 2017).

12. Mosier, N.; Wyman, C.E.; Dale, B.D.; Elander, R.T.; Lee, Y.Y.; Holtzapple, M.; Ladisch, C.M. Features of promising technologies for pretreament of lignocellulosic biomass. *Bioresour. Technol.* **2005**, *96*, 673–686. [CrossRef] [PubMed]

13. Pan, X.; Xie, D.; Gilkes, N.; Gregg, D.J.; Sadler, J.N. Strategies to enhance the enzymatic hydrolysis of pretreated softwood with high residual content. *Appl Biochem. Biotechnol.* **2005**, *124*, 1069–1079. [CrossRef]

14. Mansfield, S.D.; Mooney, C.; Saddler, J.N. Substrate and enzyme characteristics that limit cellulose hydrolysis. *Biotechnol. Prog.* **1999**, *15*, 804–816. [CrossRef] [PubMed]

15. Chandra, R.P.; Bura, R.; Mabee, W.E.; Berlin, A.; Pan, X.; Sadler, J.N. Substrate pretreatment: The key to effective enzymatic hydrolysis of lignocellulosics? *Adv. Biochem. Eng. Biotechnol.* **2007**, *108*, 67–83. [PubMed]

16. Sun, S.; Sun, S.; Cao, X.; Sun, R. The role of pretreatment in improving the enzymatic hydrolysis of lignocellulosic materials. *Bioresour. Technol.* **2016**, *199*, 49–58. [CrossRef] [PubMed]

17. Mussato, S.I.; Dragone, G.M. Biomass pretreatment, biorefineries, and potential products for a bioeconomy development. In *Biomass Fractionation Technologies for a Lignocellulosic Feedstock Biorefinery*; Mussato, S.I., Ed.; Elsevier Inc.: Amsterdam, The Netherlands, 2016; pp. 1–22.

18. Alvira, P.; Ballesteros, M.; Negro, M.J. Progress on enzymatic saccharification technologies for biofuels production. In *Biofuel Technologies: Recent Developments*; Gupta, V.K., Tuohy, G., Eds.; Springer: Berlin, Germany, 2013; pp. 145–169.

19. Jørgensen, H.; Krinstensen, J.B.; Felby, C. Enzymatic conversion of lignocellulose into fermentable sugars: Challenges and opportunities. *Biofuels Bioprod. Biorefin.* **2007**, *1*, 119–134. [CrossRef]

20. Martínez, A. How to break down crystalline cellulose. *Science* **2016**, *352*, 1050–1051. [CrossRef] [PubMed]

21. Olsson, L.; Jørgensen, H.; Krogh, K.B.R.; Roca, C. Bioethanol production from lignocellulosic material. In *Polysaccharides: Structural Diversity and Functional Versatility*; Dumitriu, S., Ed.; Marcel Dekker: New York, NY, USA, 2005; pp. 957–993.

22. Radecka, D.; Mukherjee, V.; Mateo, R.Q.; Stojiljkovic, M.; Foulquie-Moreno, M.R.; Thevelien, J.M. Looking beyond *Saccharomyces*: The potential of non-conventional yeast species for desirable traits in bioethanol fermentation. *FEMS Yeast Res.* **2015**. [CrossRef] [PubMed]

23. Jönsson, L.F.; Martín, C. Pretreatment of lignocellulose: Formation of inhibitory by-products and strategies form minimizing their effects. *Bioresour. Technol.* **2016**, *199*, 103–112. [CrossRef] [PubMed]

24. Taherzadeh, M.J.; Karimi, K. Fermentation inhibitors in ethanol processes and different strategies to reduce their effects. In *Biofuels: Alternative Feedstocks and Conversion Processes*; Pandey, A., Larroche, C., Ricke, S.C., Dussap, C.G., Gnansounou, E., Eds.; Springer: Berlin, Germany, 2011; pp. 287–311.

25. Panagiotou, G.; Olsson, L. Effect of compounds released during pretreatment of wheat straw on microbial growth and enzymatic hydrolysis rates. *Biotechnol. Bioeng.* **2006**, *96*, 250–258. [CrossRef] [PubMed]

26. Ximenes, E.; Kim, Y.; Mosier, N.; Dien, B.; Ladisch, M. Deactivation of cellulases by phenols. *Enzyme Microb. Technol.* **2011**, *48*, 54–60. [CrossRef] [PubMed]

27. Martínez, A.T.; Ruíz-Dueñas, J.; Martínez, M.J.; Del Río, J.C.; Gutiérrez, A. Enzymatic delignification of plant cell wall: From nature to mill. *Curr. Opin. Biotechnol.* **2009**, *20*, 348–357. [CrossRef] [PubMed]

28. Vanholme, R.; Demedts, B.; Morrel, K.; Ralph, J.; Boerjan, W. Lignin biosynthesis and structure. *Plant Physiol.* **2010**, *153*, 895–905. [CrossRef] [PubMed]

29. Rahikainen, J.L.; Martín-Sampedro, R.; Heikkinen, H.; Rovio, S.; Marjamaa, K.; Tamminen, T.; Rojas, O.J.; Kruus, K. Inhibitory effect of lignin during cellulose bioconversion: The effect of lignin chemistry on non-productive enzyme adsorption. *Bioresour. Technol.* **2013**, *133*, 270–278. [CrossRef] [PubMed]

30. Santos, J.I.; Martín-Sampedro, R.; Fillat, U.; Oliva, J.M.; Negro, M.J.; Ballesteros, M.; Eugenio, M.E.; Ibarra, D. Evaluating lignin-rich residues from biochemical ethanol production of wheat straw and olive tree pruning by FTIR and 2D-NMR. *Int. J. Polym. Sci.* **2015**. [CrossRef]

31. Santos, J.I.; Fillat, U.; Martín-Sampedro, R.; Ballesteros, I.; Manzanares, P.; Ballesteros, M.; Eugenio, M.E.; Ibarra, D. Lignin-enriched fermentation residues from bioethanol production of fast-growing poplar and forage sorghum. *Bioresources* **2015**, *10*, 5215–5232. [CrossRef]

32. Li, J.; Henriksson, G.; Gellerstedt, G. Lignin depolymerization/repolymerization and its critical role for delignification of aspen wood by steam-explosion. *Bioresour. Technol.* **2007**, *98*, 3061–3068. [CrossRef] [PubMed]

33. Samuel, R.; Cao, S.; Das, B.K.; Hu, F.; Pu, Y.; Ragauskas, A.J. Investigation of the fate of poplar lignin during autohydrolysis pretreatment to understand the biomass recalcitrance. *RSC Adv.* **2013**, *3*, 5305–5309. [CrossRef]

34. Sewalt, V.J.H.; Glasser, W.G.; Beauchemin, K.A. Lignin impact on fiber degradation. 3. Reversal of inhibition of enzymatic hydrolysis by chemical modification of lignin and by additives. *J. Agric. Food. Chem.* **1997**, *45*, 1823–1828. [CrossRef]

35. Helle, S.S.; Duff, S.J.; Cooper, D.G. Effect of surfactants on cellulose hydrolysis. *Biotechnol. Bioeng.* **1993**, *42*, 611–617. [CrossRef] [PubMed]

36. García-Aparicio, M.P.; Ballesteros, I.; González, A.; Oliva, J.M.; Ballesteros, M.; Negro, M.J. Effect of inhibitors released during steam-explosion pretreatment of barley straw on enzymatic hydrolysis. *Appl. Biochem. Biotechnol.* **2006**, *129*, 278–288. [CrossRef]

37. Larsson, S.; Reimann, A.; Nivelbrant, N.; Jonsson, L.J. Comparison of different methods for the detoxification of lignocellulose hydrolysates of spruce. *Appl Biochem. Biotechnol.* **1999**, *77*, 91–103. [CrossRef]

38. Wilson, J.J.; Deschatelets, L.; Nishikawa, N.K. Comparative fermentability of enzymatic and acid hydrolysates of steam pretreated aspen wood hemicellulose by *Pichia stipitis* CBS 5776. *Appl. Microbiol. Biotechnol.* **1989**, *31*, 592–596. [CrossRef]

39. Fargues, C.; Lewandowski, R.; Lameloise, M.L. Evaluation of ion-exchange and adsorbent resins for the detoxification of beet distillery effluents. *Ind. Eng. Chem. Res.* **2010**, *49*, 9248–9257. [CrossRef]

40. Rodrigues, R.; Felipe, M.G.A.; Silva, J.; Vitolo, M.; Gómez, P.V. The influence of pH, temperature and hydrolyzate concentration of volatile and nonvolatile compounds from sugarcane bagasse hemicellulosic hydrolyzate treate with charcoal before or after vacuum evaporation. *Braz. J. Chem. Eng.* **2001**, *18*, 299–311. [CrossRef]

41. Alriksson, B.; Cavka, A.; Jhonson, L.F. Improving the fermentability of enzymatic hydrolysates of lignocellulose through chemical in situ modification with reducing agents. *Bioresour. Technol.* **2011**, *102*, 1254–1263. [CrossRef] [PubMed]

42. Palmqvist, E.; Hahn-Hägerdal, B.; Szengyel, Z.; Zacchi, G.; Réczey, K. Simultaneous detoxification and enzyme production of hemicellulose hydrolysates obtained after steam pretreatment. *Enzyme Microbiol. Technol.* **1997**, *20*, 286–293. [CrossRef]

43. Okuda, N.; Soneura, M.; Ninomiya, K.; Katakura, Y.; Shioya, S. Biological detoxification of waste house wood hydrolysate using *Ureibacillus thermosphaericus* for bioethanol production. *J. Biosci. Bioeng.* **2008**, *106*, 128–133. [CrossRef] [PubMed]

44. Thomsen, M.H.; Thygesen, A.; Thomsen, A.B. Identification and characterization of fermentation inhibitors formed during hydrothermal treatment and following SSF of wheat straw. *Appl. Microbiol. Biotechonol.* **2009**, *83*, 447–455. [CrossRef] [PubMed]

45. Adeboye, P.T.; Bettiga, M.; Aldaeus, F.; Larsson, P.T.; Olsson, L. Catabolism of coniferyl aldehyde, ferulic acid and p-coumaric acid by *Saccharomyces cerevisiae* yields less toxic products. *Microb. Cell Fact.* **2015**, *14*, 149. [CrossRef] [PubMed]

46. Tomás-Pejó, E.; Ballesteros, M.; Oliva, J.M.; Olsson, L. Adaptation of the xylose fermenting yeast *Saccharomyces cerevisiae* F12 for improving ethanol production in different fed-batch SSF processes. *J. Ind. Microbiol. Biotechnol.* **2010**, *37*, 1211–1220. [CrossRef] [PubMed]

47. Larsson, S.; Nivelbrant, N.O.; Jönsson, L.J. Effect of overexpression of *Saccharomyces cerevisiae* Pad1p on the resistance to phenylacrylic acids and lignocellulose hydrolysates under aerobic and oxygenlimited conditions. *Appl. Microbiol. Biotechnol.* **2001**, *57*, 167–174. [CrossRef] [PubMed]

48. Petersson, A.; Almeida, J.R.; Modig, T.; Karhumaa, K.; Hahn-Hägerdal, B.; Gorwa-Grauslund, M.F.; Lidén, G. A 5-hydroxymethyl furfural reducing enzyme encoded by the *Saccharomyces cerevisiae ADH6* gene conveys HMF tolerance. *Yeast* **2006**, *23*, 455–464. [CrossRef] [PubMed]

49. Martínez, A.T.; Speranza, M.; Ruiz-Dueñas, F.J.; Ferreira, P.; Camarero, S.; Guillén, F.; Martínez, M.J.; Gutiérrez, A.; del Río, J.C. Biodegradation of lignocellulosics: Microbial, chemical and enzymatic aspects of fungal attack to lignin. *Int. Microbiol.* **2005**, *8*, 195–204. [PubMed]

50. Gold, M.H.; Youngs, H.L.; Gelpke, M.D. Manganese peroxidase. *Met. Ions Biol. Syst.* **2000**, *37*, 559–586. [PubMed]

51. Martínez, A.T. Molecular biology and structure-function of lignin degrading heme peroxidases. *Enzym. Microb. Tehcnol.* **2000**, *30*, 425–444. [CrossRef]

52. Martínez, M.J.; Ruiz-Dueñas, F.J.; Guillén, F.; Martínez, A.T. Purification and catalytic properties of two manganese-peroxidase isoenzymes from *Pleurotus eryngii*. *Eur. J. Biochem.* **1996**, *237*, 424–432. [CrossRef] [PubMed]

53. Ruiz-Dueñas, F.J.; Martínez, M.J.; Martínez, A.T. Molecular characterization of a novel peroxidase isolated from the ligninolytic fungus *Pleurotus eryingii*. *Mol. Microbiol.* **1999**, *31*, 223–235. [CrossRef] [PubMed]

54. Salvachúa, D.; Prieto, A.; Martínez, A.T.; Martínez, M.J. Characterization of a novel DyP-type peroxidase from *Irpex lacteus* and its application in the enzymatic hydrolysis of wheat straw. *Appl. Environ. Microbiol.* **2013**, *79*, 4316–4324. [CrossRef] [PubMed]

55. Zhang, L.H.; Li, D.; Wang, L.J.; Wang, T.-P.; Zhang, L.; Dong Chen, X.; Mao, Z.-H. Effect of steam explosion on biodegradation of lignin in wheat straw. *Bioresour. Technol.* **2008**, *99*, 8512–8515. [CrossRef] [PubMed]

56. Salvachúa, D.; Prieto, A.; López-Abelairas, M.; Lu-Chau, T.; Martínez, A.T.; Martínez, M.J. Fungal pretreatment: An alternative in second-generation ethanol form wheat straw. *Bioresour. Technol.* **2011**, *102*, 7500–7506. [CrossRef] [PubMed]

57. Yu, J.; Zhang, J.; He, J.; Liu, Z.; Yu, Z. Combinations of mild physical or chemical pretreatment with biological pretreatment for enzymatic hydrolysis of rice hull. *Bioresour. Technol.* **2009**, *100*, 903–908. [CrossRef] [PubMed]

58. Martín-Sampedro, R.; Fillat, U.; Ibarra, D.; Eugenio, M.E. Use of new endophytic fungi as pretreatment to enhance enzymatic saccharification of *Eucalyptus globulus*. *Bioresour. Technol.* **2015**, *196*, 383–390. [CrossRef] [PubMed]

59. Singh, P.; Suman, A.; Tiwari, P.; Arya, N.; Gaur, A.; Shrivastava, A.K. Biological pretreatment of sugarcane trash for its conversion to fermentable sugars. *World J. Microbiol. Biotechnol.* **2008**, *24*, 667–673. [CrossRef]

60. Yoshida, H. Chemistry of lacquer (Urishi) part I. *J. Chem. Soc.* **1883**, *43*, 472–486. [CrossRef]

61. Mayer, A.M.; Staples, R.C. Laccase: New function for an old enzyme. *Phytochemistry* **2002**, *60*, 551–565. [CrossRef]

62. Fillat, U.; Martín-Sampedro, R.; Macaya-Sanz, D.; Martín, J.A.; Ibarra, D.; Martínez, M.J.; Eugenio, M.E. Screening of eucalyptus wood endophytes for laccase activity. *Process Biochem.* **2016**, *51*, 589–598. [CrossRef]

63. Thurston, C.F. The structure and function of fungal laccases. *Microbiology* **1994**, *140*, 19–26. [CrossRef]

64. Morozova, O.; Shumakovich, G.; Gorbacheva, M.; Sheelev, S.; Yaropolov, A. "Blue laccases". *Biochemisery* **2007**, *72*, 1136–1150. [CrossRef]
65. Barreca, A.M.; Fabbrini, M.; Galli, C.; Gentili, P.; Ljunggren, S. Laccase/mediated oxidation of a lignin model for improved delignification procedures. *J. Mol. Catal. B Enzym.* **2003**, *26*, 105–110. [CrossRef]
66. Bourbonnais, R.; Paice, M.G. Oxidation of non-phenolic substrates. An expanded role for laccase in lignin biodegradation. *FEBS Lett.* **1990**, *267*, 99–102. [CrossRef]
67. Cañas, A.I.; Camarero, S. Laccases and their natural mediators: Biotechnological tools for sustainable eco-friendly processes. *Biotechnol. Adv.* **2010**, *28*, 694–705. [CrossRef] [PubMed]
68. Call, H.-P. Verfahren zur Veränderung, Abbau oder Bleichen von Lignin, Ligninhaltigen Materialien oder Ähnlichen Stoffen. WO 1994029510 A1, 22 December 1994.
69. Paice, M.G.; Bourbonnais, R.; Reid, I.; Archibald, F.S. Kraft pulp bleaching by redox enzymes. In Proceedings of the 9th International Symposium Wood Pulping Chemistry, Montreal, QC, Canada, 9–12 June 1997; pp. PL11–PL14.
70. Camarero, S.; García, O.; Vidal, T.; Colom, J.; del Río, J.C.; Gutiérrez, A.; Gras, J.M.; Monje, R.; Martínez, M.J.; Martínez, A.T. Efficient bleaching of non-wood high-quality paper pulp using laccase–mediator system. *Enzym. Microb. Technol.* **2004**, *35*, 113–120. [CrossRef]
71. Ibarra, D.; Camarero, S.; Romero, J.; Martínez, M.J.; Martínez, A.T. Integrating laccase–mediator treatment into an industrial-type sequence for totally chlorine free bleaching eucalypt kraft pulp. *J. Chem. Technol. Biotechnol.* **2006**, *81*, 1159–1165. [CrossRef]
72. Kunamneni, A.; Plou, F.J.; Ballesteros, A.; Alcalde, M. Laccases and their applications: A patent review. *Recent Pat. Biotechnol.* **2008**, *2*, 10–24. [CrossRef] [PubMed]
73. Camarero, S.; Ibarra, D.; Martínez, M.J.; Martínez, A.T. Lignin-derived compounds as efficient laccase mediators for decolorization of different types of recalcitrant dyes. *Appl. Environ. Microbiol.* **2005**, *71*, 1775–1784. [CrossRef] [PubMed]
74. Camarero, S.; Ibarra, D.; Martínez, A.T.; Romero, J.; Gutiérrez, A.; del Río, J.C. Paper pulp delignification using laccase and natural mediators. *Enzym. Microb. Technol.* **2007**, *40*, 1264–1271. [CrossRef]
75. Gutiérrez, A.; Rencoret, J.; Ibarra, D.; Molina, S.; Camarero, S.; Romero, J.; del Río, J.C.; Martínez, A.T. Removal of lipophilic extractives from paper pulp by laccase and lignin-derived phenols as natural mediators. *Environ. Sci. Technol.* **2007**, *41*, 4124–4129. [CrossRef] [PubMed]
76. Jurado, M.; Prieto, A.; Martínez-Alcalá, A.; Martínez, A.T.; Martínez, M.J. Laccase detoxification of steam-exploded wheat straw for second generation bioethanol. *Bioresour. Technol.* **2009**, *100*, 6378–6384. [CrossRef] [PubMed]
77. Kolb, M.; Sieber, V.; Amann, M.; Faulstich, M.; Schieder, D. Removal of monomer delignification products by laccase from *Trametes versicolor*. *Bioresour. Technol.* **2012**, *104*, 298–304. [CrossRef] [PubMed]
78. Camarero, S.; Caña, A.I.; Nousiainen, P.; Record, E.; Lomascolo, A.; Martínez, M.J.; Martínez, A.T. p-Hydroxycinnamics acids as natural mediators of laccase detoxification of recalcitrant compounds. *Environ. Sci. Technol.* **2008**, *42*, 6703–6709. [CrossRef] [PubMed]
79. Kalyani, D.; Dhiman, S.S.; Kim, H.; Jeya, M.; Kim, I.-W.; Lee, J.-K. Characterization of a novel laccase from the isolated *Coltricia perennis* and its application to detoxification of biomass. *Process Biochem.* **2010**, *47*, 671–678. [CrossRef]
80. Moreno, A.D.; Ibarra, D.; Fernández, J.L.; Ballesteros, M. Different laccase detoxification strategies for ethanol production from lignocellulosic biomass by the thermotolerant yeas *Kluyveromyces marxianus* CECT 10875. *Bioresour. Technol.* **2012**, *106*, 101–109. [CrossRef] [PubMed]
81. Jönsson, L.J.; Palmqvist, E.; Nivelbrant, N.-O.; Hahn-Hägerdal, B. Detoxification of wood hydrolysates with laccase and peroxidase from the white-rot fungus *Trametes versicolor*. *Appl. Microbiol. Biotechnol.* **1998**, *49*, 691–697. [CrossRef]
82. Moreno, A.D.; Ibarra, D.; Mialon, A.; Ballesteros, M. A bacterial laccase for enhancing saccharification and ethanol fermentation of steam-pretreated biomass. *Fermentation* **2016**, *2*, 11. [CrossRef]
83. Moreno, A.D.; Ibarra, D.; Ballesteros, I.; González, A.; Ballesteros, M. Comparing cell viability and ethanol fermentation of the thermotolerant yeast *Kluyveromyces marxianus* and *Saccharomyces cerevisiae* on steam-exploded biomass treated with laccase. *Bioresour. Technol.* **2013**, *135*, 239–245. [CrossRef] [PubMed]

84. Moreno, A.D.; Ibarra, D.; Alvira, P.; Tomás-Pejó, E.; Ballesteros, M. Exploring laccase and mediators behavior during saccharification and fermentation of steam-exploded wheat straw for bioethanol production. *J. Chem. Technol. Biotechnol.* **2016**, *91*, 1816–1825. [CrossRef]

85. Martín, C.; Galbe, M.; Wahlbom, C.F.; Hahn-Hägerdal, B.; Jönsson, L. Ethanol production from enzymatic hydrolysates of sugarcane bagasse using recombinant xylose-utilising *Saccharomyces cerevisiae*. *Enzym. Microb. Technol.* **2002**, *31*, 274–282. [CrossRef]

86. Fang, Z.; Liu, X.; Chen, L.; Shen, Y.; Zhang, X.; Fang, W.; Wang, X.; Bao, X.; Xiao, Y. Identification of a laccase Glac15 from *Ganoderma lucidum* 77002 and its application in bioethanol production. *Biotechnol. Biofuels* **2015**, *8*, 54. [CrossRef] [PubMed]

87. Li, J.; Lin, J.; Zhou, P.; Wu, K.; Liu, H.; Xiong, C.; Gong, Y.; Xiao, W.; Liu, Z. One-pot simultaneous saccharification and fermentation: A preliminary study of a novel configuration for cellulosic ethanol production. *Bioresour. Technol.* **2014**, *161*, 171–178. [CrossRef] [PubMed]

88. Chandel, A.K.; Kapoor, R.K.; Singh, A.; Kuhad, R.C. Detoxification of sugarcane bagasse hydrolysate improves ethanol production by *Candida shehatae* NCIM 3501. *Bioresour. Technol.* **2007**, *98*, 1947–1950. [CrossRef] [PubMed]

89. Moreno, A.D.; Ibarra, D.; Ballesteros, I.; Fernández, J.L.; Ballesteros, M. Ethanol from laccase-detoxified lignocellulose by the thermotolerant yeast *Kluyveromyces marxianus*—Effects of steam pretreatment conditions, process configurations and substrate loadings. *Biochem. Eng. J.* **2013**, *79*, 94–103. [CrossRef]

90. Moreno, A.D.; Tomás-Pejó, E.; Ibarra, D.; Ballesteros, M.; Olsson, L. In situ laccase treatement enhances the fermentability of steam-exploded wheat straw in SSCF processes at high dry matter consistencies. *Bioresour. Technol.* **2013**, *143*, 337–343. [CrossRef] [PubMed]

91. Moreno, D.; Tomás-Pejó, E.; Ibarra, D.; Ballesteros, M.; Olsson, L. Fed-batch SSCF using steam-exploded wheat straw at high dry matter consistencies and a xylose-fermenting *Saccharomyces ceresiae* strain: Effect of laccase supplementation. *Biotechnol. Biofuels* **2013**, *6*, 160. [CrossRef] [PubMed]

92. Alvira, P.; Moreno, A.D.; Ibarra, D.; Sáez, F.; Ballesteros, M. Improving the fermentation performance of *Saccharomyces cerevisiae* by laccase during ethanol production from steam-exploded wheat straw at high-substrate loadings. *Biotechnol. Prog.* **2013**, *29*, 74–82. [CrossRef] [PubMed]

93. Larsson, S.; Cassland, P.; Jönsson, L. Development of a *Saccharomyces cerevisiae* strain with enhanced resistance to phenolic fermentation inhibitors in lignocellulose hydrolysates by heterologous expression of laccase. *Appl. Environ. Microbiol.* **2001**, *67*, 1163–1170. [CrossRef] [PubMed]

94. Ludwig, D.; Amann, M.; Hirth, T.; Rupp, S.; Zibek, S. Development and optimization of single and combined detoxification processes to improve the fermentability of lignocellulose hydrolyzates. *Bioresour. Technol.* **2013**, *133*, 455–461. [CrossRef] [PubMed]

95. De La Torre, M.; Martín-Sampedro, R.; Fillat, U.; Hernández, M.; Arias, M.E.; Eugenio, M.E.; Ibarra, D. Comparison of bacterial and fungal laccases for enhancing saccharification and ethanol fermentation of steam-pretreated biomass. In Proceedings of the IX Ibero-American Congress on Pulp and Paper Research—CIADICYP, Helsinki, Finland, 5–9 September 2016; p. 29.

96. Enguita, F.J.; Matias, P.M.; Martins, L.O.; Placido, D.; Henriques, A.O.; Carrondo, M.A. Spore-coat laccase CotA from *Bacillus subtilis*: Crystallization and preliminary X-ray characterization by the MAD method. *Acta Crystallogr. D Biol. Cristallogr.* **2002**, *58*, 1490–1493. [CrossRef]

97. Moya, R.; Saastamoinen, P.; Hernández, M.; Suurnäkki, A.; Arias, E.; Mattinen, M.-L. Reactivity of bacterial and fungal laccases with lignin under alkaline conditions. *Bioresour. Technol.* **2011**, *102*, 10006–10012. [CrossRef] [PubMed]

98. Alfani, A.; Gallifuoco, F.F.; Saporosi, A.; Spera, A.; Cantarella, M. Comparison of SHF and SSF process for bioconversion of steam-exploded wheat straw. *J. Ind. Microbiol. Biotechnol.* **2000**, *25*, 184–192. [CrossRef]

99. Abdel-Banat, B.M.A.; Hoshida, H.; Ano, A.; Hoshida, H.; Nonklang, S.; Akada, R. High-temperature fermentation: How can processes for ethanol production at high temperatures become superior to the traditional process using mesophilic yeast? *Appl. Microbiol. Biotechnol.* **2010**, *85*, 861–867. [CrossRef] [PubMed]

100. Banerjee, S.; Mudliar, S.; Sen, R.; Giri, B.; Staputti, D.; Chakrrabarti, T.; Pandey, R.A. Commercializing lignocellulosic bioethanol: Technology bottlenecks and possible remedies. *Biofuels Bioprod. Biorefin.* **2009**, *4*, 77–93. [CrossRef]

101. Rosgaard, L.; Andric, P.; Dam-Johansen, K.; Pedersen, S.; Meyer, A.S. Effects of substrate loading on enzymatic hydrolysis and viscosity of pretreated barley straw to ethanol. *Appl. Biochem. Biotechnol.* **2007**, *143*, 27–40. [CrossRef] [PubMed]

102. Tabka, M.G.; Herpoel-Gimbert, I.; Monod, F.; Asther, M.; Sigoillot, J.C. Enzymatic saccharification of wheat straw for bioethanol production by a combined cellulose, xylanase and feruloyl esterase treatment. *Enzym. Microb. Technol.* **2006**, *39*, 897–902. [CrossRef]

103. Oliva-Taravilla, A.; Tomás-Pejó, E.; Demuez, M.; González-Fernández, C.; Ballesteros, M. Inhibition of cellulose enzymatic hydrolysis by laccase-derived compounds from phenols. *Biotechnol. Prog.* **2015**, *31*, 700–706. [CrossRef] [PubMed]

104. Oliva-Taravilla, A.; Tomás-Pejó, E.; Demuez, M.; González-Fernández, C.; Ballesteros, M. Phenols and lignin: Key players in reducing enzymatic hydrolysis yields of steam-pretreated biomass in presence of laccase. *J. Biotechnol.* **2016**, *218*, 94–101. [CrossRef] [PubMed]

105. Oliva-Taravilla, A.; Moreno, A.D.; Demuez, M.; Ibarra, D.; Tomás-Pejó, E.; González-Fernández, C.; Ballesteros, M. Unraveling the effects of laccase treatment on enzymatic hydrolysis of steam-exploded wheat straw. *Bioresour. Technol.* **2015**, *175*, 209–215. [CrossRef] [PubMed]

106. Kuila, A.; Mukhopadhyay, M.; Tuli, D.K.; Banerjee, R. Accessibility of enzymatically delignified Bambusa bambos for efficient hydrolysis at minimum cellulase loading: An optimization study. *Enzym. Res.* **2011**. [CrossRef] [PubMed]

107. Kuila, A.; Mukhopadhyay, M.; Tuli, D.K.; Banerjee, R. Production of ethanol from lignocellulosics: An enzymatic venture. *EXCLI J.* **2011**, *10*, 85–96. [PubMed]

108. Mukhopadhyay, M.; Kuila, A.; Tuli, D.K.; Banerjee, R. Enzymatic depolymerization of *Ricinus communis*, a potential lignocellulosic for improved saccharification. *Biomass Bioenerg.* **2011**, *35*, 3584–3591. [CrossRef]

109. Rajak, R.C.; Banerjee, R. Enzyme mediated biomass pretreatment and hydrolysis: A biotechnological venture towards bioethanol production. *RSC Adv.* **2016**, *6*, 61301–61311. [CrossRef]

110. Rencoret, J.; Pereira, A.; del Río, J.C.; Martínez, A.T.; Gutiérrez, A. Laccase-mediator pretreatment of wheat straw degrades lignin and improves saccharification. *Bioenerg. Res.* **2016**, *9*, 917–930. [CrossRef]

111. Rico, A.; Rencoret, J.; del Río, J.C.; Martínez, A.T.; Gutiérrez, A. In-depth 2D NMR study of lignin modification during pretreatment of eucalyptus wood with laccase and mediators. *Bioenerg. Res.* **2015**, *8*, 211–230. [CrossRef]

112. Singh, R.; Hu, J.; Regner, M.R.; Round, J.W.; Ralph, R.; Saddler, J.N.; Eltis, L.D. Enhanced delignification of steam-pretreated poplar by a bacterial laccase. *Sci. Rep.* **2017**, *7*, 42121. [CrossRef] [PubMed]

113. Li, J.; Sun, F.; Li, X.; Yan, Z.; Yuan, Y.; Liu, X.F. Enhanced saccharification of corn straw pretreated by alkali combining crude ligninolytic enzymes. *J. Chem. Technol. Biotechnol.* **2012**, *87*, 1687–1693. [CrossRef]

114. Yang, P.; Jiang, S.; Zheng, Z.; Luo, S.; Pan, L. Effect of alkali and laccase pretreatment of *Brassica campestris* straw: Architecture, crystallisation, and saccharification. *Polym. Renew. Resour.* **2011**, *2*, 21–34.

115. Qiu, W.; Chen, H. Enhanced the enzymatic hydrolysis efficiency of wheat straw after combined steam explosion and laccase pretreatment. *Bioresour. Technol.* **2012**, *118*, 8–12. [CrossRef] [PubMed]

116. Palonen, H.; Viikari, L. Role of oxidative enzymatic treatments on enzymatic hydrolysis of softwood. *Biotechnol. Bioeng.* **2004**, *86*, 550–557. [CrossRef] [PubMed]

117. Moilanen, U.; Kellock, M.; Galkin, S.; Viikari, L. The laccase catalyzed modification of lignin for enzymatic hydrolysis. *Enzym. Microb. Technol.* **2011**, *49*, 492–498. [CrossRef] [PubMed]

118. Moilanen, U.; Kellock, M.; Várnai, A.; Andberg, M.; Viikari, L. Mechanisms of laccase-mediator treatments improving the enzymatic hydrolysis of pre-treated spruce. *Biotechnol. Biofuels* **2014**, *7*, 177. [CrossRef] [PubMed]

119. Sitarz, A.K.; Mikkelsen, J.D.; Højrup, P.; Meyer, A.S. Identification of a laccase from *Ganoderma lucidum* CBS 229.93 having potential for enhancing cellulase catalyzed lignocellulose degradation. *Enzym. Microb. Technol.* **2013**, *53*, 378–385. [CrossRef] [PubMed]

120. Zhang, J.; Qu, Y.; Xiao, P.; Wang, X.; Wang, T.; He, F. Improved biomass saccharification by *Trichoderma reesei* through heterologous expression of *lacA* gene *Trametes* sp. AH28-2. *J. Biosci. Bioeng.* **2012**, *113*, 697–703. [CrossRef] [PubMed]

121. Davidi, L.; Moraïs, S.; Artzi, L.; Knop, D.; Hadar, Y.; Arfi, Y.; Bayer, E.A. Toward combined delignification and saccharification of wheat straw by a laccase-containing designer cellulosome. *Proc. Natl. Acad. Sci. USA* **2016**, *113*, 10854–10859. [CrossRef] [PubMed]

122. Chen, C.-H.; Hsieh, Z.-S.; Cheepudom, J.; Yang, C.-H.; Meng, M. A 24.7-kDa copper-containing oxidase, secreted by *Thermobifida fusca*, significantly increasing the xylanase/cellulase-catalyzed hydrolysis of sugarcane bagasse. *Appl. Microbiol. Biotechnol.* **2013**, *97*, 8977–8986. [CrossRef] [PubMed]

123. Barneto, G.A.; Aracri, E.; Andreu, G.; Vidal, T. Investigating the structure-effect relationships of various natural phenols used as laccase mediators in the biobleaching of kenaf and sisal pulps. *Bioresour. Technol.* **2012**, *112*, 327–335. [CrossRef] [PubMed]

124. Zanirun, Z.; Bahrin, E.K.; Lai-Yee, P.; Hassan, M.; Abd-Aziz, S. Enhancement of fermentable sugars production from oil palm empty fruit bunch by ligninolytic enzymes mediator system. *Int. Biodeterior. Biodegrad.* **2015**, *105*, 13–20. [CrossRef]

125. Lu, C.; Wang, H.; Luo, Y.; Guo, L. An efficient system for predelignification of gramineous biofuel feedstock in vitro: Application of a laccase from *Pycnoporus sanguineus* H275. *Process Biochem.* **2010**, *45*, 1141–1147. [CrossRef]

126. Al-Zuhair, S.; Abualreesh, M.; Ahmed, K.; Razak, A.A. Enzymatic delignification of biomass for enhanced fermentable sugars production. *Energy Technol.* **2015**, *3*, 1–8. [CrossRef]

127. Moniruzzaman, M.; Ono, T. Ionic liquid assisted enzymatic delignification of wood biomass: A new green and efficient approach for isolating of cellulose fibers. *Biochem. Eng. J.* **2012**, *60*, 156–160. [CrossRef]

128. Financie, R.; Moniruzzaman, M.; Uemura, Y. Enhanced enzymatic delignification of oil palm biomass with ionic liquid pretreatment. *Biochem. Eng. J.* **2016**, *110*, 1–7. [CrossRef]

129. Plácido, J.; Imam, T.; Capareda, S. Evaluation of ligninolytic enzymes, ultrasonication and liquid hot water as pretreatments for bioethanol production from cotton gin trash. *Bioresour. Technol.* **2013**, *139*, 203–208. [CrossRef] [PubMed]

130. Plácido, J.; Capareda, S. Analysis of ultrasonication pretreatment in bioethanol production from cotton gin trash using FTIR spectroscopy and principal component analysis. *Bioresour. Bioprocess.* **2014**, *1*, 1–9. [CrossRef]

131. Nagula, K.N.; Pandit, A.B. Process intensification of delignification and enzymatic hydrolysis of delignified cellulosic biomass using various process intensification techniques including cavitation. *Bioresour. Technol.* **2016**, *213*, 162–168. [CrossRef] [PubMed]

132. Gutiérrez, A.; Rencoret, J.; Cadena, E.M.; Rico, A.; Barth, D.; del Río, J.C.; Martínez, A.T. Demonstration of laccase-based removal of lignin from wood and non-wood plant feedstocks. *Bioresour. Technol.* **2012**, *119*, 114–122. [CrossRef] [PubMed]

133. Rico, A.; Rencoret, J.; del Río, J.C.; Martínez, A.; Gutiérrez, A. Pretreatment with laccase and a phenolic mediator degrades lignin and enhances saccharification of Eucalyptus feedstock. *Biotechnol. Bifuels* **2014**, *7*, 6. [CrossRef] [PubMed]

134. Al-Zuhair, S.; Ahmed, K.; Abdulrazak, A.; El-Nass, M.H. Synergistic effect of pretreatment and hydrolysis enzymes on the production of fermentable sugars from date palm lignocellulosic waste. *J. Ind. Eng. Chem.* **2013**, *19*, 413–415. [CrossRef]

135. Chen, Q.; Marshall, N.M.; Geib, S.M.; Tien, M.; Richard, T.L. Effects of laccase on lignin depolymerization and enzymatic hydrolysis of ensiled corn stover. *Bioresour. Technol.* **2012**, *117*, 186–192. [CrossRef] [PubMed]

136. Heap, L.; Green, A.; Brown, D.; Dongen, B.V.; Turner, N. Role of laccase as an enzymatic pretreatment method to improve lignocellulosic saccharification. *Catal. Sci. Technol.* **2014**, *4*, 2251–2259. [CrossRef]

137. Martín-Sampedro, R.; Capanema, E.A.; Hoeger, I.; Villar, J.C.; Rojas, O.J. Lignin changes after steam explosion and laccase-mediator treatment of eucalyptus wood chips. *J. Agric. Food. Chem.* **2011**, *59*, 8761–8769. [CrossRef] [PubMed]

138. Martín-Sampedro, R.; Eugenio, M.E.; García, J.C.; López, F.; Villar, J.C.; Diaz, M.J. Steam explosion and enzymatic pre-treatments as an approach to improve the enzymatic hydrolysis of *Eucalyptus globulus*. *Biomass Bioeng.* **2012**, *42*, 97–106. [CrossRef]

139. Milstein, O.; Haars, A.; Majchercyk, A.; Tautz, D.; Zanker, H.; Hüttermann, A. Removal of chlorophenols and chlorolignins from bleaching effluents by combining chemical and biological treatment. *Water Sci. Technol.* **1988**, *20*, 161–170.

140. Widsten, P.; Kandelbauer, A. Adhesion improvement of lignocellulosic products by enzymatic pre-treatment. *Biotechnol. Adv.* **2008**, *26*, 379–386. [CrossRef] [PubMed]

141. Chandra, R.P.; Ragauskas, A.J. Evaluating laccase-facilitated coupling of phenolic acids to high-yield kraft pulps. *Enzym. Microb. Technol.* **2002**, *30*, 855–861. [CrossRef]

142. Elegir, G.; Kindl, A.; Sadocco, P.; Orlandi, M. Development of antimicrobial cellulose packaging through laccase-mediated grafting of phenolic compounds. *Enzym. Microb. Technol.* **2008**, *43*, 84–92. [CrossRef]

143. Fillat, A.; Gallardo, O.; Vidal, T.; Pastor, F.I.J.; Díaz, P.; Roncero, M.B. Enzymatic grafting of natural phenols to flax fibres: Development of antimicrobial properties. *Carbohydr. Polym.* **2012**, *87*, 146–152. [CrossRef]

144. Roth, S.; Spiess, A.C. Laccases for biorefinery applications: A critical review on challenges and perspectives. *Bioprocess Biosyst. Eng.* **2015**, *38*, 2285–2313. [CrossRef] [PubMed]

fermentation

MDPI

Article

Process Development for Enhanced 2,3-Butanediol Production by *Paenibacillus polymyxa* DSM 365

Christopher Chukwudi Okonkwo [1], Victor C. Ujor [2,*], Pankaj K. Mishra [3] and Thaddeus Chukwuemeka Ezeji [1,*]

[1] Department of Animal Sciences, The Ohio State University, and Ohio State Agricultural Research and Development Center (OARDC), 305 Gerlaugh Hall, 1680 Madison Avenue, Wooster, OH 44691, USA; okonkwo.5@osu.edu

[2] Renewable Energy Program, Agricultural Technical Institute, The Ohio State University, 1328 Dover Road, Wooster, OH 44691, USA

[3] Crop Production Division, ICAR—Vivekananda Parvatiya Krishi Anusandhan Sansthan, Department of Agricultural Research and Education (DARE), Ministry of Agriculture and Farmers Welfare, Government of India), Almora 263601, Uttarakhand, India; Pankaj.Mishra@icar.gov.in

* Correspondence: ujor.1@osu.edu (V.C.U.); ezeji.1@osu.edu (T.C.E.); Tel.: +330-287-1268 (V.C.U.); +330-263-3796 (T.C.E.), Fax: +330-287-1205 (V.C.U.); +330-263-3949 (T.C.E.)

Academic Editor: Badal C. Saha
Received: 9 March 2017; Accepted: 2 May 2017; Published: 7 May 2017

Abstract: While chiral 2,3-Butanediol (2,3-BD) is currently receiving remarkable attention because of its numerous industrial applications in the synthetic rubber, bioplastics, cosmetics, and flavor industries, 2,3-BD-mediated feedback inhibition of *Paenibacillus polymyxa* DSM 365 limits the accumulation of higher concentrations of 2,3-BD in the bioreactor during fermentation. The Box-Behnken design, Plackett-Burman design (PBD), and response surface methodology were employed to evaluate the impacts of seven factors including tryptone, yeast extract, ammonium acetate, ammonium sulfate, glycerol concentrations, temperature, and inoculum size on 2,3-butanediol (2,3-BD) production by *Paenibacillus polymyxa* DSM 365. Results showed that three factors; tryptone, temperature, and inoculum size significantly influence 2,3-BD production ($p < 0.05$) by *P. polymyxa*. The optimal levels of tryptone, inoculum size, and temperature as determined by the Box-Behnken design and response surface methodology were 3.5 g/L, 9.5%, and 35 °C, respectively. The optimized process was validated in batch and fed-batch fermentations in a 5-L Bioflo 3000 Bioreactor, and 51.10 and 68.54 g/L 2,3-BD were obtained, respectively. Interestingly, the production of exopolysaccharides (EPS), an undesirable co-product, was reduced by 19% when compared to the control. These results underscore an interplay between medium components and fermentation conditions, leading to increased 2,3-BD production and decreased EPS production by *P. polymyxa*. Collectively, our findings demonstrate both increased 2,3-BD titer, a fundamental prerequisite to the potential commercialization of fermentative 2,3-BD production using renewable feedstocks, and reduced flux of carbons towards undesirable EPS production.

Keywords: *Paenibacillus polymyxa*; butanediol; acetoin; glycerol; optimization

1. Introduction

The compound 2,3-Butanediol (2,3-BD) is an industrial platform chemical with vast industrial applications, particularly for its potential use in the synthesis of 1,3-butadiene (1,3-BD), a monomer of synthetic rubber. Other applications of 2,3-BD include the synthesis of methyl ethyl ketone (MEK), a fuel additive with a higher heat of combustion than ethanol, and as solvents for lacquers and resins [1]. Furthermore, 2,3-BD finds applications as an antifreeze due to its low freezing temperature of −60 °C [2], an ink additive, as a chemical feedstock for the production of acetoin and diacetyl, vital flavor enhancers in the food industry [3], and as an additive in aviation fuel. Due to the finite nature of petroleum and the need to reduce society's dependence on petroleum-derived feedstocks for industrial processes, it has become imperative to develop sustainable feedstocks such as 2,3-BD from renewable resources. At present, 2,3-BD is produced from hydrocarbons by the cracking of butane and 2-butene in which the resulting product is further hydrolyzed to 2,3-BD [4,5]. Recently, the often encountered instabilities in crude oil price have re-ignited interest in fermentative 2,3-BD production from cheap renewable feedstocks. To reach this goal, multifarious research efforts are currently underway to increase the yield, titer, and productivity of microbe-derived 2,3-BD. These include metabolic engineering of producer organisms to produce and tolerate higher 2,3-BD concentrations, and the optimization of fermentation media components and conditions for maximal 2,3-BD accumulation in the broth. In this study, we sought to optimize 2,3-BD production by *Paenibacillus polymyxa* DSM 365 (hereafter referred to as *P. polymyxa*) by assessing the impacts of both medium components and fermentation conditions on 2,3-BD accumulation.

P. polymyxa was chosen for this study due to its non-pathogenicity and capacity to produce 98% levo 2,3-BD; the industrially preferred 2,3-BD isomer due to its properties, which make it amenable to important chemical reactions that generate key industrially applicable products, such as dehydration to 1,3-BD (for synthetic rubber production), dehydrogenation to acetoin or diacetyl (flavor enhancers and essential components in fragrances), ketalization to Methyl tert-butyl ether (fuel additive), and esterification to 2,3-BD diester (used as a precursor in the synthesis and compounding of cosmetics, drugs, and thermoplastic polymers [1,6].

To assess the 2,3-BD production capacity of *P. polymyxa*, we first conducted batch fermentations in 100 mL Pyrex bottles, which resulted in a maximum 2,3-BD concentration of 24 g/L [7]. Batch fermentation in the bioreactor (6-L) produced 27 g/L 2,3-BD, whereas fed-batch fermentation (in the bioreactor) resulted in 47 g/L, despite excess glucose supply [7]. Therefore, we rationalized that in addition to other factors, 2,3-BD-mediated feedback inhibition might pose a significant roadblock to the accumulation of 2,3-BD during fermentation, and this assumption was confirmed by assaying 2,3-BD toxicity against *P. polymyxa* in fermentation cultures [7]. We observed that 2,3-BD exerts a concentration-dependent toxicity on *P. polymyxa* with ~50 g/L 2,3-BD as the toxic threshold above which cell growth stalls considerably and the accumulated 2,3-BD is converted backwards to acetoin, the precursor of 2,3-BD; most plausibly to alleviate 2,3-BD-mediated toxicity [7]. In addition, a significant portion of sugar substrates are diverted to exopolysaccharides (EPS) production during 2,3-BD fermentation, thereby lowering 2,3-BD yield and complicating its recovery from the fermentation broth [7,8]. Therefore, if *P. polymyxa* 2,3-BD fermentation is to reach an industrial-scale, it is critical to determine the optimal conditions and medium components necessary for marked 2,3-BD accumulation and tolerance during fermentation. Further, cheaply available substrates such as glycerol, which is currently accumulated in excess as a by-product of biodiesel production [9,10], holds significant promise towards improving the economics of 2,3-BD fermentation, either as a sole carbon source or as a supplement to glucose or other sugars. In fact, glycerol has been shown to support 2,3-BD production by *Klebsiella pneumoniae* as a sole carbon source [11,12]. Thus, we investigated the optimal conditions and medium components for high 2,3-BD production by *P. polymyxa* using a glycerol-supplemented medium. In addition to lowering the overall substrate cost, glycerol catabolism furnishes the cell with additional NADH [13,14], which supplies extra reducing power for 2,3-BD dehydrogenase, the final enzyme of the 2,3-BD pathway, which consumes NADH during the conversion of acetoin to 2,3-BD [15].

Previous optimization studies focused largely on enhancing 2,3-BD production. These studies either targeted medium components only, or fermentation conditions without a holistic evaluation of both parameters (medium components and fermentation conditions; [8,16]). Medium components and fermentation conditions such as temperature, inoculum size, pH, and aeration rate most reasonably interact during fermentation to engender 2,3-BD production. Therefore, in this study, select medium components and fermentation conditions were assessed collectively for their capacity to enhance 2,3-BD production by employing various optimization strategies. Plackett-Burman experimental design, path of steepest ascent method, Box-Behnken experimental design, and response surface methodology strategies were employed to optimize 2,3-BD production by *P. polymyxa*. The medium components tested in this study include yeast extract, tryptone, ammonium acetate, ammonium sulfate, and crude glycerol; whereas the fermentation conditions that were extensively investigated include temperature and inoculum size. These factors were shown to influence 2,3-BD production by *P. polymyxa* from our one-factor-at-a-time experiments.

2. Methodology

2.1. Experimental Methods

2.1.1. Microorganism and Culture Preparation

P. polymyxa was obtained from the German Collection of Microorganisms and Cell Culture, Braunschweig, Germany (DSMZ—Deutsche Sammlung von Mikroorganismen und Zellkulturen). Lyophilized cells were reactivated by inoculating into Luria bertani (LB) broth, grown overnight (12 h), and then stored as glycerol (50% sterile glycerol) stock at −80 °C until further use. Glucose, yeast extract, and tryptone were prepared and sterilized separately at 121 °C for 15 min followed by cooling to 50 °C prior to mixing with filter-sterilized components (buffer and trace element solution), and this mixture forms the final pre-culture medium. For inoculum preparation, 1 mL of *P. polymyxa* glycerol stock was inoculated into 30 mL of pre-culture medium. The pre-culture medium contained (g/L); 20.0 glucose, 5.0 yeast extract (YE; Sigma-Aldrich, St. Louis, MO, USA), 5.0 tryptone (Sigma-Aldrich, St. Louis, MO, USA), 0.2 $MgSO_4$, and 3.0 $(NH_4)_2SO_4$. The pre-culture was supplemented with 0.9 mL of phosphate buffer (pH 6.5) and 0.09 mL of trace element solution. The phosphate buffer (pH 6.5) contained (g/L); 3.5 KH_2PO_4, 2.75 K_2HPO_4, while the trace element solution was prepared by dissolving 0.4 g $FeSO_4$ in 3 mL 25% HCl, followed by the addition of 500 mL double-distilled H_2O and (g); 0.8 H_3BO_3, 0.04 $CuSO_4 \cdot 5H_2O$, 0.04 $NaMoO_4 \cdot 2H_2O$, 5.0 $MnCl_2 \cdot 4H_2O$, 0.1 $ZnSO_4 \cdot 7H_2O$, 0.08 Co $(NO_3)_2 \cdot 6H_2O$, 1.0 $CaCl_2 \cdot 2H_2O$, and 0.01 biotin. The trace element solution was made up to 1 L with double-distilled H_2O. The pre-culture was incubated aerobically at 37 °C and 200 rpm for 10–12 h in an incubator shaker (New Brunswick Scientific, Edison, NJ, USA). When the optical density (OD_{600nm}) of the pre-culture reached 1.0–1.2, 30 mL (10 mL each) of actively growing cells were distributed in three 250 mL flasks containing 90 mL sterile pre-culture medium each and incubated for another 2–3 h until the OD_{600nm} reached 1.0–1.2, and then the pre-culture was transferred to production medium. Phosphate buffer and trace element solution were prepared separately and filter-sterilized using a 0.22 μm PES filter (Corning Incorporated, Corning, NY, USA).

2.1.2. Batch and Fed-Batch Fermentations

Batch and fed-batch fermentations were conducted in a 5 L Bioflo 3000 Bioreactor (New Brunswick Scientific, Edison, NJ, USA) with a 2 L starting volume. The bioreactor was equipped with sensors for measuring pH and temperature and stirrers for medium agitation. The medium was continuously stirred by means of 2 Rushton impellers (3-plate). Fermentations were conducted aerobically by sparging sterile air into the medium at a flow rate of 150 mL/min through a 0.2 μm PTFE Acro®50 sterile filter (Pall Corporation, Ann Arbor, MI, USA) using a Masterflex L/S® Pump (Cole-Parmer Instrument Company, Vernon Hills, IL, USA) through the top of the bioreactor into the fermentation medium. In addition to the concentrations of the medium components studied, the production medium contained (g/L): 120 glucose, 3.5 KH_2PO_4, 2.75 K_2HPO_4, 0.2 $MgSO_4$, 0.05 $CoCl_2$, 10.0 3-(*N*-morpholino) propanesulfonic acid (MOPS), and 6 mL of trace element solution (described above: microorganism and culture preparation). All medium components were prepared separately and then mixed under aseptic conditions. Fermentation was started at an initial pH of 6.5 ± 0.1 and the pH was externally controlled with 12.5% NH_4OH or 6.5 normal H_3PO_4 when the pH dropped below 6.0 ± 0.1 or increased above 6.5 ± 0.1. The fermentation medium was stirred at 300 rpm and the culture was fed when the broth glucose concentration fell below 15 g/L for fed-batch fermentations. Each round of sugar feeding was accompanied by the addition of half-strength of the other medium components (buffer and trace element solutions).

2.1.3. Analytical Methods

Cell growth was determined by measuring optical density (OD_{600}) in a DU® Spectrophotometer (Beckman Coulter Inc., Brea, CA, USA). Changes in pH were measured using an Acumen® Basic pH meter (Fischer Scientific, Pittsburgh, PA, USA). The concentrations of 2,3-BD, acetoin, ethanol, and acetic acid were determined using a 7890A Agilent gas chromatograph (Agilent Technologies Inc., Wilmington, DE, USA) equipped with a flame ionization detector (FID) and a J $\times$ W 19091 N-213 capillary column [30 m (length) $\times$ 320 µm (internal diameter) $\times$ 0.5 µm (HP-Innowax film thickness)]. The carrier gas was nitrogen, and the inlet and detector temperatures were maintained at 250 and 300 °C, respectively. The oven temperature was programmed to span from 60 to 200 °C with 20 °C min^{-1} increments, and a 5 min hold at 200 °C. Samples (1 µL) were injected with a split ratio of 10:1.

Glucose concentration was determined by HPLC using a Waters 2796 Bioseparations Module equipped with an Evaporative Light Scattering Detector (ELSD; Waters, Milford, MA, USA) and a 9 µm Aminex HPX-87P column; 300 mm (length) $\times$ 7.8 mm (internal diameter) connected in series to a 4.6 mm (internal diameter) $\times$ 3 cm (length) Aminex deashing guard column (Bio-Rad, Hercules, CA, USA). The column temperature was maintained at 65 °C. The mobile phase was HPLC-grade water maintained at a flow rate of 0.6 mL/min. The amounts of EPS produced were measured using a previously described method [7].

2.2. Experimental Design and Data Analysis

Other authors have previously reported that yeast extract, ammonium acetate, ammonium sulfate, glycerol, and tryptone, as well as temperature and inoculum size influence microbial 2,3-BD production [16–20]. Some of these studies concluded that using high concentrations of expensive yeast extract (up 60 g/L) was crucial for optimal 2,3-BD production. Therefore, we first conducted one-factor-at-a-time experiments to evaluate the degrees of effect exerted by these factors on the 2,3-BD production capacity of *P. polymyxa*. The one-factor-at-a-time experiments underlined the effects of yeast extract, ammonium acetate, ammonium sulfate, glycerol, and tryptone, as well as temperature and inoculum size on 2,3-BD production by *P. polymyxa*; hence, we employed the Plackett-Burman design, path of steepest ascent, Box-Behnken, and response surface methodology for further optimization studies.

2.2.1. Plackett-Burman Design

Placket-Burman design allowed for the evaluation of important factors that influence 2,3-BD production based on the assumption that the selected factors do not interact. In this design, each factor was defined at two levels; a high (+1) and a low (−1), which represent two different concentrations or condition set points. The actual experimental values were defined according to the equation;

$$X_i = x_i - x_0/\Delta x_i \ (I = 1, 2, 3, \dots, k) \tag{1}$$

From the equation above, X_i is the defined value of an independent factor such as inoculum size, temperature, glycerol, tryptone, yeast extract, ammonium acetate, and ammonium sulfate, x_i is the real value of an independent factor, and x_0 represents the real value of an independent factor at a center point value. Furthermore, Δx_i is the difference between the real value at the center point (x_0) and the real values at the lower or upper point of an independent factor. The data obtained from the design were fitted to a first-order model for 2,3-BD production as shown in the equation below;

$$Y = \beta_0 + \sum \beta_i X_i \tag{2}$$

Y is the concentration of 2,3-BD obtained from each experimental run, while β_0 is the intercept, and X_i represents the ith factor (X_1–X_7; see Table 1) and β_i is the regression coefficient of each factor (X_1–X_7; see Table 1) [21]. The resulting data were then analyzed and fitted to a linear regression using the Design Expert software package (version 10.0, Stat-Ease, Inc. Minneapolis, MN, USA). Analysis of variance (ANOVA) at 95% confidence interval ($p < 0.05$) was used to determine the significant factors. The significant factors were chosen for the path of steepest ascent experiment.

Table 1. Statistical analysis of Plackett-Burman design results showing the effect of medium components and fermentation conditions on 2,3-BD production by *P. polymyxa*.

Factor	Low Level (−1)	High Level (+1)	% Contribution	*t* Value	*p* Value
X_1: Inoculum size (%)	6	10	39.56	10.29	0.0260 *
X_2: Temperature (°C)	35	37	23.23	−7.88	0.0348 *
X_3: CH_3COONH_4 (g/L)	3	5	4.87	3.61	0.0753
X_4: $(NH_4)_2SO_4$ (g/L)	2	4	9.30	4.99	0.0595
X_5: Glycerol (g/L)	5	10	0.29	0.89	0.2112
X_6: Yeast extract (g/L)	5	7	4.51	3.47	0.1174
X_7: Tryptone (g/L)	5	7	18.23	−6.99	0.0377 *

$R^2 = 0.9993$, Adj $R^2 = 0.9951$ * Statistical significance.

2.2.2. Path of Steepest Ascent

The path of steepest ascent method enables the determination of three optimal levels at which to further optimize each of the significant factors obtained from the Plackett-Burman design. This is carried out by moving the center point value of each factor sequentially along the path of steepest ascent until no further increase in 2,3-BD is obtained; i.e., the center point value of each selected factor is either increased or decreased until the maximum 2,3-BD achieved begins to decline [22,23].

2.2.3. Box-Behnken Design and Response Surface Methodology

The optimal nutrient concentrations and fermentation conditions for maximum 2,3-BD production were determined by employing the Box-Behnken design and response surface methodology. The three significant factors—tryptone, temperature, and inoculum size—selected from the Plackett-Burman design were varied at three levels. These factors were shown to exert the most significant effects on 2,3-BD production. The Box-Behnken design, when integrated with response surface methodology quantifies the relationship between the independent input factors and the obtained response surfaces [24]. In this study, the association between the responses and the three important factors were determined according to the second order polynomial function:

$$Y = \beta_0 + \sum \beta_i X_i + \sum \beta_{ii} X_i^2 + \sum \beta_{ij} X_i X_j \ (i,j = 1,2 \dots ,k) \tag{3}$$

In the equation above, Y is the calculated 2,3-BD response function and β_0 is the estimated regression coefficient of the fitted response at the center point of the design, while X_i represents the corresponding actual value factors for inoculum size, temperature, and tryptone. The regression coefficient for the linear terms is represented by β_i, whereas β_{ij} is the interaction effects and β_{ii} is the quadratic effect. The Design Expert software package (Version 10.0, Stat-Ease Inc., Minneapolis, MN, USA) was used to calculate and analyze the second-order polynomial coefficients. Analysis of variance (ANOVA) was used to test the significance of independent factors (tryptone, inoculum size, temperature) and their interactions at an alpha (α) level of 0.05.

3. Results and Discussion

3.1. Plackett-Burman Design

To obtain an optimized medium and fermentation conditions for improved 2,3-BD production by *P. polymyxa*, medium components and fermentation conditions were evaluated. Nutrient components including tryptone, glycerol, ammonium sulfate, and other fermentation parameters (temperature and inoculum size) were studied to determine the extent to which these factors impact 2,3-BD production by *P. polymyxa*. All media contained crude glycerol. Crude glycerol served a dual purpose in the medium—a source of carbon and an additional source of NADH, as glycerol catabolism generates two additional molecules of NADH, relative to glucose on a molar basis [13,14]. The reducing equivalent furnished by NADH is critical for 2,3-BD biosynthesis, as NADH is required for the reduction of acetoin to 2,3-BD [25,26].

Tryptone was used as an organic nitrogen source in addition to yeast extract, whereas ammonium acetate and ammonium sulfate served as inorganic sources of nitrogen. For fermentation conditions, temperature and inoculum size were selected for investigation. Like most microbial processes, 2,3-BD biosynthesis is enzyme-controlled, therefore, fermentation temperature impacts substrate consumption and 2,3-BD production, because enzyme activity is temperature-dependent [19,20]. Inoculum size has been reported to increase substrate utilization with improved 2,3-BD production and yield [19,27]. To our knowledge, this is the first report of combined optimization of fermentation nutrients and conditions for enhanced 2,3-BD production by *P. polymyxa*. To determine the factors with significant influence on 2,3-BD production, the Plackett-Burman design was employed to test each of the factors at two levels. Our preliminary experiments involving each factor (one-factor-at-a-time experiment) showed that the concentrations of yeast extract, tryptone, ammonium acetate, ammonium sulfate, and glycerol that had significant influence on 2,3-BD production were approximately 5.0, 5.0, 4.0, 3.0, and 7.0 g/L, respectively (Figures S1–S5). Furthermore, the approximate temperature and inoculum size that exerted marked effects on 2,3-BD accumulation from the one-factor-at-a-time experiments were 36 °C and 8%, respectively (Figures S6 and S7).

The results obtained from our one-factor-at-a-time experiments informed the selection of two levels for each factor, which were then tested using the Plackett-Burman design. The two levels for each of the factors were determined by employing Equation (1) above; designated as the low and high levels, respectively, as shown in Table 1. The ANOVA generated from the experimental runs using the Plackett-Burman design is shown in Table 1. The linear regression coefficient of the model, R^2, was 0.9993 and the adjusted determination coefficient, *Adj* R^2, was 0.9951, which are both significantly close to unity, indicating the robustness of the model for further studies. R^2 measures variations in 2,3-BD response that are explained by the tested factors for a linear regression model, whereas *Adj* R^2 is the measure of goodness-of-fit for the model. As shown in Table 1, the *p*-values for tryptone, temperature, and inoculum size were 0.0377, 0.0348, and 0.0260, respectively, indicating that tryptone, temperature, and inoculum size were the most important factors that influenced 2,3-BD production by *P. polymyxa* among the factors studied, at a 95% confidence interval.

Among the medium components tested, tryptone exerted the most significant effect on 2,3-BD production by *P. polymyxa*. However, it was observed that further increase in the concentration of tryptone in the fermentation medium increased biomass formation without increasing 2,3-BD production as shown by the t-value of tryptone (Table 1). The pattern observed with tryptone was not unusual considering that tryptone supplies amino acids for protein (including enzymes) biosynthesis, as well as serves as a source of nitrogen for the biosynthesis of nucleic acids. In light of this, we speculated that a concentration threshold may exist for tryptone, within which both biomass and 2,3-BD accumulations by *P. polymyxa* occur optimally. On the other hand, increasing the temperature of the fermentation medium negatively impacted 2,3-BD production by *P. polymxa*, as revealed by the t-value of temperature in Table 1. Temperature regulation is essential for cells to function optimally, and when temperature falls below, or exceeds the optimum range for an organism, cellular metabolism is impeded, which in this case, adversely affects 2,3-BD production [6]. Increasing inoculum size in the fermentation medium was found to positively influence 2,3-BD production by *P. polymyxa* as shown by the t-value of inoculum (Table 1). Inoculum size determines the population of viable cells in the fermentation medium at time zero, and the greater the number of cells in the medium at time zero, the shorter the lag phase of growth, which ultimately translates to a faster conversion of substrates to 2,3-BD. For instance, the use of high inoculum size has been shown to decrease the duration of bacterial acclimation to culture medium with increased productivity as opposed to when a low inoculum volume is used [28]. A study conducted using *Bacillus licheniformis* showed that increasing inoculum size from 0.5 to 10 g/L increased volumetric 2,3-BD productivity and 2,3-BD yield from ~0.04 to 0.35 g/L/h and 0.11 to 0.35 g/g, respectively [19]. Furthermore, Jyothi et al. [29] obtained higher glutamic acid concentrations when the inoculum size of *Brevibacterium divaricatum* was increased from 3% to 7% with concomitant higher glutamic acid yields.

3.2. Path of Steepest Ascent Design

The path of steepest ascent method was used to determine the optimum levels for each of the three significant factors obtained by the Plackett-Burman design. The optimum level for each of the factors is critical for the Box-Behnken design and response surface methodology. For the path of steepest ascent study, the center point (value) between the low and high levels in the Plackett-Burman design was employed. The center point value for each factor was moved along a path that ensures an increase in 2,3-BD production. The direction for which the center point of each factor is moved was informed by the t-values shown in Table 1. The effects of tryptone and temperature were negative (−6.99 and −7.88, respectively), whereas that of inoculum size was positive (+10.29). These imply that to increase 2,3-BD production by *P. polymyxa*, the center point of each factor with a positive effect needs to be increased sequentially while that with a negative effect is to be decreased until no further increase in 2,3-BD production is observed. The center points for tryptone, temperature, and inoculum size were 5 g/L, 36 °C and 8%, respectively. Consequently, tryptone and temperature were decreased sequentially from 5 g/L and 36 °C to 2.5 g/L and 32 °C, respectively, while inoculum size was sequentially increased from 8% to 10.5% until no further increase in 2,3-BD production was observed (Table 2). From Table 2, the best three experiments in terms of 2,3-BD concentration were experiments 2, 3, and 4. The levels of each factor corresponding to experiments 2, 3, and 4 in the path of steepest ascent were selected for further optimization using the Box-Behnken design and response surface methodology. Based on these, the low, center, and high values, respectively, selected for temperature were 33, 34, and 35 °C, and those of inoculum size were 8.5%, 9.0%, and 9.5%, whereas the selected concentrations for tryptone were 3.5, 4.0, and 4.5 g/L.

Table 2. The path of steepest ascent experimental design and 2,3-BD production by *P. polymyxa*.

Run	Factors			2,3-BD (g/L)
	Inoculum Size (%)	Temperature (°C)	Tryptone (g/L)	
1	8.0	36	5.0	46.56
2	8.5	35	4.5	47.87
3	9.0	34	4.0	51.96
4	9.5	33	3.5	48.26
5	10	32	3.0	45.50

3.3. Box-Behnken Design and Response Surface Methodology

Based on the results obtained from the Plackett-Burman design and the path of steepest ascent method, the Box-Behnken design was used to conduct 15 experimental runs to further optimize the levels of temperature, tryptone, and inoculum size as shown in Table 3. Data obtained from the design matrix were analyzed using multiple regression and a second order polynomial equation model was obtained as shown below:

$$Y = 51.97 + 0.64X_1 - 0.16X_2 + 1.17X_7 + 2.36X_1X_2 - 0.79X_1X_7 + 0.64X_2X_7 - 2.31X_1^2 - 0.71X_2^2 - 0.059X_7^2 \quad (4)$$

where Y was the predicted response, and X_1, X_2, and X_7 were the defined values of inoculum size, temperature, and tryptone, respectively.

Table 3. Box-Behnken design and response results for 2,3-BD production.

Run	Coded Values			Actual Values			2,3-BD (g/L)
	X_1	X_2	X_7	X_1 (%)	X_2 (°C)	X_7 (g/L)	
1	−1	0	+1	8.5	34	4.5	50.65
2	+1	0	+1	9.5	34	4.5	50.99
3	−1	+1	0	8.5	35	4.0	46.20
4	0	0	0	9.0	34	4.0	52.56
5	0	−1	−1	9.0	33	3.5	50.98
6	−1	−1	0	8.5	33	4.0	51.04
7	0	0	0	9.0	34	4.0	51.98
8	+1	−1	0	9.5	33	4.0	46.99
9	0	+1	+1	9.0	35	4.5	52.70
10	+1	0	−1	9.5	34	3.5	50.13
11	+1	+1	0	9.5	35	4.0	51.58
12	0	+1	−1	9.0	35	3.5	49.18
13	0	0	0	9.0	34	4.0	51.37
14	0	−1	+1	9.0	33	4.5	51.94
15	−1	0	−1	8.5	34	3.5	46.65

X_1, inoculum size; X_2, temperature; X_7, tryptone. Error bars show standard deviations of the means ($n = 3$).

Analysis of variance (ANOVA) was used to test the statistical significance of the model as shown in Table 4. The model had a p-value of 0.0017 which is far less than 0.05 (indicative of significance). The regression coefficient, R^2, of the model was 0.9750 and the adjusted determination coefficient, *Adj* R^2, of the model was 0.9300, implying that 93% of variation in the response can be explained by the model. The lack of fit p-value was 0.5904, which indicates that the lack of fit was not significant, which confirms that the model was adequate for predicting 2,3-BD production. The lack of fit test is used to compare residual errors to the pure errors and gives an F-value for the model [30]. The F-value of the model was 21.65, which is low, thereby confirming that the model is significantly robust.

Table 4. ANOVA for 2,3-BD production by *P. polymyxa* according to the response surface quadratic model (lack of fit is not significant).

Factors	Sum of Squares	Degree of Freedom	Mean of Squares	F-Value	p-Value
Model	61.55	9	6.84	21.65	0.0017 *
X_1	3.32	1	3.32	10.50	0.0230 *
X_2	0.21	1	0.21	0.66	0.4539
X_7	10.90	1	10.90	34.52	0.0020 *
$X_1 X_2$	22.23	1	22.23	70.38	0.0004 *
$X_1 X_7$	2.46	1	2.46	7.80	0.0383 *
$X_2 X_7$	1.64	1	1.64	5.19	0.0718
X_1^2	19.64	1	19.64	62.18	0.0005 *
X_2^2	1.87	1	1.87	5.91	0.0592
X_7^2	0.013	1	0.013	0.040	0.8487
Residual	1.58	5	0.32		
Lack of fit	0.87	3	0.29	0.82	0.5904
Pure error	0.71	2	0.35		
Cor. Total	63.13	14			

$R^2 = 0.9750$, Adj $R^2 = 0.9300$; *Cor.* = Corrected; * Statistical significance.

To further evaluate the optimal levels of the individual factors, the significance of each factor and their interaction terms were tested using F-test, and the corresponding p-values for each of the model terms are shown in Table 4. Model terms with a p-value less than 0.05 shows that the terms are significant. The model terms for inoculum size, tryptone, inoculum size and temperature (inoculum size x temperature), inoculum size and tryptone (inoculum size x tryptone), and the quadratic term, inoculum size and inoculum size (inoculum size)^2 were all found to be significant. As shown in Table 4, tryptone was found to exert the largest effect on 2,3-BD production by *P. polymyxa* amongst the individual terms studied, whereas inoculum size and temperature (inoculum size x temperature) exhibited the largest effect when the individual and interaction terms were compared.

The interaction effects of factors were also evaluated by response surface methodology. The response surface is a three-dimensional plot that graphically represents the regression equation and shows relationship between the response and the independent factors [31,32]. Concave or convex response surfaces show that the maximum or minimum response is located within the experimental region, whereas a saddled surface shows a relative maximum and a relative minimum response, respectively [16,33]. Furthermore, the contour plots are two-dimensional representations of the response surface, which enhance the visual interpretation of the response surface [32,33]. Plots showing elliptical contours indicate significant interactions between the independent factors and the center of the smallest ellipse refers to a point of maximum or minimum response [31]. Also, plots with circular contours show that the interactions between the independent factors are negligible [32].

In the present study, the interaction between temperature and inoculum size when tryptone is maintained at the center value (4.0 g/L) shows a concave surface (Figure 1a), suggesting the presence of an apparent optimum condition. The corresponding elliptical contour plot shows that the interaction between temperature and inoculum size has significant effect on 2,3-BD production by *P. polymyxa* (Figure 1b). Additionally, the interaction between inoculum size and tryptone when the temperature is kept at the center point value also shows a concave surface (Figure 1c) with an elliptical contour (Figure 1d), thereby indicating a significant interaction. Conversely, the interaction between temperature and tryptone shows a contour that is not elliptical, therefore, not significant (Figure 2b). The optimum levels for the factors where maximum 2,3-BD production by *P. polymyxa* is predicted was obtained from the elliptical contour plot of Figure 1b, where strong interactions was observed. The maximum levels of inoculum size and temperature were indicated at the point of intersection between the major and minor axes confined by the smallest ellipse in Figure 1b [16,34,35]. The optimum conditions for maximum predicted 2,3-BD production was calculated when the coordinates of the important points (from Figure 1b) were inserted into Equation (3), and the partial derivatives were set to zero. The maximum predicted 2,3-BD was 51.52 g/L, which corresponds to a temperature of 34.98 °C and an inoculum size of 9.45%. The concentration of tryptone at this maximum predicted 2,3-BD was 3.5 g/L. The contour plots showing interactions between inoculum size and tryptone and between temperature and tryptone were not fully elliptical. The lack of a perfect elliptical contour is an indication that little or minimal interaction exists between the factors under evaluation. Thus, the optimized fermentation medium and conditions obtained in this study were 9.5% inoculum size, 120 g/L glucose, 3.5 g/L tryptone, and a temperature of 35 °C with the addition of yeast extract, 5 g/L; ammonium acetate, 4 g/L; NH_4SO_4, 3 g/L, crude glycerol, 7 g/L; KH_2PO_4, 3.5 g/L; K_2HPO_4, 2.75 g/L; $CoCl_2$, 0.05 g/L, $MgSO_4$, 0.2 g/L; MOPS, 10 g/L; and trace element solution.

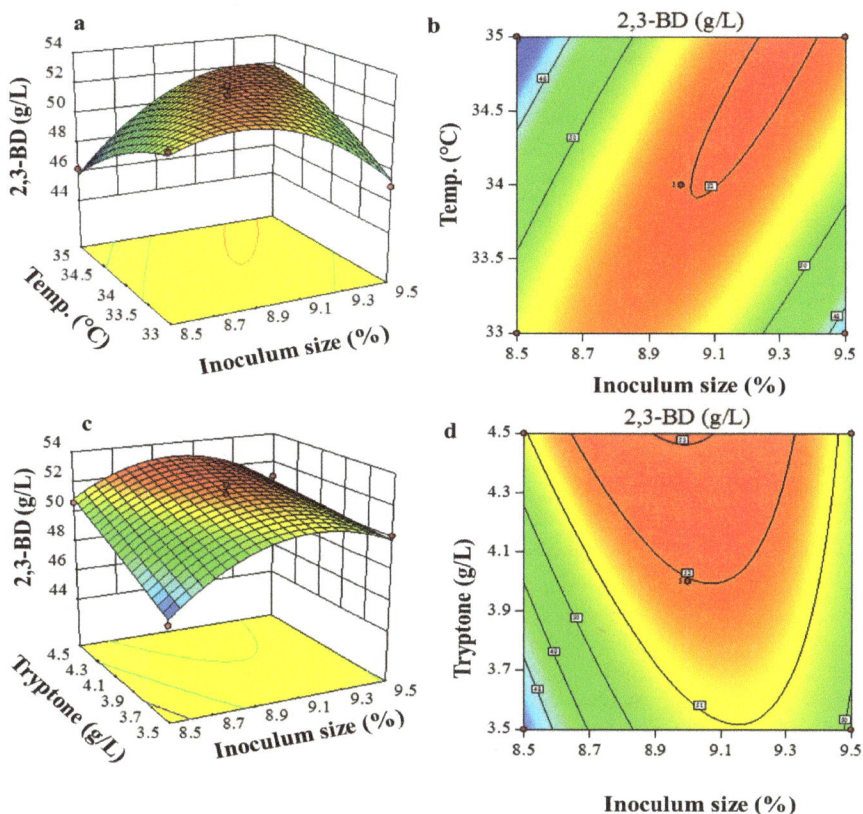

Figure 1. Contour and response surface plots. (**a**) The response surface plot; (**b**) the resultant contour plot showing the effects of temperature and inoculum size on 2,3-BD production by *P. polymyxa* with tryptone fixed at 4 g/L; (**c**) response surface plot and (**d**) the resultant contour plot depicting the effects of tryptone and inoculum size on 2,3-BD production by *P. polymyxa* with the temperature fixed at 34 °C.

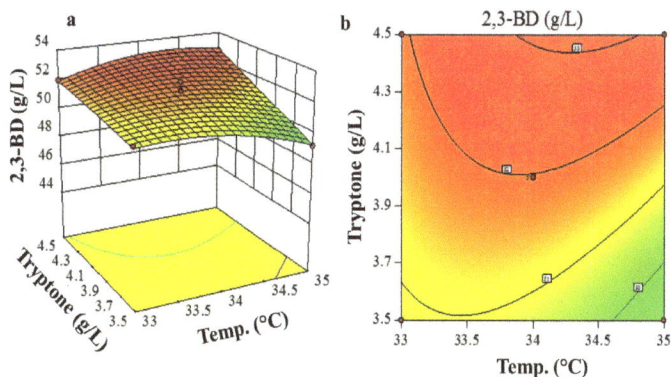

Figure 2. Contour and response surface plots. (**a**) The response surface plot and (**b**) the resultant contour plot (**b**) showing the effects of tryptone and temperature on 2,3-BD production by *P. polymyxa* with the inoculum size fixed at 9%.

3.4. Experimental Validation of the Optimized Medium and Conditions in Batch and Fed-Batch Fermentations

The optimized fermentation medium and conditions obtained from the analyses above were then used to conduct batch and fed-batch fermentations in a 5 L bioreactor to validate the fermentation medium and conditions. In each case, fermentation was conducted with a starting volume of 2 L. As shown above, apart from inoculum size, temperature, and tryptone, all the other factors that were tested by the Plackett-Burman experimental design were kept at their center point values. Batch and fed-batch fermentations were conducted in triplicate. The concentration of 2,3-BD obtained from the mean of three biological replicates using the optimized medium and conditions was 51.10 g/L, which was 99% of the predicted maximum 2,3-BD concentration of 51.52 g/L by response surface methodology. The yield and productivity of 2,3-BD obtained in the batch fermentation were 0.42 and 1.70, respectively (Table 5). The batch fermentation profile in Figure 3a shows complete glucose utilization, which is an indication of efficient glucose conversion to 2,3-BD with minimal formation of competing products. Thus, in addition to increased 2,3-BD production, the concentrations of ethanol, acetic acid, EPS, and acetoin were considerably diminished (Table 5 and Figure S8), when compared to non-optimized fermentations [7]. Due to reduced diversion of carbon to the EPS and ethanol biosynthesis pathways, and most likely, increased carbon flux from acetoin and acetic acid to 2,3-BD production during batch fermentation by *P. polymyxa*, the 2,3-BD yield increased from 0.32 g/g glucose under non-optimized fermentation conditions [7] to 0.42 g/g glucose under optimized conditions, accounting for a 31% increase in 2,3-BD yield.

To further determine the maximum 2,3-BD that *P. polymyxa* can accumulate using the optimized fermentation medium and conditions, a fed-batch fermentation was conducted. During the fed-batch process, glucose was intermittently replenished in the culture, accompanied by the addition of half-strength of the other nutrient components until no further increase in 2,3-BD production or glucose consumption was observed. The maximum 2,3-BD obtained from the mean of three independent fed-batch fermentations was 68.54 g/L with a yield and productivity of 0.34 and 0.70, respectively (Table 5; Figure 3b). Similarly, the concentrations of competing products, namely acetoin, ethanol, and acetic acid were considerably reduced, indicating that the optimized process enabled the efficient conversion of substrate carbon to the desired product, 2,3-BD. Nonetheless, acetoin was observed to increase towards the end of the fermentation, at which point, a corresponding decline in 2,3-BD was observed; a mechanism that *P. polymyxa* is thought to adopt for reducing the toxicity of 2,3-BD at an elevated concentration [7]. Notably, the yield and productivity of 2,3-BD reduced as the fermentation mode was switched from batch to fed-batch (Table 5). This is not unusual considering that greater amounts of glucose are consumed in the fed-batch cultures, some of which is funneled to cell maintenance and growth. In fact, the cell dilution effect resulting from intermittent feeding of glucose and other nutrients into the broth engenders a lag phase; albeit transient, thereby triggering transient cell growth following glucose supplementation, which diverts glucose away from 2,3-BD biosynthesis, momentarily. On the other hand, the concentration of acetoin in the fed-batch fermentation increased 3-fold whereas those of ethanol and acetic acid exhibited 1.2-fold increases, when compared to the batch fermentation (Table 5). A portion of the additional glucose fed into the fed-batch cultures were further converted to ethanol and acetic acid, which were observed to increase relative to the batch cultures that were not fed additional glucose. Interestingly, the optimized culture medium and conditions resulted in a 19% reduction in EPS production in both the batch and fed-batch fermentations (Table 5; Figure S8).

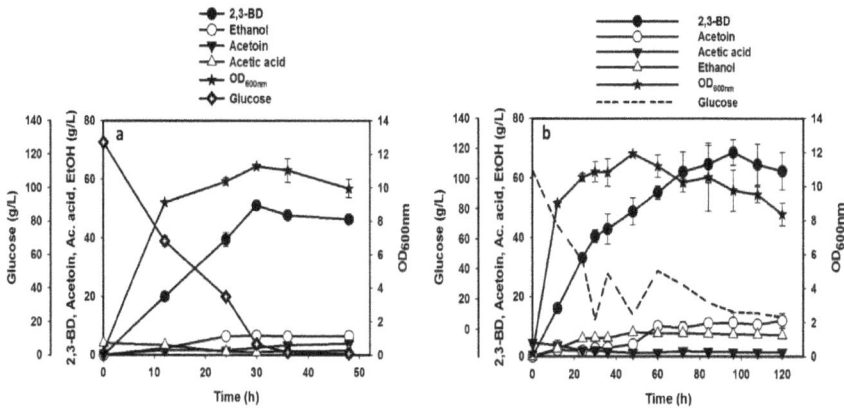

Figure 3. The fermentation profiles of *P. polymyxa* using optimized culture medium and conditions. (a) Batch and fed-batch (b) fermentations.

Table 5. The product profiles of *P. polymyxa* grown in batch and fed-batch fermentations under optimized conditions.

Product Profile	Batch Fermentation			Fed-Batch Fermentation		
	Max. conc. (g/L)	Yield (g/g)	Productivity (g/L/h)	Max. conc. (g/L)	Yield (g/g)	Productivity (g/L/h)
2,3-BD	51.10 ± 0.61	0.42 ± 0.01	1.70 ± 0.02	68.54 ± 4.25	0.34 ± 0.03	0.70 ± 0.04
Ethanol	6.64 ± 0.19	0.06 ± 0.00	0.22 ± 0.01	8.15 ± 0.51	0.06 ± 0.00	0.17 ± 0.01
Acetoin	3.96 ± 0.08	0.03 ± 0.00	0.08 ± 0.00	12.01 ± 2.43	0.05 ± 0.01	0.10 ± 0.02
Acetic acid	1.51 ± 0.07	0.01 ± 0.00	0.03 ± 0.00	1.82 ± 0.71	0.01 ± 0.00	0.03 ± 0.01
EPS	4.97 ± 0.15	0.04 ± 0.00	0.41 ± 0.01	4.69 ± 0.02	0.02 ± 0.00	0.39 ± 0.00
Glucose consumed	120.54 ± 1.55	N/A	N/A	199.97 ± 8.53	N/A	N/A

N/A: Not applicable. Error bars show standard deviations of the means (*n* = 3).

A comparison of the 2,3-BD concentration obtained in this study to those from other studies is summarized in Table 6. Different strains of *P. polymyxa* have been shown to possess the enzymatic repertoire for metabolizing several carbon sources to 2,3-BD, with yields ranging from 0.33 to 0.51 g/g glucose (Table 6). Notably, Häßler et al. [8] reported the production of 111 g/L 2,3-BD using a complex fermentation medium containing 60 g/L yeast extract. The use of 60 g/L yeast extract in large-scale 2,3-BD fermentation would increase the cost of production astronomically. Hence, an important achievement of this study is the development of an inexpensive fermentation medium for 2,3-BD fermentation. To assess the degree of impact exerted on 2,3-BD production by tryptone and yeast extract in this study, each of the nitrogen sources (yeast extract and tryptone) were incorporated in the growth medium up to 7 g/L (0, 3, 5, 7 g/L) prior to optimization studies (one-factor-at-a-time experiments). As depicted in Figures S4 and S5, the addition of yeast extract and tryptone in the fermentation medium resulted in increased 2,3-BD production in a concentration dependent manner. However, the increase in 2,3-BD accumulation in the fermentation broth diminished at tryptone and yeast extract concentrations above 5 g/L (that is, at 7 g/L), suggesting that increasing tryptone and yeast extract concentrations above 5 g/L may not engender a further increase in 2,3-BD production. In light of this, we rationalized that lower tryptone and yeast extract concentrations may be ideal for 2,3-BD production by *P. polymyxa*; an economic benefit for large-scale fermentation. Indeed, optimization studies demonstrated that 5 and 3.5 g/L yeast extract and tryptone, respectively, resulted in increased 2,3-BD production from ~27 and ~47 (in un-optimized) to 51 and 68.5 g/L 2,3-BD (in optimized) batch and fed-batch fermentations, respectively.

It is worth mentioning that Häßler et al. [8] used varying fermentation parameters, namely 500 rpm, 0.2 vvm, 37 °C, and sucrose as a carbon source whereas 300 rpm, 0.075 vvm, 35 °C, and glucose (a more readily available sugar from different biomass materials) were used in the present study. However, it is unlikely that interactions between medium components and fermentation conditions contributed to such high levels of 2,3-BD production (111 g/L). Indeed, Okonkwo et al. [7] demonstrated recently that 2,3-BD-mediated toxicity on *P. polymyxa* increased remarkably when the total 2,3-BD concentration in the fermentation broth exceeded the toxic threshold (48 g/L), thereby necessitating backward conversion to acetoin (acetoin ↔ 2,3-butanediol), possibly to alleviate 2,3-BD toxicity. Overall, the 2,3-BD yield of 0.42 and productivity of 1.70 obtained in the batch fermentation in this study compares favorably to those reported by other authors. It is worth noting that the yield and productivity obtained in this study were achieved using lower amounts of organic nitrogen sources in the forms of tryptone and yeast extract when compared to the other studies; a critical economic consideration for large-scale operations.

Table 6. Comparison of 2,3-BD concentrations obtained in this study to those of other studies using *P. polymyxa*.

Carbon Source	2,3-BD (g/L)	2,3-BD Yield (g/g)	2,3-BD Prod. (g/L/h)	GO	ONS (g/L)	PP	FM	Ref.
Glucose	51.10 ± 0.61	0.42 ± 0.01	1.70 ± 0.02	11.27 ± 0.06	YE, 5; tryp., 3.5	PP DSM 365	B	This work
Glucose	68.54 ± 4.25	0.34 ± 0.03	0.70 ± 0.04	11.92 ± 0.11	YE, 5; tryp., 3.5	PP DSM 365	FB	This work
Raw inulin extract from Jerusalem artichoke tubers	37.57 ± 0.32	0.51	0.89	26.79 ± 0.35 *	YE, 3; pep, 2	PP ZJ-9	B	[16]
Glucose	71.71	ND	1.33	13	YE, 10	PP CJX518	FB	[36]
Sucrose	111	ND	2.06	23 [†]	YE, 60	PP DSM 365	FB	[8]
Glucose	16.50	0.33	2.01	9.5	YE, 15	PP ICGEB2008	B	[37]
Inulin	51.3	ND	ND	NS	YE, 6; Pep., 3	PP ZJ-9 (XG-1)	FB	[38]
Inulin	36.8	ND	ND	11 [†]	YE, 6; Pep., 3	PP ZJ-9 (XG-1)	B	[38]

* Unit in g/L; [†] Determined at OD_{660nm}; NS—Not shown; ND—not detected; GO—Growth (OD_{600nm}); ONS—Organic nitrogen source; PP—*Paenibacillus polymyxa*; FM—Fermentation mode; B—batch; FB—Fed-batch; YE—yeast extract; Pep.—peptone; tryp.—tryptone; Ref.—references. The data shown are maximum product concentrations and cell growth achieved during fermentation.

4. Conclusions

Based on the results from the Box-Behnken design and response surface methodology, an optimized medium (7 g/L crude glycerol included) and culture conditions for enhanced 2,3-BD production by *P. polymyxa* were developed. The optimized conditions were validated in batch and fed-batch fermentations, leading to the production of 51.10 and 68.54 g/L, respectively, of 2,3-BD. These account for 47% and 31% increases in 2,3-BD production in batch and fed-batch cultures, respectively, with attendant diminished generation of competing co-products, especially EPS, relative to the non-optimized fermentations. The results presented here underline the interplay between medium components, culture conditions, and product-mediated toxicity (feedback inhibition), as the earlier determined toxic threshold of 2,3-BD (50 g/L) on *P. polymyxa* in a non-optimized medium [7] was significantly exceeded in this work (68.54 g/L). However, it is worth mentioning that glycerol was incorporated in the fermentation medium used in this study, which may contribute to 2,3-BD biosynthesis via improved NADH regeneration, especially in the optimized medium, relative to the un-optimized control medium. Collectively, we demonstrate that lower amounts of the expensive organic nitrogen sources, tryptone and yeast extract, can be used for optimal 2,3-BD production. This represents a significant reduction in operating costs in the efforts to commercialize biological production of 2,3-BD.

Supplementary Materials: The following are available online at http://www.mdpi.com/2311-5637/3/2/18/s1.

Acknowledgments: Salaries and research support were provided in part by State funds appropriated to the Ohio Agricultural Research and Development Center (OARDC), OARDC interdisciplinary grant, and the Hatch grant (Project No. OHO01222). We would like to also acknowledge Peloton Technologies LLC for financial support.

Author Contributions: Thaddeus Chukwuemeka Ezeji conceived the experiments; Thaddeus Chukwuemeka Ezeji, Victor C. Ujor, Christopher Chukwudi Okonkwo and Pankaj K. Mishra designed the experiments; Christopher Chukwudi Okonkwo and Pankaj K. Mishra conducted the experiments; Thaddeus Chukwuemeka Ezeji, Victor C. Ujor and Christopher Chukwudi Okonkwo interpreted the results; Christopher Chukwudi Okonkwo and Victor C. Ujor wrote the manuscript.

Conflicts of Interest: The authors declare no conflict of interest.

References

1. Celinska, E.; Grajet, W. Biotechnological production of 2,3-butanediol-current states and prospects. *Biotechnol. Adv.* **2009**, *27*, 715–725. [CrossRef] [PubMed]
2. Soltys, K.A.; Batta, A.K.; Koneru, B. Successful nonfreezing, subzero preservation of rat liver with 2,3-butanediol and type I antifreeze protein. *J. Surg. Res.* **2001**, *96*, 30–34. [CrossRef] [PubMed]
3. Bartowsky, E.J.; Henschke, P.A. The 'buttery' attribute of wine–diacetyl–desirability, spoilage and beyond. *Int. J. Food Microbiol.* **2004**, *96*, 235–252. [CrossRef] [PubMed]
4. Mitchell, R.L.; Robertson, N.C. Oxidation of Hydrocarbons. U.S. Patent US2689253 A, 1954.
5. O'Neil, M.J. *The Merck Index-An Encyclopedia of Chemicals, Drugs, and Biologicals*, 13th ed.; Merck and Co., Inc.: Whitehouse Station, NJ, USA, 2001.
6. Garg, S.K.; Jain, A. Fermentative production of 2,3-butanediol: A review. *Bioresour. Technol.* **1995**, *51*, 103–109. [CrossRef]
7. Okonkwo, C.C.; Ujor, V.; Ezeji, T.C. Investigation of relationships between 2,3-butanediol toxicity and production during growth of *Paenibacillus polymyxa*. *New Biotechnol.* **2017**, *34*, 23–31. [CrossRef] [PubMed]
8. Häßler, T.; Schieder, D.; Pfaller, R.; Faulstich, M.; Sieber, V. Enhanced fed-batch fermentation of 2,3-butanediol by *Paenibacillus polymyxa* DSM 365. *Bioresour. Technol.* **2012**, *124*, 237–244. [CrossRef] [PubMed]
9. Ripoll, V.; de Vicente, G.; Morán, B.; Rojas, A.; Segarra, S.; Montesinos, A.; Tortajada, M.; Ramón, D.; Ladero, M.; Santos, V.E. Novel biocatalysts for glycerol conversion into 2,3-butanediol. *Process Biochem.* **2016**, *51*, 740–748. [CrossRef]
10. Kim, T.; Cho, S.; Woo, H.M.; Lee, S.M.; Lee, J.; Um, Y.; Seo, J.H. High production of 2,3-butanediol from glycerol without 1,3-propanediol formation by *Raoultella ornithinolytica* B6. *Appl. Microbiol. Biotechnol.* **2017**, *101*, 2821–2830. [CrossRef] [PubMed]
11. Petrov, K.; Petrova, P. Enhanced production of 2,3-butanediol from glycerol by forced pH fluctuations. *Appl. Microbiol. Biotechnol.* **2010**, *87*, 943–949. [CrossRef] [PubMed]
12. Biebl, H.; Zeng, A.P.; Menzel, K.; Deckwer, W.D. Fermentation of glycerol to 1,3-propanediol and 2,3-butanediol by *Klebsiella pneumoniae*. *Appl. Microbiol. Biotechnol.* **1998**, *50*, 24–29. [CrossRef] [PubMed]
13. Lin, E.C.C. Glycerol dissimilation and its regulation in bacteria. *Annu. Rev. Microbiol.* **1976**, *30*, 535–578. [CrossRef] [PubMed]
14. Neijssel, O.M.; Hueting, S.; Crabbendam, K.J.; Tempest, D. Dual pathways of glycerol assimilation in *Klebsiella aerogenes* NCIB 418. *Arch. Microbiol.* **1975**, *104*, 83–87. [CrossRef] [PubMed]
15. Voloch, M.; Ladish, M.R.; Rodwell, V.W.; Tsao, G.T. Reduction of acetoin to 2,3-butanediol in *Klebsiella pneumoniae*: A new model. *Biotechnol. Bioeng.* **1983**, *26*, 173–183. [CrossRef] [PubMed]
16. Gao, J.; Xu, H.; Li, Q.-J.; Feng, X.-H.; Li, S. Optimization of medium for one-step fermentation of inulin extract from Jerusalem artichoke tubers using *Paenibacillus polymyxa* ZJ-9 to produce R,R-2,3-butanediol. *Bioresour. Technol.* **2010**, *101*, 7076–7082. [CrossRef] [PubMed]
17. Anvari, M.; Motlagh, R.S. Enhancement of 2,3-butanediol production by *Klebsiella oxytoca* PTCC 1402. *BioMed Res. Int.* Available online: https://www.mysciencework.com/publication/show/16bdb9aba87406e5a516be552043648a (accessed on 8 January 2017).
18. Zhang, L.; Yang, Y.; Sun, J.; Shen, Y.; Wei, D.; Zhu, J.; Chu, J. Microbial production of 2,3-butanediol by a mutagenized strain of *Serratia marcescens* H30. *Bioresour. Technol.* **2010**, *101*, 1961–1967. [CrossRef] [PubMed]

19. Perego, P.; Converti, A.; del Borghi, M. Effects of temperature, inoculum size and starch hydrolyzate concentration on butanediol production by *Bacillus licheniformis*. *Bioresour. Technol.* **2003**, *89*, 25–131. [CrossRef]

20. Marwoto, B.; Nakashimada, Y.; Kakizono, T.; Nishio, N. Enhancement of (R,R)-2,3-butanediol production from xylose by *Paenibacillus polymyxa* at elevated temperatures. *Biotechnol. Lett.* **2002**, *24*, 109–114. [CrossRef]

21. Khuri, A.I.; Mukhopadhyay, S. Response surface methodology. *Comp. Stat.* **2010**, *2*, 128–149. [CrossRef]

22. Box, G.E.P.; Hunter, J.S.; Hunter, W.G. *Statistics for Experimenters*; John Wiley and Sons: New York, NY, USA, 1978.

23. Chang, Y.; Huang, J.; Lee, C.; Shih, I.; Tzeng, Y. Use of response surface methodology to optimize culture medium for production of lovastatin by *Monascus ruber*. *Enzyme Microbial Technol.* **2002**, *30*, 889–894. [CrossRef]

24. Kwak, J.S. Application of Taguchi and response surface methodologies for geometric error in surface grinding process. *Int. J. Machine Tools Manuf.* **2005**, *45*, 327–334. [CrossRef]

25. Xiao, Z.; Xu, P. Acetoin metabolism in bacteria. *Crit. Rev. Microbiol.* **2007**, *33*, 127–140. [CrossRef] [PubMed]

26. Li, J.; Wang, W.; Ma, Y.; Zheng, A. Medium optimization and proteome analysis of (R,R)-2,3-butanediol production by *Paenibacillus polymyxa* ATCC 12321. *Appl. Microbiol. Biotechnol.* **2013**, *97*, 585–597. [CrossRef] [PubMed]

27. Nilegaonkar, S.; Bhosale, S.B.; Kshirsagar, D.C.; Kapidi, A.H. Production of 2,3-butanediol from glucose by *Bacillus licheniformis*. *World J. Microbiol. Biotechnol.* **1992**, *8*, 378–381. [CrossRef] [PubMed]

28. Yu, E.K.; Saddler, J.N. Fed-batch approach to production of 2,3-butanediol by *Klebsiella pneumoniae* grown on high substrate concentrations. *Appl. Environ. Microbiol.* **1983**, *46*, 630–635. [PubMed]

29. Jyothi, A.N.; Sasikiran, K.; Nambisan, B.; Balagopalan, C. Optimisation of glutamic acid production from cassava starch factory residues using Brevibacterium divaricatum. *Process Biochem.* **2005**, *40*, 3576–3579. [CrossRef]

30. Muthukumar, M.; Mohan, D.; Rajendran, M. Optimization of mix proportions of mineral aggregates using Box Behnken designs of experiments. *Cem. Concr. Compos.* **2003**, *25*, 751–758. [CrossRef]

31. Bas, D.; Boyaci, I.H. Modeling and optimization I: Usability of response surface methodology. *J. Food Eng.* **2007**, *78*, 836–845. [CrossRef]

32. Liu, J.; Luo, J.; Ye, H.; Sun, Y.; Lu, Z.; Zeng, X. Medium optimization and structural characterization of exopolysaccharides from endophytic bacterium *Paenibacillus polymyxa* EJS-3. *Carbohydr. Polym.* **2010**, *79*, 206–213. [CrossRef]

33. Bezerra, M.A.; Santelli, R.E.; Oliveira, E.P.; Villar, L.S.; Escaleira, L.A. Response surface methodology (RSM) as a tool for optimization in analytical chemistry. *Talanta* **2008**, *76*, 965–977. [CrossRef] [PubMed]

34. Tanyildizi, M.S.; Ozer, D.; Elibol, M. Optimization of α-amylase by *Bacillus* sp. using response surface methodology. *Process Biochem.* **2005**, *40*, 2291–2296. [CrossRef]

35. Myers, R.H.; Montgomery, D.C.; Anderson-Cook, C.M. *Response Surface Methodology: Process and Product Optimization Using Designed Experiments*; John Wiley & Sons, Inc.: New York, NY, USA, 2009.

36. Dai, J.-J.; Cheng, J.-S.; Liang, Y.-Q.; Jiang, T.; Yuan, Y.-J. Regulation of extracellular oxidoreduction potential enhanced (R,R)-2,3-butanediol production by *Paenibacillus polymyxa* CJX518. *Bioresour. Technol.* **2014**, *167*, 433–440. [CrossRef] [PubMed]

37. Adlakha, N.; Yazdani, S.S. Efficient production of (R,R)-2,3-butanediol from cellulosic hydrolysate using *Paenibacillus polymyxa* ICGEB2008. *J. Ind. Microbiol. Biotechnol.* **2015**, *42*, 21–28. [CrossRef] [PubMed]

38. Zhang, L.; Xu, Y.; Gao, J.; Xu, H.; Cao, C.; Xue, F.; Ding, G.; Peng, Y. Introduction of the exogenous NADH coenzyme regeneration system and its influence on intracellular metabolic flux of *Paenibacillus polymyxa*. *Bioresour. Technol.* **2016**, *201*, 319–328. [CrossRef] [PubMed]

fermentation

MDPI

Article

Kinetics of Bioethanol Production from Waste Sorghum Leaves Using *Saccharomyces cerevisiae* BY4743

Daneal C. S. Rorke and Evariste Bosco Gueguim Kana *

School of Life Sciences, University of KwaZulu-Natal, Pietermaritzburg 3201, South Africa;
danealrorke@gmail.com
* Correspondence: kanag@ukzn.ac.za; Tel.: +27-33-260-5527

Academic Editor: Thaddeus Ezeji
Received: 16 January 2017; Accepted: 2 May 2017; Published: 8 May 2017

Abstract: Kinetic models for bioethanol production from waste sorghum leaves by *Saccharomyces cerevisiae* BY4743 are presented. Fermentation processes were carried out at varied initial glucose concentrations (12.5–30.0 g/L). Experimental data on cell growth and substrate utilisation fit the Monod kinetic model with a coefficient of determination (R^2) of 0.95. A maximum specific growth rate (μ_{max}) and Monod constant (K_S) of 0.176 h^{-1} and 10.11 g/L, respectively, were obtained. The bioethanol production data fit the modified Gompertz model with an R^2 value of 0.98. A maximum bioethanol production rate ($r_{p,m}$) of 0.52 g/L/h, maximum potential bioethanol concentration (P_m) of 17.15 g/L, and a bioethanol production lag time (t_L) of 6.31 h were observed. The obtained Monod and modified Gompertz coefficients indicated that waste sorghum leaves can serve as an efficient substrate for bioethanol production. These models with high accuracy are suitable for the scale-up development of bioethanol production from lignocellulosic feedstocks such as sorghum leaves.

Keywords: Monod equation; modified Gompertz equation; bioethanol; sorghum leaves

---

1. Introduction

Ideal crops for commercial bioethanol production in South Africa include maize, grain sorghum, and sugar cane [1]; however, in order to completely utilise these materials, post-harvest field waste should be employed for biofuel production. Sweet sorghum (*Sorghum bicolor* (L.) Moench), in particular, yields significant amounts of biomass (leaves and pressed stalks) and sugar (found in stalks) [2]. Bioconversion of lignocellulosic material to renewable fuels is currently receiving great interest since it does not impact food security [2]. Several studies on the enhancement of fermentable sugar release from lignocellulosic substrates have been reported [3–5]. Microwave-assisted pre-treatment has received increased attention due to its lower energy demand and shorter process times [6]. Microwave radiation alters the structure of lignocellulose by emitting electromagnetic radiation, which results in the formation of thermal pockets. These pockets ultimately explode due to an increase in heat, leading to the relocation of crystalline structures within the lignocellulosic material [6]. Gabhane et al. [7] studied the individual and interactive effects of acid and alkali pre-treatments using an autoclave, microwave, and ultrasonicator, and obtained a maximal reducing sugar yield of 36.84% from acid pre-treated banana waste by using microwave radiation. Despite the vast information available on lignocellulosic pre-treatment, a significant knowledge gap exists between this and the kinetic assessment of the fermentation efficiency of pre-treated lignocellulosic substrates for biofuel production.

Bioethanol is one such fuel which exhibits several advantages over conventional fossil fuels. This includes its renewable nature, ease of storage, higher oxygen content, higher octane number, the fact that it is free of sulfur, and contributes less to global warming and air pollution [8,9]. In recent times, the application of bioethanol as a fuel replacement has become more appealing [9]. Subsequent to lignocellulosic pre-treatment, the fermentable sugars released are converted to bioethanol by the exploitation of microbial metabolism, with the simultaneous release of carbon dioxide. Numerous studies have focused on the production of bioethanol from sorghum [10–12]. Suryaningsih and Irhas [10] obtained an optimal ethanol concentration of 40 g/L after 48 h of fermentation using sorghum grain, while an ethanol content of 12.4 g/L was obtained by Massoud et al. [11] by using the lignocellulosic hydrolysate of pressed sorghum stalks. However, very few have determined the efficiency of bioethanol production from sorghum leaves alone. Globally, efforts are being made to further expedite the use of renewable fuel sources as an alternative. These efforts are being challenged by a significant increase in the cost of production [13]. This suggests that further modelling and optimisation studies are required for the development of biofuel from lignocellulosic substrates.

Kinetic modelling refers to a mathematical description of the changes in the properties of a system in which biochemical reactions take place [14]. These models assist in the design of a production process by representing the complex biochemistry of cells. Kinetic models can be used to understand, predict, and evaluate the effects of altering the components of a fermentation process [15]. Most commonly, these models are used to increase yield and productivity as well as minimise the formation of undesired by-products, ensuring that the product is of high quality [15]. Models capable of describing the kinetics of microbial growth, substrate utilisation, and product formation play a fundamental role in process optimisation and control [16] by providing a basis for process design, control, and scale-up [17].

Monod kinetics models are commonly used to describe biomass growth and product formation with respect to the limiting substrate [18], while the modified Gompertz models are used to determine production lag time, maximum production rate, and maximum product concentration on a given substrate [8,19]. The original Gompertz function has been applied in a wide range of research areas, such as ecology, marketing, actuarial sciences, medicine, and biology [20]. Although the modified Gompertz equation has been used in many studies for ethanol and hydrogen production [19,21,22], very few studies have reported on bioethanol fermentation kinetics using lignocellulosic biomass as a feedstock [2,17,18]. These studies include feedstocks such as populus hydrolysate [17], sweet sorghum stalks [2], and rice hulls [18]. Despite this, there is a scarcity of information regarding the fermentation kinetics of this fuel using waste sorghum leaves.

Knowledge from fermentation kinetic studies on waste sorghum leaves will provide fundamental information on process characteristics and behaviour. Furthermore, decisions involving process control and improvement can be made with relative ease when a bioprocess is fully understood, advancing its commercial application. In this study, the Monod and modified Gompertz models were used to assess the kinetic behaviour of a bioethanol fermentation process (in a batch system) using waste sorghum leaves.

2. Materials and Methods

2.1. Feedstock Preparation and Pre-Treatment

Sorghum leaves used in this study were harvested from Ukulinga Research Farm, Pietermaritzburg, South Africa (29°67′ E, 30°40′ S). Approximately five to eight sorghum leaves were cut off at the leaf collar of mature (approximately 100–120 days) plants. They were immediately oven-dried at 70 °C for 48 h and milled to particle sizes of 1–2 mm using a centrifugal miller (Retsch ZM-1, Durban, South Africa). Milled leaves were treated under previously optimised conditions [23]; i.e., a 3.83% (*v*/*v*) HCl (Merck, Durban, South Africa) solution at a solid-to-liquid (S:L) ratio of 16.66% for 2 min at 600 W in a 1000 W capacity microwave oven (Samsung, Model: ME9114S1, Durban, South Africa).

2.2. Enzymatic Hydrolysis

Pre-treated biomass was rinsed with distilled water until a pH of 4.0 was achieved. The biomass was then oven-dried at 60 °C overnight and enzymatically hydrolysed using powdered cellulase enzyme, Onozuka R-10 (Merck, Durban, South Africa) in 500 mL Erlenmeyer flasks. A solid loading rate of 20 g dry biomass in 200 mL 0.05 M citrate buffer, with an enzyme loading rate of 50 mg/g of dry biomass was employed. The pH during enzymatic hydrolysis was 4.8, and the temperature was maintained at 50 °C using a water bath at 120 rpm for 72 h. The hydrolysate was filtered, and the filtrate was analysed for glucose concentration.

2.3. Fermentation Medium Formulation

A mineral salt solution (pH 4.5) containing (in g/L); yeast extract, 1.0; $(NH_4)_2SO_4$, 2.0 and $MgSO_4$, 1.0 was autoclaved at 121 °C for 15 min. All reagents purchased from Merck (Durban, South Africa). Filter-sterilised enzymatic hydrolysate was then added to the mineral salts. Initial glucose concentrations within a range of 12.5–30.0 g/L were obtained by diluting or—where needed—supplementing with pure glucose.

2.4. Microorganism and Inoculum Preparation

The *S. cerevisiae* BY4743 used in this study was obtained from the Department of Genetics, University of KwaZulu-Natal, Pietermaritzburg, South Africa. A single flask containing 100 mL Yeast-Peptone-Dextrose (YPD) medium was inoculated with a single colony and grown at 150 rpm, 30 °C overnight, until the exponential growth phase was reached. This culture was inoculated (10%) into the prepared fermentation medium (working volume of 100 mL) containing an initial glucose concentration of 12.5 g/L. The culture was then grown under the same conditions as previously described and then used as a starter culture for subsequent fermentation processes.

2.5. Fermentation Process and Analytical Methods

Fermentation processes were carried out in sterilised 250 mL flasks with a working volume of 100 mL. Aliquots of 10 mL (10% inoculation) *S. cerevisiae* were aseptically added to the fermentation flasks, and the cultures were incubated at 30 °C at 120 rpm for 24 h, or until glucose concentrations were depleted. Fermentations were aseptically sampled every two h and assessed for biomass concentration, sugar content, and bioethanol content.

The sugar content of filtered enzymatic hydrolysate and fermentation media was determined using a YSI 2700 Model Biochemical Analyser (YSI, Yellow Springs, OH, USA). Ethanol content was determined in the gas phase of the fermentation process, using an ethanol vapour sensor (ETH-BTA, Vernier Software and Technology, Beaverton, OR, USA). The absorbance of the culture broth was measured using a spectrophotometer (UV-Vis Spectrophotometer, UVmini-1240, Shimadzu, Kyoto, Japan) at 650 nm. Cell biomass quantification was achieved by using absorbance as a function of the concentration of yeast cells. A standard curve was prepared by determining the dry weights and corresponding absorbance values of yeast biomass at varied dilutions of a 24 h *S. cerevisiae* culture, grown in fermentation media containing 12.5 g/L glucose. Dry weights were determined by centrifuging 5 mL of each dilution (1, 1/2, 1/4, 1/8, and 1/10) for 10 min at 5000 rpm. The supernatant was removed, and the remaining biomass was dried at 60 °C until a constant mass was obtained.

2.6. Calculations of Kinetic Model Constants

The average specific growth rates (μ) of fermentation processes carried out in duplicate were calculated using Equation (1). The specific growth rate values (μ) and the substrate concentration data were subsequently used to estimate the maximum specific growth rate (μ_{max}) and Monod constant (K_S) by the double reciprocal Lineweaver–Burk plot. The Lineweaver–Burk plot has the possibility of distorting the error structure of the data, but it is still used for representation of kinetic data because the double reciprocal plot usually automatically and conveniently provides a considerably improved weighting for linear graphs of most kinetic parameters as a function of substrate concentration [24].

$$\mu = \frac{lnX_2 - lnX_1}{t_2 - t_1} \tag{1}$$

where X_2 and X_1 are biomass concentrations (g/L) at time instants t_2 and t_1, respectively.

The linear form of this equation is as follows:

$$\frac{1}{\mu} = \frac{1}{\mu_{max}} + \frac{K_s}{\mu_{max}}\left(\frac{1}{S}\right) \tag{2}$$

where S represents substrate concentration. In addition, experimental data on bioethanol production over time were used to fit the modified Gompertz model (Equation (3)) using the least squares method, (CurveExpert V1.5.5), which showed the lag time, maximum bioethanol production rate, and the potential maximum product concentration.

$$P = P_m \cdot exp\left\{-exp\left[\frac{r_{p,m} \cdot exp(1)}{P_m}\right] \cdot (t_L - t) + 1\right\}. \tag{3}$$

where P is bioethanol concentration (g/L), P_m is potential maximum bioethanol concentration (g/L), $r_{p,m}$ is maximum bioethanol production rate (g/L/h), and t_L is the time from the beginning of fermentation to exponential bioethanol production (h).

Sugar utilisation, ethanol yield, ethanol productivity, and fermentation efficiency were calculated using the following Equations (4)–(7) respectively [25]:

$$Sugar\ utilisation\ (\%) = \frac{Original\ sugar\ content - Residual\ sugar\ content}{Original\ sugar\ content} \times 100 \tag{4}$$

$$Ethanol\ yield\left[\frac{g(ethanol)}{g(glucose)}\right] = \frac{Maximum\ ethanol\ concentration\ (g/L)}{Utilised\ glucose\ (g/L)} \tag{5}$$

$$Ethanol\ productivity\ (g/L/h) = \frac{Maximum\ ethanol\ concentration\ (g/L)}{Fermentation\ time\ (h)} \tag{6}$$

$$Fermentation\ efficiency\ (\%) = \frac{Actual\ ethanol\ yield\ (g/L)}{Theoretical\ ethanol\ yield\ (g/L)} \times 100 \tag{7}$$

3. Results and Discussion

3.1. Monod Kinetic Model of S. cerevisiae on Waste Sorghum Leaves

The microwave-assisted acid pre-treatment and subsequent enzymatic hydrolysis of the biomass of sorghum leaves resulted in a glucose yield of 0.153 g/g sorghum leaves. Cell biomass, bioethanol production, and glucose consumption were monitored throughout the fermentation process. The correlation between absorbance and dry weight of yeast biomass was determined by linear regression, which gave a correlation coefficient (r) of 0.96. The specific growth rate (μ) values were calculated using the exponential (log) phase of microbial growth. The values obtained were 0.096, 0.104, 0.114, 0.122, and 0.123 h^{-1} at initial substrate concentrations of 12.5, 13.3, 19.4, 21.8, and 23.1 g/L, respectively (Figure 1). In comparison, Echegaray et al. [26] obtained a range of specific growth rates between 0.019 and 0.240 h^{-1} using diluted sugarcane molasses as a substrate (170–270 g/L total reducing sugar range) under anaerobic cultivation of *S. cerevisiae*. In addition, an increase in μ values from 0.096 to 0.123 h^{-1} was observed when the initial glucose concentration increased from 12.5 to 23.0 g/L. A similar trend was reported by Laopaiboon et al. [27], whereby an increase in glucose concentration from 10 to 150 g/L resulted in an increase of μ value from 0.43 to 0.49 h^{-1}. These findings suggest that the specific growth rate of a culture increases with increasing substrate concentration, until substrate saturation is reached [28].

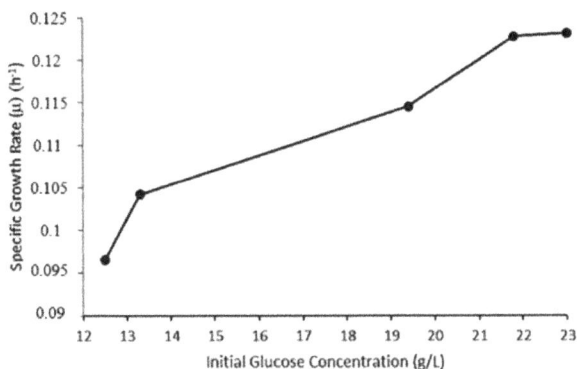

Figure 1. Specific growth rates (μ) of *S. cerevisiae* BY4743 at varied initial glucose concentrations.

Data on the specific growth rate (μ) values and initial substrate concentrations were used to estimate K_S and μ_{max} (Figure 2). A maximum specific growth rate (μ_{max}) value of 0.176 h^{-1} was obtained, which was close to the value of 0.169 h^{-1} previously reported by Dodić et al. [19] using *S. cerevisiae* cells grown on raw sugar beet juice. As cell growth rate is largely dependent on substrate concentration, it is expected that a higher initial sugar concentration will result in higher Monod coefficients [28]. The K_S value obtained (10.11 g/L) was in line with values previously reported from several studies on lignocellulosic substrates (Table 1). Using citrus pulp waste as a substrate, a K_S value of 10.690 g/L was reported by Raposo et al. [29], while Srimachai et al. [16] obtained a K_S value of 10.210 g/L using oil palm frond juice. These observations imply that *S. cerevisiae* has a similar affinity ($1/K_S$) to sorghum leaves as oil palm frond juice, glucose, and citrus waste pulp. In contrast to this, Ariyajaroenwong et al. [2] reported a Monod constant (K_S) of 47.510 g/L when using sweet sorghum juice as a substrate. This decreased affinity may be due to the presence of more than one type of sugar in sweet sorghum juice [2]. Singh and Sharma [30] reported a K_S value of 3.700 g/L using glucose, which is much lower than the range observed in previous studies; however, this corresponds to a higher affinity constant, which is expected as glucose is metabolised with ease.

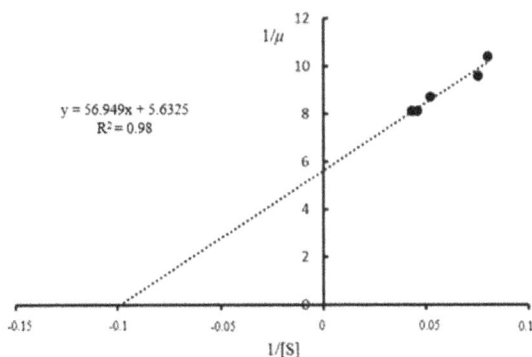

Figure 2. Lineweaver–Burk plot used to estimate Monod constants for batch ethanol production from waste sorghum leaves (SL).

Variations in K_S values (from 3.7 to 213.6 g/L) can be attributed to substrate type and concentration, strains of yeast employed, or the fermentation process itself [2]. These data demonstrate that the suitability of waste sorghum leaves as a substrate for *S. cerevisiae* growth is similar to that of raw sugar beet juice and oil palm frond juice. Furthermore, the fermentation volume size may impact the K_S value. This is illustrated by the vast differences in substrate affinity for glucose obtained by Shafaghat et al. [22] using a working volume of less than 250 mL and Ahmad et al. [9] with a working volume of 8 L. The differences observed between the aforementioned studies may be attributed to additional process challenges encountered in large volume, such as poor agitation, low mass transfer, and inhomogeneity.

Table 1. Comparison of the obtained Monod model coefficients with previous studies.

Substrate	μ_{max} (h^{-1})	K_S (g/L)	Reference
Sorghum leaves	0.176	10.110	This study
Oil palm frond juice (10–20 years)	0.150	10.210	Srimachai et al. [25]
Sugar beet raw juice	0.169	ND	Dodić et al. [19]
Sweet sorghum juice	0.313	47.510	Ariyajaroenwong et al. [2]
Glucose	0.291	ND	Govindaswamy et al. [31]
Banana peels	1.500	25.000	Manikandan et al. [32]
Glucose	0.084	213.60	Ahmad et al. [9]
Glucose	0.650	11.390	Shafaghat et al. [22]
Citrus waste pulp	0.350	10.690	Raposo et al. [29]
Glucose	0.133	3.700	Singh and Sharma [30]

ND: Not determined.

3.2. Bioethanol Production

The bioethanol production trend of *S. cerevisiae* cultivated on fermentation medium prepared from sorghum leaves is shown in Figure 3. A rapid depletion of glucose was observed from 0 to 32 h. A lag phase in bioethanol production of 6 h was obtained. This corresponds to cell adaptation and synthesis of key nutrients required for biomass or product (bioethanol) formation [14]. Ardestani and Shafiei [33] reported exponential growth of *S. cerevisiae* after 7 h of incubation. A rapid increase in ethanol concentration was observed from 6 to 28 h, corresponding to the exponential stage (Figure 4). This is expected, as ethanol is a primary metabolite and is therefore produced during the exponential phase of cell growth. A similar observation was reported by Lin et al. [34], where a steady increase in ethanol was observed over a duration of 48 h at 30 and 40 °C. An average ethanol yield of 0.49 g-ethanol/g-glucose was obtained, corresponding to a 96% fermentation efficiency during this period. Fermentation efficiencies between 72.78% and 78.43% have been reported by Srimachai et al. [25] using oil palm frond juice as a substrate, whilst ethanol yields between 0.40 and 0.49 g/g have been obtained from raw sugar beet juice [19]. Waste sorghum leaves show excellent potential for lignocellulosic bioethanol production. A productivity of 0.345 g/L/h was observed in this study. Ethanol productivities on other lignocellulosic substrates in the range of 0.25 to 1.01 g/L/h have been reported [35–38], further pointing to the relative higher potential of waste sorghum leaves for bioethanol production.

The modified Gompertz model was fitted to the experimental data, and kinetic coefficients were determined (Equation (8)).

$$P = 17.15 exp \left\{ -exp \left[\frac{0.52 \, exp(1)}{17.15} \right] \cdot (6.31 - t) + 1 \right\} \tag{8}$$

The fitted regression curve exhibited an R^2 value of 0.98 and a correlation coefficient (r) of 0.99, suggesting that this model is able to efficiently describe bioethanol production during the fermentation of sorghum leaf wastes. The Gompertz coefficients for maximum potential bioethanol concentration (P_m), maximum bioethanol production rate ($r_{p,m}$), and lag time were 17.15 g/L, 0.52 g/L/h, and 6.31 h, respectively, from waste sorghum leaves. Very few studies have reported a lag time of longer than one hour [19]. This suggests that a duration of at least 6 h was required for yeast cells to adapt to fermentation medium derived from waste sorghum leaves. Additionally, the maximum potential bioethanol concentration of 17.15 g/L—which corresponds to 2.17% (*v*/*v*)—illustrates that the impact of ethanol concentration within the medium may have a slight effect on the specific growth rate of *S. cerevisiae*. This is supported by an earlier study by Dinh et al. [39], which showed that a higher initial ethanol concentration within fermentation media resulted in an increase in the time required for cells to reach the optimal bioethanol production rate as well as a reduction in the maximum ethanol concentration.

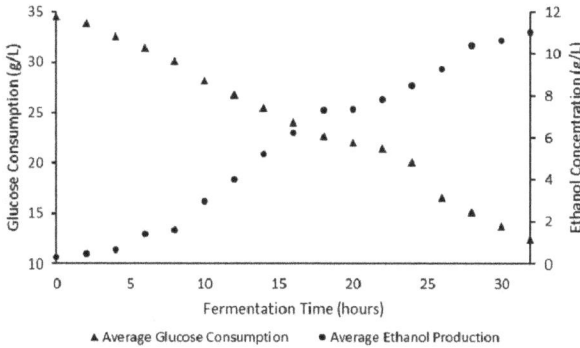

Figure 3. Average glucose utilisation and ethanol formation during batch fermentation by *S. cerevisiae* BY4743.

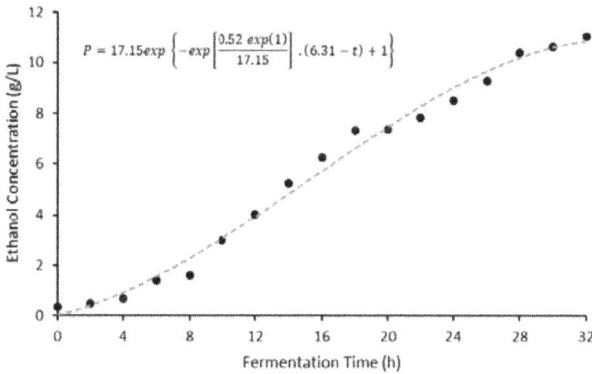

$$P = 17.15exp\left\{-exp\left[\frac{0.52\ exp(1)}{17.15}\right].(6.31-t)+1\right\}$$

Figure 4. Plot illustrating the fitted modified Gompertz curve.

Table 2 shows a comparison of the Gompertz coefficients obtained from this study using sorghum leaves and those reported from oil palm frond juice and sugar beet raw juice. From sorghum leaves, a higher maximum potential bioethanol concentration was achieved. In addition to this, an observed bioethanol production rate of 0.52 g/L/h was two times that achieved by Srimachai et al. [25] from oil palm frond juice. This illustrates the higher potential of waste sorghum leaves to accommodate a higher production rate.

Table 2. Comparison of modified Gompertz model parameters with previous studies.

Substrate	P_m (g/L)	$r_{p,m}$ (g/L/h)	t_L (h)	Reference
Sorghum leaves	17.15	0.52	6.31	This study
Oil palm frond juice (10–20 years)	3.79	0.08	0.77	Srimachai et al. [25]
Oil palm frond juice (3–4 years)	11.50	0.24	0.12	Srimachai et al. [25]
Sugar beet raw juice	73.31	4.39	1.04	Dodić et al. [19]

4. Conclusions

This study developed two kinetic models to describe the growth of *S. cerevisiae* BY4743 on pre-treated waste sorghum leaves for bioethanol production. Experimental data fitted the Monod and modified Gompertz model with high accuracy giving R^2 values of 0.95 and 0.98, respectively. From the Monod model, a maximum specific growth rate and Monod constant of 0.176 h^{-1} and 10.11 g/L were obtained, respectively. These findings show that waste sorghum leaves have a greater potential for bioethanol production with a higher production rate and productivity than several lignocellulosic substrates. Furthermore, a maximum yield of 0.49 g-ethanol/g-glucose was achieved after 32 h of fermentation. The generated kinetic knowledge of *S. cerevisiae* growth on sorghum leaves and bioethanol formation in this study is of high importance for process optimisation and scale-up towards commercialisation of this fuel.

Acknowledgments: The financial assistance of the National Research Foundation of South Africa (NRF) towards this research is hereby acknowledged.

Author Contributions: Daneal C. S. Rorke and Evariste Bosco Gueguim Kana conceived and designed the experiments; Daneal C. S. Rorke performed the experiments and analysed the data; Daneal C. S. Rorke wrote the article with the help and guidance of Evariste Bosco Gueguim Kana.

Conflicts of Interest: The authors declare no conflict of interest.

References

1. DoE. Biofuels Pricing and Manufacturing Economics. Department of Energy: South Africa, 2012. Available online: http://www.energy.gov.za/files/esources/renewables/BiofuelsPricingAndManufacturingEconomics.png (accessed on 10 November 2016).

2. Ariyajaroenwong, P.; Laopaiboon, P.; Salakkam, A.; Srinophakun, P.; Laopaiboon, L. Kinetic models for batch and continuous ethanol fermentation from sweet sorghum juice by yeast immobilized on sweet sorghum stalks. *J. Taiwan Inst. Chem. Eng.* **2016**, *66*, 210–216. [CrossRef]

3. Diaz, A.B.; de Souza Moretti, M.M.; Bezerra-Bussoli, C.; da Costa Carreira Nunes, C.; Blandino, A.; da Silva, R.; Gomes, E. Evaluation of microwave-assisted pretreatment of lignocellulosic biomass immersed in alkaline glycerol for fermentable sugars production. *Bioresour. Technol.* **2015**, *185*, 316–323. [CrossRef] [PubMed]

4. Mai, N.L.; Ha, S.H.; Koo, Y. Efficient pretreatment of lignocellulose in ionic liquids/co-solvent for enzymatic hydrolysis enhancement into fermentable sugars. *Process Biochem.* **2014**, *49*, 1144–1151. [CrossRef]

5. Vani, S.; Sukumaran, R.K.; Savithri, S. Prediction of sugar yields during hydrolysis of lignocellulosic biomass using artificial neural network modeling. *Bioresour. Technol.* **2015**, *188*, 128–135. [CrossRef] [PubMed]

6. Aguilar-Reynosa, A.; Romaní, A.; Rodríguez-Jasso, R.M.; Aguilar, C.N.; Garrote, G.; Ruiz, H.A. Microwave heating processing as alternative of pretreatment in second-generation biorefinery: An overview. *Energy Convers. Manag.* **2017**, *136*, 50–65. [CrossRef]

7. Gabhane, J.; Prince William, S.P.M.; Gadhe, A.; Rath, R.; Vaidya, A.N.; Wate, S. Pretreatment of banana agricultural waste for bio-ethanol production: Individual and interactive effects of acid and alkali pretreatments with autoclaving, microwave heating and ultrasonication. *Waste Manag.* **2014**, *34*, 498–503. [CrossRef] [PubMed]

8. Putra, M.D.; Abasaeed, A.E.; Atiyeh, H.K.; Al-Zahrani, S.M.; Gaily, M.H.; Sulieman, A.K.; Zeinelabdeen, M.A. Kinetic Modeling and Enhanced Production of Fructose and Ethanol from Date Fruit Extract. *Chem. Eng. Commun.* **2015**, *202*, 1618–1627. [CrossRef]

9. Ahmad, F.; Jameel, A.T.; Kamarudin, M.H.; Mel, M. Study of growth kinetic modeling of ethanol production by *Saccharomyces cerevisiae*. *Afr. J. Biotechnol.* **2011**, *16*, 18842–18846.
10. Massoud, M.I.; El-Razek, A.M.A. Suitability of *Sorghum bicolor* L. stalks and grains for bioproduction of ethanol. *Ann. Agric. Sci.* **2011**, *56*, 83–87. [CrossRef]
11. Suryaningsih, R. Bioenergy Plants in Indonesia: Sorghum for Producing Bioethanol as an Alternative Energy Substitute of Fossil Fuels. *Energy Procedia* **2014**, *47*, 211–216. [CrossRef]
12. Shen, F.; Zeng, Y.; Deng, S.; Liu, R. Bioethanol production from sweet sorghum stalk juice with immobilized yeast. *Procedia Environ. Sci.* **2011**, *11*, 782–789. [CrossRef]
13. Martins, F.; Gay, J.C. Biofuels: From Boom to Bust? Bain Brief, Bain and Company, 17 September 2014. Available online: http://www.bain.com/publications/articles/biofuels-from-boom-to-bust.aspx# (accessed on 23 October 2016).
14. Lee, S.Y. Kinetic Modeling and Simulation. *Encycl. Syst. Biol.* **2013**. [CrossRef]
15. Almquist, J.; Cvijovic, M.; Hatzimanikatis, V.; Nielsen, J.; Jirstrand, M. Kinetic models in industrial biotechnology—Improving cell factory performance. *Metab. Eng.* **2014**, *24*, 38–60. [CrossRef] [PubMed]
16. Ordoñez, M.C.; Raftery, J.P.; Jaladi, T.; Chen, X.; Kao, K.; Karim, M.N. Modelling of batch kinetics of aerobic carotenoid production using *Saccharomyces cerevisiae*. *Biochem. Eng. J.* **2016**, *114*, 226–236. [CrossRef]
17. Linville, J.L.; Rodriguez, M., Jr.; Mielenz, J.R.; Cox, C.D. Kinetic modeling of batch fermentation for *Populus* hydrolysate tolerant mutant and wild type strains of *Clostridium thermocellum*. *Bioresour. Technol.* **2013**, *147*, 605–613. [CrossRef] [PubMed]
18. Imamoglu, E.; Sukan, F.V. Scale-up and kinetic modeling for bioethanol production. *Bioresour. Technol.* **2013**, *144*, 311–320. [CrossRef] [PubMed]
19. Dodić, J.M.; Vučurović, D.G.; Dodić, S.N.; Grahovac, J.A.; Popov, S.D.; Nedeljković, N.M. Kinetic modelling of batch ethanol production from sugar beet raw juice. *Appl. Energy* **2012**, *99*, 192–197. [CrossRef]
20. Jukić, D.; Kralik, G.; Scitovski, R. Least-squares fitting Gompertz curve. *J. Comput. Appl. Math.* **2004**, *169*, 359–375. [CrossRef]
21. Mu, Y.; Wang, G.; Yu, H.Q. Kinetic modelling of batch hydrogen production process by mixed anaerobic cultures. *Bioresour. Technol.* **2006**, *97*, 1302–1307. [CrossRef] [PubMed]
22. Shafaghat, H.; Najafpour, G.D.; Rezaei, P.S.; Sharifzadeh, M. Growth Kinetics and Ethanol Productivity of *Saccharomyces cerevisiae* PTCC 24860 on Various Carbon Sources. *World Appl. Sci. J.* **2009**, *7*, 140–144.
23. Rorke, D.C.S.; Suinyuy, T.N.; Gueguim Kana, E.B. Microwave-assisted chemical pre-treatment of waste sorghum leaves: Process optimization and development of an intelligent model for determination of volatile compound fractions. *Bioresour. Technol.* **2017**, *224*, 590–600. [CrossRef] [PubMed]
24. Burk, D. Enzyme kinetic constants: The double reciprocal plot. *Trends Biochem. Sci.* **1984**, *9*, 202–204. [CrossRef]
25. Srimachai, T.; Nuithitikul, K.; O-thong, S.; Kongjan, P.; Panpong, K. Optimization and Kinetic Modeling of Ethanol Production from Oil Palm Frond Juice in Batch Fermentation. *Energy Procedia* **2015**, *79*, 111–118. [CrossRef]
26. Echegaray, O.F.; Carvalho, J.C.M.; Fernandes, A.N.R.; Satos, S.; Aquarone, E.; Vitolo, M. Fed-batch culture of *Saccharomyces cerevisiae* in sugar-cane blackstrap molasses: Invertase activity of intact cells in ethanol fermentation. *Biomass Bioenergy* **2000**, *19*, 39–50. [CrossRef]
27. Laopaiboon, L.; Nuanpeng, S.; Srinophakun, P.; Klanrit, P.; Laopaiboon, P. Selection of *Saccharomyces cerevisiae* and Investigation of its Performance for Very High Gravity Ethanol Fermentation. *Biotechnology* **2008**, *7*, 493–498. [CrossRef]
28. Okpokwasili, G.C.; Nweke, C.O. Microbial growth and substrate utilization kinetics. *Afr. J. Biotechnol.* **2005**, *5*, 305–317.
29. Raposo, S.; Pardão, J.M.; Díaz, I.; Lima-Costa, M.E. Kinetic modelling of bioethanol production using agro-industrial by-products. *Int. J. Energy Environ.* **2009**, *1*, 1–8.
30. Singh, J.; Sharma, R. Growth kinetic and modeling of ethanol production by wilds and mutant *Saccharomyces cerevisiae* MTCC 170. *Eur. J. Exp. Biol.* **2015**, *5*, 1–6.
31. Govindaswamy, S.; Vane, L.M. Kinetics of growth and ethanol production on different carbon substrates using genetically engineered xylose-fermenting yeast. *Bioresour. Technol.* **2007**, *98*, 677–685. [CrossRef] [PubMed]

32. Manikandan, K.; Saravanan, V.; Viruthagiri, T. Kinetics studies on ethanol production from banana peel waste using mutant strain of *Saccharomyces cerevisiae*. *Indian J. Biotechnol.* **2008**, *7*, 83–88.

33. Ardestani, F.; Shafiei, S. Non-Structured Kinetic Model for the Cell Growth of *Saccharomyces cerevisiae* in a Batch Culture. *Iran. J. Energy Environ.* **2014**, *5*, 8–12. [CrossRef]

34. Lin, Y.; Zhang, W.; Li, C.; Sakakibara, K.; Tanaka, S.; Kong, H. Factors affecting ethanol fermentation using *Saccharomyces cerevisiae* BY4742. *Biomass Bioenergy* **2012**, *47*, 395–401. [CrossRef]

35. Abdullah, S.S.S.; Shirai, Y.; Bahrin, E.K.; Hassan, M.A. Fresh oil palm frond juice as a renewable, non-food, non-cellulosic and complete medium for direct bioethanol production. *Ind. Crops Prod.* **2015**, *63*, 357–361. [CrossRef]

36. Ramos, C.L.; Duarte, W.F.; Freire, A.L.; Dias, D.R.; Eleutherio, E.C.; Schwan, R.F. Evaluation of stress tolerance and fermentative behaviour of indigenous *Saccharomyces cerevisiae*. *Braz. J. Microbiol.* **2013**, *44*, 935–944. [CrossRef] [PubMed]

37. Wu, X.; Staggenborg, S.; Propheter, J.L.; Rooney, W.L.; Yu, J.; Wang, D. Features of sweet sorghum juice and their performance in ethanol fermentation. *Ind. Crops Prod.* **2010**, *31*, 164–170. [CrossRef]

38. Pavlečič, M.; Vrana, I.; Vibovec, K.; Šantek, M.I.; Horvat, P.; Šantek, B. Ethanol Production from Different Intermediates of Sugar Beet Processing. *Food Technol. Biotechnol.* **2010**, *48*, 362–367.

39. Dinh, T.N.; Nagahisa, K.; Hirasawa, T.; Furusawa, C.; Shimizu, H. Adaptation of *Saccharomyces cerevisiae* Cells to High Ethanol Concentration and Changes in Fatty Acid Composition of Membrane and Cell Size. *PLoS ONE* **2008**, *3*, e2623. [CrossRef] [PubMed]

fermentation

MDPI

Article

Continuous Ethanol Production from Synthesis Gas by *Clostridium ragsdalei* in a Trickle-Bed Reactor

Mamatha Devarapalli [1], Randy S. Lewis [2] and Hasan K. Atiyeh [1,*]

[1] Department of Biosystems and Agricultural Engineering, Oklahoma State University,
    Stillwater, OK 74078, USA; mamathd@ostatemail.okstate.edu
[2] Department of Chemical Engineering, Brigham Young University, Provo, UT 84602, USA;
    randy.lewis@byu.edu
*   Correspondence: hasan.atiyeh@okstate.edu; Tel.: +1-405-744-8397

Academic Editors: Thaddeus Ezeji and Badal C. Saha
Received: 27 March 2017; Accepted: 18 May 2017; Published: 24 May 2017

Abstract: A trickle-bed reactor (TBR) when operated in a trickle flow regime reduces liquid resistance to mass transfer because a very thin liquid film is in contact with the gas phase and results in improved gas–liquid mass transfer compared to continuous stirred tank reactors (CSTRs). In the present study, continuous syngas fermentation was performed in a 1-L TBR for ethanol production by *Clostridium ragsdalei*. The effects of dilution and gas flow rates on product formation, productivity, gas uptakes and conversion efficiencies were examined. Results showed that CO and H_2 conversion efficiencies reached over 90% when the gas flow rate was maintained between 1.5 and 2.8 standard cubic centimeters per minute (sccm) at a dilution rate of 0.009 h^{-1}. A 4:1 molar ratio of ethanol to acetic acid was achieved in co-current continuous mode with both gas and liquid entered the TBR at the top and exited from the bottom at dilution rates of 0.009 and 0.012 h^{-1}, and gas flow rates from 10.1 to 12.2 sccm and 15.9 to 18.9 sccm, respectively.

Keywords: continuous syngas fermentation; *Clostridium ragsdalei*; ethanol; trickle bed reactor

---

1. Introduction

Syngas fermentation is part of the hybrid conversion technology for the conversion of renewable feedstocks or gas waste streams containing CO, CO_2 and H_2 to biofuels and chemicals. *Clostridium ljungdahlii*, *Clostridium carboxidivorans*, *Clostridium ragsdalei*, and *Alkalibaculum bacchi* are among the microorganisms that metabolize CO, CO_2 and H_2 via the reductive acetyl-CoA pathway to produce ethanol, acetic acid and cell carbon [1–4]. One major advantage of the hybrid conversion process is the ability to utilize feedstocks such as municipal solid wastes, industrial fuel gases and biomass [5]. However, challenges for this technology include mass transfer limitations, enzyme inhibition, low cell concentration and low ethanol productivity.

Ethanol has been reported to be a non-growth associated product of gas fermentation by certain *Clostridium* species [5,6]. Many researchers focused on improving ethanol productivity by optimizing media components, adding reducing agents, adjusting pH, adding nanoparticles and optimizing the bioreactor design to improve the mass transfer of CO and H_2 in fermentation medium [7–13]. *C. ljungdahlii* is one of the most extensively studied microorganisms for ethanol production using syngas fermentation. Tenfold (from 5 to 48 g/L) and over threefold (from 0.4 to 1.5 g/L) increases in ethanol and cell mass concentrations, respectively, were achieved in a continuous stirred tank reactor (CSTR) with cells recycled using *C. ljungdahlii* by designing a defined production medium and controlling pH at 4.5 [14].

Syngas fermentation bioreactors must maximize gas–liquid mass transfer, while achieving high cell densities to promote fast reaction [10]. Bioreactors such as air-lift reactors, continuous stirred tank reactors (CSTRs), trickle-bed reactors (TBRs) and hollow fiber membrane (HFM) reactors have been characterized for their capabilities for CO mass transfer into fermentation medium [12,15–19]. Further, improved ethanol production over batch bottle fermentations was reported when fermentations were performed in various bioreactors that provided larger working volume, greater cell recycling, continuous addition of nutrients and syngas, and better control of operating parameters [1,16,20–23]. For example, *C. carboxidivorans* produced only 0.9 g/L ethanol in batch bottles [24] compared to 1.6 g/L ethanol in bubble column [21]. About 19 g/L ethanol was produced by *C. ljungdahlii* in a two stage CSTRs with cell recycle [22] compared to about 1 g/L ethanol in batch bottles [25]. *C. ragsdalei* produced about 1.5 g/L ethanol in bottles [8] compared to 2 g/L in two stage CSTRs with partial cell recycle [20]. However, 1.7 g/L ethanol was produced by *A. bacchi* in bottles compared 6 g/L ethanol in a CSTR with cell recycling.

In our previous study [19], the TBR was reported to provide greater mass transfer capabilities compared to a CSTR. Further, in semi-continuous fermentation, formation of a biofilm in the TBR improved the H_2 uptake by decreasing the CO inhibition on hydrogenase because CO is consumed by cells as it flows through the TBR [18]. However, higher acetic acid production was observed in the semi-continuous fermentations due to repetitive medium replacement that provided a growth supporting environment. During batch and semi-continuous fermentations, cells undergo lysis as the nutrient levels depleted, causing the fermentation to cease. Production of ethanol in a batch process is time- and labor-intensive due to the long doubling times of syngas-fermenting microbes [22]. During continuous fermentation, high cell concentrations and productivity can be maintained for a longer period. Further, a continuous supply of fresh medium would maintain the cell's activity and adapt cells in the biofilm to produce more solvent by controlling fermentation parameters such as dilution rate, pH and gas flow rate. The focus of the present study is to improve ethanol production in a TBR during continuous fermentation. Supply of nutrients to a TBR was controlled by altering the dilution rate. The effect of dilution rate and gas flow rate on gas conversion, gas uptake, product concentrations, yields and productivities in both counter-current and co-current modes of operation were studied.

2. Materials and Methods

2.1. Microorganism and Medium Preparation

Clostridium ragsdalei (ATCC-PTA-7826) was maintained and grown on standard yeast extract medium. The medium contained 0.5 g/L yeast extract (YE), 10 g/L 2-(*N*-morpholino)ethanesulfonic acid (MES) as the buffer, 25 mL/L mineral solution without NaCl, 10 mL/L vitamin solution, 10 mL/L metal solution and 10 mL/L of 4% (*w/v*) cysteine sulfide. The detailed medium composition was previously reported [4]. *C. ragsdalei* stock culture was passaged three times (i.e., inoculum was transferred to fresh medium three times for cells' adaptation) prior to inoculating the TBR to reduce the lag phase. Detailed inoculum preparation was reported previously [18].

2.2. Fermentation Experimental Setup

A schematic of the continuous syngas fermentation setup is shown in Figure 1. The TBR designed in house was made of a borosilicate glass column of 5.1 cm diameter and 61 cm long. The detailed reactor design was reported earlier [18]. The packing material was 6-mm soda lime glass beads. The TBR liquid outlet was connected to a 500 mL Pyrex glass bottle, which was used as a sump to hold 500 mL of medium. It was operated both in counter-current and co-current modes. A peristaltic pump circulated the liquid at a desired flow rate. The pH and ORP probes (Cole-Parmer, Vernon, IL, USA) were placed in line in the recirculation loop. A liquid sample port was placed in the recirculation loop. Fresh medium from the feed tank was pumped to the reactor through a 6.4 mm T-connector placed in the recirculation loop before the medium enters the top of the TBR using a Bioflo pump and controller (New Brunswick Scientific Co., Edison, NJ, USA). The product stream was connected to another Bioflo pump that pumped out the product to a tank to maintain a constant amount of liquid in the reactor and recirculation loop. A port for acid and base addition was placed in the recirculation loop after the sampling port and was connected to the Bioflo controller for pH control. N_2 was purged continuously through the feed and product tanks at 20 standard cubic centimeters per minute (sccm) to maintain anaerobic conditions. A one-way valve was connected at the gas outlet of both tanks to make sure the gas flowed out and air did not flow back into the tanks. In counter-current mode of operation, the gas entered at the bottom of the TBR. The exhaust gas from the TBR was fed into the sump headspace and then out to the sump gas exit line. In co-current operation, both gas and liquid entered the TBR at the top and exited through the same exit line to the sump. The sump acted as a gas–liquid separator. Further, a back pressure regulator was connected to the sump gas exit line to ensure that a 115 kPa pressure was maintained in the TBR. A pressure gauge was connected at the TBR gas exit line to measure the pressure in the TBR. An additional gas exit line was connected to the sump as a safety exhaust line with a pressure switch and a solenoid valve to vent the excess pressure in the TBR. A bubbler was placed after the pressure regulator to minimize losses of products exiting with the gas.

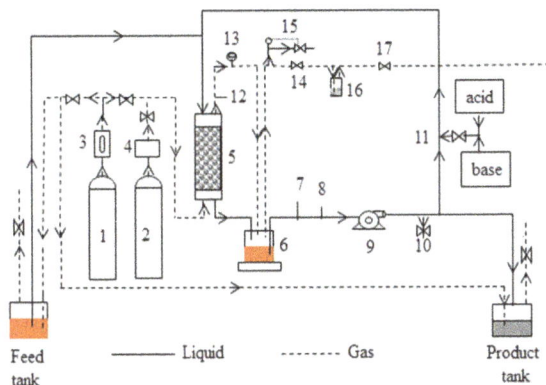

Figure 1. Continuous syngas fermentation in trickle bed reactor (TBR) setup for counter-current flow. (1) Nitrogen cylinder; (2) Syngas cylinder; (3) Rotameter; (4) mass flow controller; (5) TBR; (6) Sump to hold medium; (7) ORP probe; (8) pH probe; (9) Masterflex pump; (10) Liquid sample port; (11) Acid/base addition port in liquid circulation loop; (12) TBR gas sample port; (13) Pressure gauge; (14) Back pressure regulator; (15) By pass safety line with solenoid valve and pressure switch; (16) Gas bubbler; (17) Ball valve to exhaust line. Dashed lines indicate gas lines and solid lines indicate liquid lines. For co-current flow, gas was fed with the liquid medium from the top of the TBR and exited to the sump. Dashed lines with arrows point in feed and product tanks contain N_2 to ensure anaerobic conditions.

2.3. Continuous Fermentation Procedure

The TBR column, sump, tubing and liquid medium were sterilized in an autoclave (Primus Sterilizer Co., Inc., Omaha, NE, USA) at 121 °C for 20 min. After sterilization, the TBR was setup and purged with N_2 for 5 h. Then, 200 mL of fresh sterile medium was added into the TBR and purged with N_2 for 8 h. Next, the gas was switched to syngas with 38% CO, 5% N_2, 28.5% CO_2 and 28.5% H_2 (by volume) (Stillwater Steel and Supply Company, Stillwater, OK, USA), which is similar to the composition of coal derived syngas [26]. A 60% (*v/v*) inoculum was aseptically added into the TBR through the liquid sample port. The temperature of the TBR was maintained at 37 °C. The liquid recirculation rate was set at 200 mL/min. At the beginning of the fermentation, the gas flow rate was set at 1.5 sccm. Initially, the TBR was operated in semi-continuous mode. After the CO and H_2 conversion efficiencies reached about 90%, the TBR was switched to continuous mode by turning on the fresh medium and product pumps at a desired flow rate. Effects of three dilution rates of 0.006, 0.009 and 0.012 h^{-1} on product formation and gas conversion efficiency were examined. The conversion efficiency of each gas during fermentation was estimated based on the amount of gas converted by *C. ragsdalei* relative to the amount of gas fed to the TBR. Dilution rate equals to the feed rate divided by the medium volume in TBR, sump and recirculation loop. At each dilution rate, the effect of gas flow rate on cell growth, gas conversion, product formation and yields was examined. The gas flow rate in the TBR was gradually increased until CO conversion efficiency dropped below 40%. Then, the gas flow rate was decreased and a new dilution rate was applied. Gas and liquid samples were aseptically withdrawn from the TBR periodically. To avoid flooding of the TBR by cell debris, the recirculation rate was increased from 200 to 500 mL/min for about 10 min at every sampling time to remove cell cells debris between the packing materials.

2.4. Sample Analysis

The cell optical density of the fermentation medium from the liquid sample port in the circulation loop was measured at 660 nm (OD$_{660}$) with a UV spectrophotometer (Cole Parmer, Vernon, Hills, IL, USA). The total cell optical density of the attached cells was measured at the end of the fermentation as described previously [18]. The pH measurements were logged into a computer using Biocommand software (New Brunswick Scientific Co.). Fermentation samples were analyzed for ethanol and acetic acid using a DB-FFAP capillary column gas chromatography with flame ionization detector (GC-FID). Gas samples were analyzed in a 6890N gas chromatography with thermal conductivity detector (TCD) (Agilent Technologies, Wilmington, DE, USA). More details of the methods used to analyze gas and liquid samples were described previously [18].

3. Results

3.1. Cell Growth and pH

C. ragsdalei cell OD$_{660}$ and pH profiles in the TBR when continuously operated for 3200 h in counter-current and co-current modes are shown in Figure 2. *C. ragsdalei* started to grow after 174 h of lag phase. The cell OD$_{660}$ was 0.35 at 197 h when the TBR was switched to continuous operation with a dilution rate of 0.012 h^{-1} and a gas flow rate of 1.9 sccm. Cell OD$_{660}$ further increased to 0.53 at 207 h and remained constant until 224 h. However, the cell OD$_{660}$ started decreasing slowly to 0.20 at 305 h, most likely due to cell washout. At this point, the dilution rate was decreased by 50% (D = 0.006 h^{-1}) which resulted in an increase in the cell OD$_{660}$ to 0.30 by 357 h.

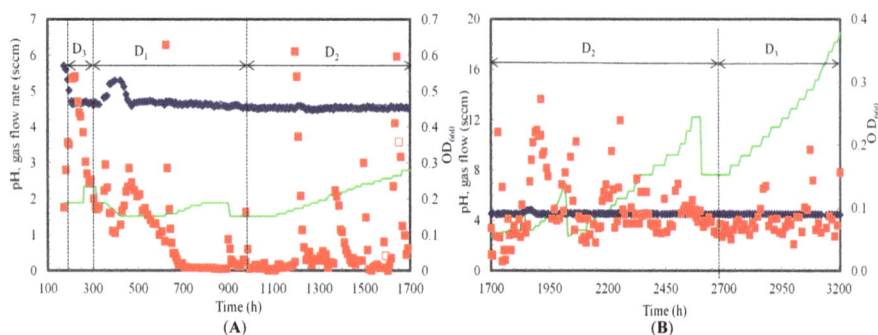

Figure 2. pH and cell mass optical density (OD$_{660}$) profiles during continuous syngas fermentation in (**A**) Counter-current and (**B**) Co-current flow modes at various dilution rates (D$_1$, D$_2$ and D$_3$ of 0.006, 0.009 and 0.012 h^{-1}, respectively); (♦) pH (■) OD$_{660}$ (—) Gas flow rate (Open symbols indicate flooded TBR; 0 to 174 h: lag phase resulted in no data).

A brief power interruption between 329 and 351 h resulted in no gas flow to the TBR. This caused a decrease in cell activity (i.e., decrease in CO and H_2 gas uptake rates) up to 398 h. The fermentation slowly recovered when the gas flow and dilution rates were reset to 1.5 sccm and 0.006 h^{-1}, respectively. The cell OD_{660} increased from 0.10 at 398 h to 0.29 at 461 h. The cell OD_{660} in the liquid medium dropped to approximately zero around 700 h. However, the gas uptake rates were maintained indicating continued cell activity due to a biofilm in TBR rather than suspended cells. This was also confirmed by measuring the total cell mass concentration at the end of the TBR run, which showed a much higher cell mass concentration attached to the TBR than was suspended in the liquid medium as discussed below. Formation of biofilm refers to all cell mass excluding suspended cells. During counter-current flow mode, cells from the biofilm were resuspended into the medium when the pressure was released to clear the medium between the beads in the flooded TBR at 627, 901 and 909 h. This resulted in a sudden increase in the measured cell OD_{660}. To avoid flooding issues in counter-current mode, the liquid recirculation rate was intermittently increased from 200 to 500 mL/min for 10 min at various sampling times of 1197, 1371, 1498, 1628, 1643 and 1652 h. Unlike in counter-current mode, cell OD_{660} was between 0.05 and 0.3 during co-current mode from 1700 to 3200 h (Figure 2B).

The TBR and glass beads were washed with DI water to calculate the total amount of cells attached to the beads in the TBR after 3200 h of continuous fermentation. The beads were collected in a tub and washed three times with 1 L DI water. The column was also washed with 1 L of DI water to account of cells attached to the column walls. The cell OD_{660} in the beads from wash-1, wash-2, wash-3 and column-wash were 8.86, 0.41, 0.13 and 1.97, respectively. Based on this analysis, the estimated overall dry cell weight in the TBR at the end of the fermentation was 4.24 g.

During cell growth, the medium pH decreased from 5.7 at 174 h to 4.7 at 207 h. The pH of the medium was then maintained at 4.6 by addition of about 0.5 to 1 mL of 2 N KOH after every sampling time. After the power interruption between 329 and 351 h, the pH was increased to 5.2 to maintain a pH slightly favorable to cell growth conditions to recover fermentation activity. The pH dropped from 5.2 to 4.7 as cell OD_{660} increased between 422 to 461 h. After 461 h, the pH was maintained between 4.5 and 4.6.

3.2. Gas Conversion

The CO and H_2 conversion efficiencies in the TBR are estimated as the amount utilized divided by the amount flowing into the TBR. The CO and H_2 conversion efficiencies by *C. ragsdalei* were 92% and 72%, respectively, when the fermentation was switched from semi-continuous to continuous mode at 197 h (Figure 3A).

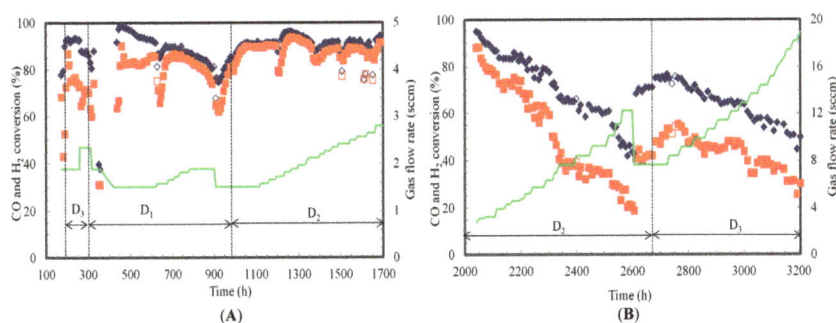

Figure 3. Gas conversion efficiencies during continuous syngas fermentation in TBR (**A**) Counter-current and (**B**) Co-current flow modes at various dilution rates (D_1, D_2 and D_3 are 0.006, 0.009 and 0.012 h^{-1}, respectively); (◆) CO (■) H_2 (—) Gas flow rate (Open symbols indicate flooded TBR; 0 h to 174 h: lag phase resulted in no data; 1700 to 2042 h: gas leak resulted in no data).

In counter-current mode, the TBR was operated at dilution rates of 0.012 h^{-1} (D_3), 0.006 h^{-1} (D_1) and 0.009 h^{-1} (D_2) from 197 to 305 h, 305 to 989 h and 989 to 1700 h, respectively. The CO and H_2 conversion efficiencies were 93% and 74%, respectively, at 0.012 h^{-1} and gas flow rate of 1.9 sccm. However, when the gas flow rate was increased to 2.3 sccm at the same dilution rate, the CO and H_2 conversion efficiencies dropped slightly to 88% and 71%, respectively.

CO and H_2 conversions efficiencies continued to decrease to 81% and 60%, respectively, when the gas flow rate and dilution rate were 2.3 sccm and 0.006 h^{-1}, respectively, at 305 h. However, the CO and H_2 conversion efficiencies increased to 88% and 74%, respectively, when gas flow rate was reduced to 1.9 sccm at 319 h. The conversion efficiencies of CO and H_2 decreased to 40% and 31%, respectively, due to power shutdown from 329 to 351 h that hindered the fermentation. The fermentation recovered slowly after 398 h with CO and H_2 conversion efficiencies reached to 92% and 86%, respectively at 620 h. As the gas flow rate was increased from 1.5 to 1.9 sccm, the liquid medium flooded the TBR (at 627, 901 and 909 h), which decreased the CO and H_2 conversion efficiencies to about 65% at 909 h (Figure 3A). The TBR flooding caused gas bypass from the bottom of the TBR to the headspace sump and decreased the availability of syngas to cells in the TBR. The gas flow rate was decreased from 1.9 to 1.5 sccm at 909 h to avoid further flooding. CO and H_2 conversion efficiencies recovered to 85% and 81%, respectively, at 981 h before increasing the dilution rate to 0.009 h^{-1}.

Gas conversion efficiencies of 91% CO and 90% H_2 were achieved between 989 and 1115 h at 0.009 h^{-1}. While the CO conversion efficiency was about the same at both 0.006 and 0.009 h^{-1}, the H_2 conversion efficiency was 5% higher at 0.009 h^{-1} than at 0.006 h^{-1} at the same gas flow rate. The increase in gas uptake is due to higher cells' activity due to availability of more nutrients at higher dilution rate. A decrease in CO and H_2 conversion efficiencies was observed at various sample points (1197, 1371, 1498 and 1628 h) when the liquid recirculation rate was increased for 10 min to clear the cell debris in the TBR. During the period between 1197 and 1628 h, the gas flow rate was increased from 1.7 sccm to 2.6 sccm. The combination of the increase in gas supply and possible removal of active cells with removal of cell debris likely contributed to the decrease in gas conversion. The TBR operation at 0.009 h^{-1} from 989 to 1700 h with an increase in gas flow rate from 1.5 sccm to 2.8 sccm resulted in CO and H_2 conversion efficiencies of about 91% and 89%, respectively.

The TBR was operated in co-current mode at dilution rates of 0.009 h^{-1} (D_2) and 0.012 h^{-1} (D_3) from 1700 to 2672 h and from 2672 to 3200 h, respectively, with a gradual increase in gas flow rate from 2.8 sccm to 18.9 sccm (Figure 3B). The TBR gas inlet leaked from 1700 to 2042 h, which resulted in inaccurate gas flow rate measurements. No gas data was obtained during this time period. During the TBR operation at 0.009 h^{-1}, the increase in gas flow rate from 2.8 sccm at 2042 h to 12.2 sccm at 2607 h resulted in a decrease in CO and H_2 conversion efficiencies from 95% and 88% to 43% and 19%, respectively. A decrease in conversion is expected since the length of time the gas is in the reactor decreases with increasing flow rate. The gas flow rate was reduced to 7.6 sccm at 2607 h to increase CO and H_2 conversion before a new dilution rate of 0.012 h^{-1} was used. CO and H_2 conversion efficiencies increased to 71% and 42%, respectively. The gas conversion efficiencies at 2660 h were slightly higher than those obtained at 2375 h and the same operating conditions.

The dilution rate was increased to 0.012 h^{-1} (D_3) at 2672 h with a gas flow rate of 7.6 sccm (Figure 3B). CO and H_2 conversion efficiencies reached 77% and 53%, respectively, at 2725 h. These gas conversion efficiencies at 0.012 h^{-1} and the same gas flow were 8% higher for CO and 21% higher for H_2 than at 0.009 h^{-1}. This is due to the increase in cells' activity with additional nutrients at the higher dilution rate. Further, when the gas flow rate was increased from 8.4 sccm at 2732 h to 18.9 sccm at 3200 h, the CO and H_2 conversion efficiencies slowly dropped to 50% CO and 30% H_2, respectively. The differences between CO and H_2 conversion efficiencies at dilution rate of 0.012 h^{-1} was lower than at 0.009 h^{-1} indicating an increase in gas uptake at higher dilution rates (Figure 4B). It can also be observed from Figure 3B that the decrease in CO and H_2 gas conversion efficiencies were lower at 0.012 h^{-1} than at 0.009 h^{-1} indicating higher cells' activity at 0.012 h^{-1}. In two stages of continuous syngas fermentation with *C. ljungdahlii* in a CSTR followed by a bubble column with gas and cells recycling, the increase in dilution rate from 0.01 to 0.016 h^{-1} was reported to increase the cell OD$_{600}$ from 9.9 to 17.8 due to supply of more nutrients [22]. The same study reported CO and H$_2$ conversion efficiencies in the CSTR were 46% and 49%, respectively, at 23 sccm compared to CO and H$_2$ conversion efficiencies of 86% and 82% in the bubble column at 121 sccm. The high gas conversion efficiency in the second stage is attributed to the high cell OD$_{600}$ of 17.8 that was achieved with cell recycling.

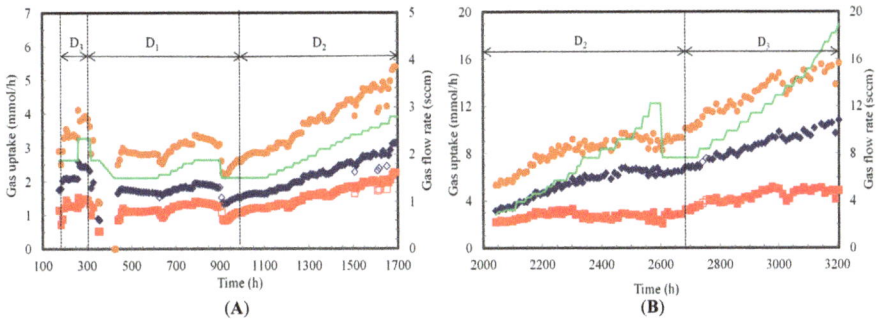

Figure 4. Gas uptake rates during continuous syngas fermentation in TBR (**A**) Counter-current and (**B**) Co-current flow modes at various dilution rates (D$_1$, D$_2$ and D$_3$ are 0.006, 0.009 and 0.012 h^{-1}, respectively); (◆) CO (■) H$_2$ (●) CO+H$_2$ (—) Gas flow rate (Open symbols indicate flooded TBR; 0 h to 174 h: lag phase resulted in no data; 1700 to 2042 h: gas leak resulted in no data).

3.3. Gas Uptake Profiles

The gas uptake profiles during continuous syngas fermentation by *C. ragsdalei* in the TBR are shown in Figure 4. The specific gas uptake rates in mmol/gcell·h were not calculated because the cell mass concentration in the biofilm during operation was not known and difficult to quantify. Hence, the gas uptakes are described only in terms of mmol/h. CO and H$_2$ uptake rates at the start of the continuous fermentation at 197 h, 0.012 h^{-1} and 1.9 sccm were 2.0 and 1.2 mmol/h, respectively. When the gas flow rate was increased to 2.3 sccm at 261 h, CO and H$_2$ uptake rates slightly increased. However, cell OD$_{660}$ decreased to 0.20 at 305 h due to cell washout (Figure 2A). Therefore, the dilution rate was decreased to 0.006 h^{-1} at 305 h. CO and H$_2$ uptake rates at 319 h decreased to 1.8 and 1.0 mmol/h, respectively.

To increase gas consumption, the gas flow rate was reduced to 1.9 sccm at 319 h. CO and H$_2$ uptake rates at 329 h recovered back to 2.0 and 1.2 mmol/h, respectively. However, the cell activity decreased when a power failure occurred between 329 and 351 h. The fermentation slowly recovered with CO and H$_2$ uptake rates of 1.7 and 1.1 mmol/h between 375 and 620 h.

The gas flow rate was increased from 1.5 to 1.9 sccm between 620 and 787 h, which resulted in a slight increase in CO and H$_2$ uptake rates to 2.0 and 1.4 mmol/h at 787 h, respectively. These gas uptakes were maintained up to 900 h. However, flooding at 901 h and 909 h resulted in a decline in CO and H$_2$ uptake rates to 1.5 and 1.0 mmol/h, respectively. Hence, the gas flow rate was decreased to 1.5 sccm at 909 h and was maintained at this flow rate until 981 h. At 981 h, the CO and H$_2$ uptake rates were essentially still the same as at 909 h. However, since the same gas uptake rates were achieved at a lower gas flow rate, CO and H$_2$ conversion efficiencies increased (Figure 3). The dilution rate was maintained at 0.009 h^{-1} during counter-current operation between 989 and 1700 h. A step increment increase in gas flow rate by 5–10% from 1.5 sccm at 989 h to 2.8 sccm at 1700 h resulted in an increase of gas uptake rate to 3.1 mmol/h of CO and 2.1 mmol/h of H$_2$.

The TBR was switched at 1700 h to co-current mode due to frequent flooding issues. However, there was gas leak in the inlet to the TBR from 1700 to 2042 h, which resulted in inaccurate gas flow rate measurements and no gas uptake data. The gas flow rate was gradually increased from 2.8 to 6.3 sccm in between 2042 and 2313 h, which increased the gas uptake rates to 5.9 mmol/h CO and 3.3 mmol/h H_2 at 2313 h (Figure 4B). A further increase in the gas flow rate from 6.3 to 12.2 sccm from 2313 to 2672 h resulted in a decrease of H_2 uptake rate to between 2.4 and 3.0 mmol/h, while the CO uptake rate increased between 6.0 and 6.7 mmol/h.

The average total CO and H_2 gas uptake rate between 2313 and 2672 h was 8.5 mmol/h. It can be observed that the increase in the dilution rate from 0.009 to 0.012 h^{-1} and gas flow rate from 2.8 to 18.9 sccm increased the overall CO and H_2 uptake rates. In co-current flow, it was also observed that the increase in dilution rate by 36% (0.009 h^{-1} to 0.012 h^{-1}) resulted in an increase in total CO and H_2 uptake rate by 47%. The gas uptake rates in co-current mode were 2.5 fold higher than in counter-current mode. This was attributed to the ability to operate the TBR in co-current mode at higher gas flow rates.

In the previous study with semi-continuous fermentation in co-current mode TBR, the maximum CO and H_2 conversion efficiencies at 4.6 sccm were 80% (CO uptake rate of 4.4 mmol/h) and 55% (H_2 uptake rate of 2.2 mmol/h), respectively [18]. In the present study during co-current continuous fermentation at 4.6 sccm and 0.009 h^{-1}, gas conversion efficiencies of 82% CO (CO uptake rate of 4.4 mmol/h) and 72% H_2 (H_2 uptake rate of 2.74 mmol/h) were achieved. The high gas conversion efficiency and uptake rates are due to high cells' activity with continuous addition of nutrients during the fermentation.

3.4. Product Profiles

At the beginning of continuous fermentation (197 h), ethanol and acetic acid concentrations were 0.8 g/L and 2.4 g/L, respectively (Figure 5A). Ethanol and acetic acid concentrations increased to 2.0 g/L and 5.0 g/L, respectively between 197 and 305 h at a dilution rate of 0.0012 h^{-1}. A slight increase in ethanol and acetic acid concentrations was observed when dilution rate was decreased to 0.006 h^{-1}. However, ethanol and acetic acid concentrations decreased between 329 h and 398 h. This decrease was associated with the power shutdown and washout cells and products.

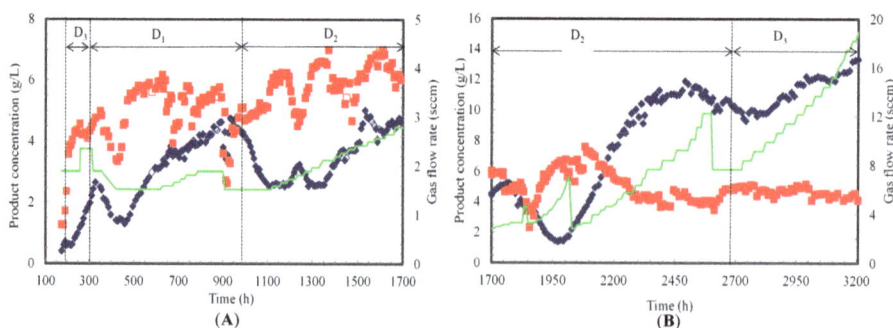

Figure 5. Product concentrations during continuous syngas fermentation in TBR. (**A**) Counter-current and (**B**) Co-current flow modes at various dilution rates (D_1, D_2 and D_3 are 0.006 h^{-1}, 0.009 h^{-1} and 0.012 h^{-1}, respectively); (◆) Ethanol (■) Acetic acid (—) Gas flow rate (Open symbols indicate flooded TBR; 0 h to 174 h: lag phase resulted in no data).

The fermentation slowly recovered and product concentrations were observed to be stable from 398 to 454 h after which ethanol and acetic acid concentrations at 627 h reached 3.2 and 6.2 g/L, respectively, when the gas flow rate was increased by 20% from 1.5 sccm between 627 and 787 h, ethanol concentration increased by 20% while the acetic acid concentration decreased by 20%. Ethanol concentration slowly increased to 4.3 g/L while acetic acid concentration remained at 5.0 g/L between 787 and 909 h. Due to flooding at 901 and 909 h, the gas flow rate was reduced from 1.9 to 1.5 sccm to recover the fermentation in the TBR.

The dilution rate was increased from 0.006 to 0.009 h^{-1} between 989 and 1700 h. This increased the product removal rate from the TBR, which decreased the ethanol concentration by about 40% to 2.5 g/L at 1115 h. However, the acetic acid concentration increased by 15% to around 5.9 g/L at 1115 h. The increase in acetic acid concentration was due to increase in cells' activity and concentration in the biofilm. The increase in cell concentration is associated with acetic acid production and ATP generation as a high amount of energy is required for cell maintenance [1].

The gas flow rate was increased from 1.5 to 2.8 sccm in a step increment of 5–10% every 24 to 36 h between 1115 and 1700 h (Figure 5A). Ethanol and acetic acid concentrations were stable at 2.5 and 6.2 g/L, respectively, when the gas flow rate was increased from 1.5 to 1.7 sccm from 1115 to 1197 h. The liquid recirculation rate was increased from 200 to 500 mL/min for 10 min to clear the cell debris at 1197 h. This resulted in a slow increase in the ethanol concentration to 3.2 g/L and a decrease in the acetic acid concentration to 4.7 g/L at 1245 h. This intermittent increase in liquid recirculation rate could have cleared the cell debris from the packing and caused better gas uptake resulting in a positive effect on ethanol production.

To test the positive effect of the intermittent increase of liquid recirculation rate on ethanol production, the liquid flow rate was again increased to 500 mL/min for 10 min at 1371 h. Ethanol concentration slowly increased to 3.7 g/L while acetic acid remained at 4.7 g/L at 1474 h. Since increasing the liquid recirculation rate intermittently had a positive effect on ethanol production, it was performed when the cell OD$_{660}$ in the medium decreased to zero. The intermittent increase in liquid recirculation rate and gradual increase in gas flow rate to 2.4 sscm resulted in production of 5.0 g/L ethanol and 6.1 g/L acetic acid at 1532 h. The gas flow rate was further increased to 2.8 sccm between 1532 and 1700 h. Ethanol and acetic acid concentrations were 4.4 and 5.8 g/L, respectively at 1700 h.

In co-current mode, the gas flow rate was gradually increased from 2.8 to 18.9 sccm (Figure 5B). Similar to the counter-current mode, the gradual increase in gas flow rate with an increase in liquid recirculation rate from 200 to 500 mL/min for 10 min at every sampling time increased ethanol production. Ethanol and acetic acid concentrations were 2.7 and 6.7 g/L, respectively, at 2052 h and 3.1 sccm. An increasing trend in ethanol production and a decreasing trend in acetic acid production were observed as the gas flow rate was increased from 3.1 to 9.2 sccm between 2052 and 2493 h. Ethanol and acetic acid concentrations at 2493 h were 11.9 and 4.6 g/L, respectively. A further increase in the gas flow rate from 9.2 to 11.1 sccm at 2542 h did not increase ethanol production. Additionally, the increase in gas flow rate from 11.1 to 12.2 sccm at 2566 h slightly decreased ethanol and acetic acid concentrations to 10.5 and 4.1 g/L, respectively. This indicates that beyond a gas flow rate of 9.2 sccm, cells reached a kinetic limitation and were not able to process more gas even when more gas was provided (Figure 4B).

Further, when the gas flow rate was decreased from 12.2 to 7.6 sccm at 2607 h, ethanol and acetic acid concentrations were stable at 10.8 and 3.8 g/L, respectively, until 2672 h. When the dilution rate was increased from 0.009 to 0.012 h^{-1} at 2510 h, the ethanol concentration dropped slowly to 9.9 g/L. However, the acetic acid concentration slightly increased to 5.0 g/L at 2551 h and 7.6 sccm. Further increase in the gas flow rate from 7.6 to 18.9 sccm resulted in 13.2 g/L ethanol and 4.3 g/L acetic acid at 3200 h. The gas uptake rate at 0.012 h^{-1} was higher than at 0.009 h^{-1} due to higher cells' activity, which resulted in more ethanol production at 0.012 h^{-1}.

3.5. Productivity and Yields

Ethanol and acetic acid productivities were estimated by multiplying the dilution rate by the product concentration. Ethanol and acetic acid yields were estimated based on CO consumed as previously reported [27]. One mole of ethanol is formed from six moles of CO and one mole of acetic acid is produced from four moles of CO.

During counter-current operation, the highest ethanol productivity of 45 mg/L·h was obtained during operation at 0.009 h^{-1} and 1556 h. However, the highest acetic acid productivity was 63 mg/L·h at 0.009 h^{-1} and 1611 h. During counter-current mode, acetic acid productivity was always higher than ethanol productivity. However, ethanol productivity was higher during co-current mode. The maximum ethanol productivity during co-current operation was 158 mg/L·h at 0.012 h^{-1} and 3200 h, while a maximum acetic acid productivity of 68 mg/L·h was obtained at 0.009 h^{-1} at 2083 h. The ethanol productivity achieved in the present study with continuous syngas fermentation in the TBR was over four times higher than reported during semi-continuous fermentation (37 mg/L·h) in the TBR [18].

Moreover, the molar ratio of ethanol to acetic acid produced in the present study during continuous fermentation at 0.012 h^{-1} and 18.9 sccm in the TBR was 4:1, which was higher than in semi-continuous TBR fermentation (1:2). In semi-continuous fermentation, as nutrients were depleted from the medium, the gas conversion efficiencies and uptake rates decreased. Replacement of the medium in semi-continuous fermentations resulted in a nutrient-rich environment at pH 5.8 that promoted cell growth and thus more acetic acid production. However, during continuous fermentation the nutrients levels were maintained by altering the dilution rate and the pH was maintained at 4.5 that favored ethanol production. This clearly shows the advantages of the continuous syngas fermentation process.

During counter-current mode, ethanol yield was 22% while acetic acid yield was 42% at 197 h and 1.9 sccm (Figure 6). However, acetic acid yield slowly dropped to 15% while ethanol yield increased to 58% at 294 h. At dilution rate of 0.006 h^{-1} from 305 h to 989 h, the ethanol yield increased from 28% at 461 h to 85% at 850 h, while acetic acid yield decreased from 38% at 461 h to 13% at 850 h. At a dilution rate of 0.009 h^{-1} between 989 and 1700 h, the average ethanol yield was about 85%, while the average acetic acid yield was about 20%. As the gas flow rate was increased and the pH was maintained at 4.5, ethanol yields increased due to the availability of more reductants (CO and H$_2$) and pH values that favored solvent production conditions.

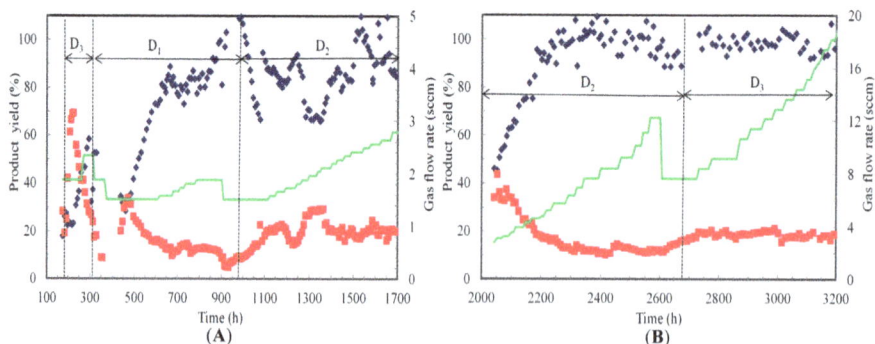

Figure 6. Product yields based on CO consumed during continuous syngas fermentation in TBR. (**A**) Counter-current and (**B**) Co-current flow modes at various dilution rates (D$_1$, D$_2$ and D$_3$ are 0.006, 0.009 and 0.012 h^{-1}, respectively); (◆) Ethanol (■) Acetic acid (—) Gas flow rate. (0 to 174 h: lag phase resulted in no data. 1700 to 2042 h: gas leak resulted in no data.

In co-current operation at 0.009 h^{-1}, ethanol yield increased from 46% at 2042 h to 100% at 2232 h. Ethanol yield remained close to 100% during fermentation from 2232 to 2607 h. However, the acetic acid yield decreased from 34% at 2042 h to 16% at 2232 h. It remained close to about 13% from 2232 to 2607 h. Ethanol and acetic acid yields were about 100% and 20%, respectively, during operation at 0.012 h^{-1}. The higher ethanol yield in co-current mode was due to higher cell activity that processed more gas.

4. Discussion

As discussed in Section 3.2, high dilution rates provided more nutrients to cells, which increased the cells' activity. The increase in the gas flow rate increased CO and H_2 transfer rates into the medium, which supported ethanol production. Ethanol produced in the present study (13.2 g/L) was higher than that reported in a CSTR with cell recycle using *A. bacchi* (6 g/L), a bubble column reactor (1.6 g/L) and a monolithic biofilm reactor (4.9 g/L) using *C. carboxidivorans* [1,5,28]. However, 19 g/L ethanol was reported in a two-stage continuous syngas fermentation in a CSTR followed by a bubble column with gas and cell recycling [23] and 48 g/L ethanol was reported in a CSTR with cell recycle using *C. ljungdahlii* [14]. Up to 24 g/L ethanol production was reported with hollow fiber membrane biofilm reactor (HFM-BR) using *C. carboxidivorans* [29]. However in addition to syngas, the presence of 10 g/L of fructose in ATCC 1745 PETC medium as previously reported [29] could have contributed to more ethanol production.

Compared to other microorganisms and reactor designs, the maximum ethanol productivity of 158 mg/L·h achieved by *C. ragsdalei* in the TBR in the present study was higher than ethanol productivity of 140 mg/L·h reported for *C. carboxidivorans* in HFM-BR, 110 mg/L·h reported for *C. ljungdahlii* in CSTR without cell recycle or for *A. bacchi* in a CSTR (70 mg/L·h) [1,29,30]. However, ethanol productivity in the present study was lower than the 301 mg/L·h ethanol productivity reported for *C. ljungdahlii* in a two stage CSTR and bubble column with gas and cell recycling [23].

The results also showed the many advantages of using continuous syngas fermentation in a TBR compared to other reactors. The cells' activity in the TBR was recovered after power shutdown and multiple flooding issues occurred during counter-current flow. Intermittent increase in liquid recirculation rate cleared cell debris in the TBR and improved gas uptake and ethanol production. However, further improvements in TBR performance are expected by utilizing better packing material for immobilization of cells and increasing the H_2:CO ratio in the syngas. Glass beads used in this study have a void fraction of 0.38, which is lower than the void fraction provided by other packing materials such as intalox saddles (0.6 to 0.9) and pall rings (0.9) [31]. Low void fraction reduces the availability of free space for gas-liquid mass transfer and decreases the reactive holdup volume. Further, use of cell immobilization techniques [32,33] such as covalent coupling using cross linking agents, entrapment, and adsorption on packing with rough surfaces can reduce the time of the biofilm formation and improve the TBR performance. Additionally, there is a need to grow more cells in the TBR, recycle unused gas, perform two-stage reactors using TBRs or in combination with other reactors to increase syngas utilization and productivity, which warrant further investigation.

5. Conclusions

To our knowledge, this is the first study on continuous operation of syngas fermentations in TBR for ethanol and acetic acid production, and this report highlighted the operational constraints and challenges of continuous syngas fermentation in TBR, and how the bioreactor operation can be restarted after major accidents such as flooding and power shutdown. The highest ethanol concentration, productivity and ethanol to acetic acid molar ratio of 13.2 g/L, 158 mg/L·h and 4:1, respectively, were obtained during co-current continuous syngas fermentation at a dilution rate of 0.012 h^{-1}. In co-current mode, the total gas uptake rates more than doubled and ethanol productivity increased over fivefold with the increase in the gas flow rate from 2.8 to 18.9 sccm and dilution rate from 0.009 to 0.012 h^{-1}. Operating TBR in co-current mode avoided flooding issues that occurred during counter-current mode and allowed production of over twofold more ethanol than in counter current mode.

Acknowledgments: This research was supported by the Sun Grant Initiative through the U.S. Department of Transportation, USDA-NIFA Project No. OKL03005 and Oklahoma Agricultural Experiment Station.

Author Contributions: Mamatha Devarapalli, Randy S. Lewis and Hasan K. Atiyeh conceived and designed the experiments; Mamatha Devarapalli performed the experiments; Mamatha Devarapalli, Randy S. Lewis and Hasan K. Atiyeh analyzed the data and wrote the paper.

Conflicts of Interest: The authors declare no conflict of interest.

References

1. Liu, K.; Atiyeh, H.K.; Stevenson, B.S.; Tanner, R.S.; Wilkins, M.R.; Huhnke, R.L. Continuous syngas fermentation for the production of ethanol, n-propanol and n-butanol. *Bioresour. Technol.* **2014**, *151*, 69–77. [CrossRef] [PubMed]
2. Phillips, J.R.; Clausen, E.C.; Gaddy, J.L. Synthesis gas as substrate for the biological production of fuels and chemicals. *Appl. Biochem. Biotechnol.* **1994**, *45*, 145–157. [CrossRef]
3. Ukpong, M.N.; Atiyeh, H.K.; de Lorme, M.J.; Liu, K.; Zhu, X.; Tanner, R.S.; Wilkins, M.R.; Stevenson, B.S. Physiological response of *Clostridium carboxidivorans* during conversion of synthesis gas to solvents in a gas-fed bioreactor. *Biotechnol. Bioeng.* **2012**, *109*, 2720–2728. [CrossRef] [PubMed]
4. Maddipati, P.; Atiyeh, H.K.; Bellmer, D.D.; Huhnke, R.L. Ethanol production from syngas by *clostridium* strain p11 using corn steep liquor as a nutrient replacement to yeast extract. *Bioresour. Technol.* **2011**, *102*, 6494–6501. [CrossRef] [PubMed]
5. Datar, R.P.; Shenkman, R.M.; Cateni, B.G.; Huhnke, R.L.; Lewis, R.S. Fermentation of biomass-generated producer gas to ethanol. *Biotechnol. Bioeng.* **2004**, *86*, 587–594. [CrossRef] [PubMed]
6. Cotter, J.L.; Chinn, M.S.; Grunden, A.M. Ethanol and acetate production by *Clostridium ljungdahlii* and *Clostridium autoethanogenum* using resting cells. *Bioprocess Biosyst. Eng.* **2009**, *32*, 369–380. [CrossRef] [PubMed]
7. Babu, B.K.; Atiyeh, H.; Wilkins, M.; Huhnke, R. Effect of the reducing agent dithiothreitol on ethanol and acetic acid production by *clostridium* strain p11 using simulated biomass-based syngas. *Biol. Eng.* **2010**, *3*, 19–35. [CrossRef]
8. Gao, J.; Atiyeh, H.K.; Phillips, J.R.; Wilkins, M.R.; Huhnke, R.L. Development of low cost medium for ethanol production from syngas by *Clostridium ragsdalei*. *Bioresour. Technol.* **2013**, *147*, 508–515. [CrossRef] [PubMed]
9. Hu, P. Thermodynamic, Sulfide, Redox Potential and pH Effects on Syngas Fermentation. Dissertation, Bringham Young University, Provo, UT, USA, 2011.
10. Klasson, K.; Ackerson, M.; Clausen, E.; Gaddy, J. Bioreactor design for synthesis gas fermentations. *Fuel* **1991**, *70*, 605–614. [CrossRef]
11. Saxena, J.; Tanner, R.S. Effect of trace metals on ethanol production from synthesis gas by the ethanologenic acetogen, *Clostridium ragsdalei*. *J. Ind. Microbiol. Biotechnol.* **2011**, *38*, 513–521. [CrossRef] [PubMed]
12. Yasin, M.; Park, S.; Jeong, Y.; Lee, E.Y.; Lee, J.; Chang, I.S. Effect of internal pressure and gas/liquid interface area on the co mass transfer coefficient using hollow fibre membranes as a high mass transfer gas diffusing system for microbial syngas fermentation. *Bioresour. Technol.* **2014**, *169*, 637–643. [CrossRef] [PubMed]
13. Kim, Y.-K.; Lee, H. Use of magnetic nanoparticles to enhance bioethanol production in syngas fermentation. *Bioresour. Technol.* **2016**, *204*, 139–144. [CrossRef] [PubMed]
14. Phillips, J.R.; Klasson, T.K.; Clausen, E.C.; Gaddy, J.L. Biological production of ethanol from coal synthesis gas. *Appl. Biochem. Biotechnol.* **1993**, *39–40*, 559–571. [CrossRef]
15. Munasinghe, P.C.; Khanal, S.K. Syngas fermentation to biofuel: Evaluation of carbon monoxide mass transfer coefficient (kLa) in different reactor configurations. *Biotechnol. Prog.* **2010**, *26*, 1616–1621. [CrossRef] [PubMed]
16. Lee, P.-H.; Ni, S.-Q.; Chang, S.-Y.; Sung, S.; Kim, S.-H. Enhancement of carbon monoxide mass transfer using an innovative external hollow fiber membrane (HFM) diffuser for syngas fermentation: Experimental studies and model development. *Chem. Eng. J.* **2012**, *184*, 268–277. [CrossRef]
17. Ungerman, A.J.; Heindel, T.J. Carbon monoxide mass transfer for syngas fermentation in a stirred tank reactor with dual impeller configurations. *Biotechnol. Prog.* **2007**, *23*, 613–620. [CrossRef] [PubMed]
18. Devarapalli, M.; Atiyeh, H.K.; Phillips, J.R.; Lewis, R.S.; Huhnke, R.L. Ethanol production during semi-continuous syngas fermentation in a trickle bed reactor using *Clostridium ragsdalei*. *Bioresour. Technol.* **2016**, *209*, 56–65. [CrossRef] [PubMed]

19. Orgill, J.J.; Atiyeh, H.K.; Devarapalli, M.; Phillips, J.R.; Lewis, R.S.; Huhnke, R.L. A comparison of mass transfer coefficients between trickle-bed, hollow fiber membrane and stirred tank reactors. *Bioresour. Technol.* **2013**, *133*, 340–346. [CrossRef] [PubMed]

20. Kundiyana, D.K.; Huhnke, R.L.; Wilkins, M.R. Effect of nutrient limitation and two-stage continuous fermentor design on productivities during "*Clostridium ragsdalei*" syngas fermentation. *Bioresour. Technol.* **2011**, *102*, 6058–6064. [CrossRef] [PubMed]

21. Rajagopalan, S.; Datar, R.P.; Lewis, R.S. Formation of ethanol from carbon monoxide via a new microbial catalyst. *Biomass Bioenerg.* **2002**, *23*, 487–493. [CrossRef]

22. Richter, H.; Martin, M.; Angenent, L. A two-stage continuous fermentation system for conversion of syngas into ethanol. *Energies* **2013**, *6*, 3987–4000. [CrossRef]

23. Martin, M.E.; Richter, H.; Saha, S.; Angenent, L.T. Traits of selected clostridium strains for syngas fermentation to ethanol. *Biotechnol. Bioeng.* **2016**, *113*, 531–539. [CrossRef] [PubMed]

24. Hurst, K.M.; Lewis, R.S. Carbon monoxide partial pressure effects on the metabolic process of syngas fermentation. *Biochem. Eng. J.* **2010**, *48*, 159–165. [CrossRef]

25. Köpke, M.; Held, C.; Hujer, S.; Liesegang, H.; Wiezer, A.; Wollherr, A.; Ehrenreich, A.; Liebl, W.; Gottschalk, G.; Dürre, P. Clostridium ljungdahlii represents a microbial production platform based on syngas. *Proc. Natl. Acad. Sci. USA* **2010**, *107*, 13087–13092. [CrossRef] [PubMed]

26. Klasson, T.K.; Ackerson, M.D.; Clausen, E.C.; Gaddy, J.L. Biological conversion of coal and coal-derived synthesis gas. *Fuel* **1993**, *72*, 1673–1678. [CrossRef]

27. Liu, K.; Atiyeh, H.K.; Tanner, R.S.; Wilkins, M.R.; Huhnke, R.L. Fermentative production of ethanol from syngas using novel moderately alkaliphilic strains of *Alkalibaculum bacchi*. *Bioresour. Technol.* **2012**, *104*, 336–341. [CrossRef] [PubMed]

28. Shen, Y.; Brown, R.; Wen, Z. Enhancing mass transfer and ethanol production in syngas fermentation of *Clostridium carboxidivorans* P7 through a monolithic biofilm reactor. *Appl. Energy* **2014**, *136*, 68–76. [CrossRef]

29. Shen, Y.; Brown, R.; Wen, Z. Syngas fermentation of *Clostridium carboxidivoran* P7 in a hollow fiber membrane biofilm reactor: Evaluating the mass transfer coefficient and ethanol production performance. *Biochem. Eng. J.* **2014**, *85*, 21–29. [CrossRef]

30. Mohammadi, M.; Younesi, H.; Najafpour, G.; Mohamed, A.R. Sustainable ethanol fermentation from synthesis gas by *Clostridium ljungdahlii* in a continuous stirred tank bioreactor. *J. Chem. Technol. Biotechnol.* **2012**, *87*, 837–843. [CrossRef]

31. Coulson, J.; Richardson, J.; Sinnot, R. Chemical engineering. In *An Introduction to Chemical Engineering Design*; Pergamon Press: New York, NY, USA, 1999; p. 838.

32. Klein, J.; Ziehr, H. Immobilization of microbial cells by adsorption. *J. Biotechnol.* **1990**, *16*, 1–15. [CrossRef]

33. Núñez, M.; Lema, J. Cell immobilization: Application to alcohol production. *Enzym. Microb. Technol.* **1987**, *9*, 642–651. [CrossRef]

fermentation

MDPI

Article

Production of Bioethanol from Agricultural Wastes Using Residual Thermal Energy of a Cogeneration Plant in the Distillation Phase

Raffaela Cutzu and Laura Bardi *

CREA-IT Council for Agricultural Research and Economics,- Research Centre for Engineering and Agro-Food Processing, USSA Turin, Area di Ricerca di Torino, Strada delle Cacce, 73, 10135 Turin, Italy; raffaela.cutzu@gmail.com
* Correspondence: laura.bardi@crea.gov.it; Tel.: +39-011-346288

Academic Editor: Thaddeus Ezeji
Received: 27 March 2017; Accepted: 21 May 2017; Published: 25 May 2017

Abstract: Alcoholic fermentations were performed, adapting the technology to exploit the residual thermal energy (hot water at 83–85 °C) of a cogeneration plant and to valorize agricultural wastes. Substrates were apple, kiwifruit, and peaches wastes; and corn threshing residue (CTR). *Saccharomyces bayanus* was chosen as starter yeast. The fruits, fresh or blanched, were mashed; CTR was gelatinized and liquefied by adding Liquozyme® SC DS (Novozymes, Dittingen, Switzerland); saccharification simultaneous to fermentation was carried out using the enzyme Spirizyme® Ultra (Novozymes, Dittingen, Switzerland). Lab-scale static fermentations were carried out at 28 °C and 35 °C, using raw fruits, blanched fruits and CTR, monitoring the ethanol production. The highest ethanol production was reached with CTR (10.22% (v/v) and among fruits with apple (8.71% (v/v)). Distillations at low temperatures and under vacuum, to exploit warm water from a cogeneration plant, were tested. Vacuum simple batch distillation by rotary evaporation at lab scale at 80 °C (heating bath) and 200 mbar or 400 mbar allowed to recover 93.35% (v/v) and 89.59% (v/v) of ethanol, respectively. These results support a fermentation process coupled to a cogeneration plant, fed with apple wastes and with CTR when apple wastes are not available, where hot water from cogeneration plant is used in blanching and distillation phases. The scale up in a pilot plant was also carried out.

Keywords: bioethanol; fruits; corn threshing residue; fermentation; distillation

1. Introduction

The rising demand for renewable energy sources induced the development of new technologies to produce biofuels [1,2]. Among them, microbial biotechnologies have been largely developed, allowing the development and production of several different biofuels, also using effluents and wastes as substrates: by this way, the costs of the processes are reduced, improving their economical competitiveness and simultaneously reducing the environmental load for wastes disposal [3,4].

Bioethanol can be used as a fuel, either pure or blended with gasoline (gasohol). In the United States, it is used as 10% solution in gasoline (E-10) while in Brazil it is used both blended (24% ethanol, 76% gasoline) and hydrated in flexible-fuel vehicles [5]. Others mixtures are E-15 (15% ethanol, 85% gasoline) and E-85 (85% ethanol e 15% gasoline). Bioethanol can also replace other additives, as octane boosters, in gasoline fuel, and ethanol–gasoline blend provides the highest brake power [6]. Other benefits come from using bioethanol as biofuel: it is totally biodegradable and sulphur free, and the products from its incomplete oxidation (acetic acid and acetaldehyde) are less toxic in comparison to other alcohols [7].

The raw materials that can be used for alcoholic fermentations are sugar crops (sugar cane, sugar beet and sorghum, fruits), starchy crops (corn, wheat and barley), and cellulosic crops (stems, leaves, trunks, branches, husks), the latter needing a pre-treatment to make fermentation possible. They vary in relation to geographic areas: corn is generally used in USA and China, while in tropical areas (India, Brazil, Colombia) sugar cane is more diffused [8]. Nowadays, the use of ligno-cellulosic biomasses, as forest management residues, food industry wastes, or specific plants, is expanding [9].

Among fruits, grape fermentation is well known worldwide; other fruit fermentation is typical of peculiar areas. Apple (*Malus domestica*) fermentation product is called cider and it is typical of the United Kingdom, France, Spain, Germany, Ireland, The Netherlands, Finland, and Switzerland. The alcoholic fermentation of peach (*Prunus persica*) and kiwifruit (*Actinidia chinensis*) is extremely rare [10,11], but possible due to the sugar content of these fruits. Until now, the fermentation of these fruits has mainly been addressed for nutritional uses; bioethanol destination is less diffused, due to ethical and economical considerations; however, much of fruit residue that is disposed as waste could become low cost substrates for bio ethanol production.

Starch from starchy crops, such as cereals, to become fermentable needs a pre-treatment composed of three steps: gelatinization, to allow the starch to lose its crystallinity and become an amorphous gel; liquefaction, where starch is hydrolyzed to dextrins by an alfa-amylase and viscosity is reduced; and saccharification, where a gluco-amylase is added to convert dextrins to glucose [12–14]. Saccharification can be managed to be simultaneous to fermentation: this makes the glucose gradually available to microorganisms and reduces contamination risks, process duration, and costs [15,16].

The microorganisms chosen for alcoholic fermentation are usually yeasts, mainly belonging to *Saccharomyces* genus. The preferred characteristics for industrial bioethanol production are: high ethanol yield; high ethanol tolerance; high ethanol productivity (>5.0 g/L/h); aptitude to grow in simple, inexpensive, and undiluted media; aptitude to grow in presence of inhibitors, at low pH, or high temperature [17]. *S. cerevisiae* is usually considered the typical yeast of wine and cider fermentations; among other species of the genus *Saccharomyces*, *S. bayanus*—characterized by high ethanol tolerance—is used for the production of wine, sparkling wines, and cider, and it can also be used in industrial applications for bioethanol production [18].

The temperature is a fundamental parameter of the fermentation process. According to some authors, the ethanol production increases with increasing temperature [14], under a limit above which the production rate decreases, because high temperatures can become a stress factor for micro-organisms. The temperatures that allow a good microbial growth and a good ethanol yield generally range between 20 and 35 °C. As fermentation is an exergonic process, particular attention is required for fermentation temperature control [5]. Then, to maximize the ethanol yield, yeast strains resistant to high temperatures should be chosen [19]; this is the best choice for bioethanol production, allowing high yields and low costs, while it could be unsuitable for fermentations aimed to reach alcoholic beverages, because sensory properties could be compromised [20,21].

The fermentation duration must also be chosen to obtain an adequate microbial growth and ethanol yield, taking into account that the shorter the duration is, the lower the costs are. The study of the microbial growth and ethanol production kinetics in relation to the substrate allows the identification of the correct duration of the fermentation process, also in consideration of volumetric productivity (g/L/h), both in batch and in continuous fermentations.

The distillation of ethanol formed during fermentation from ethanol-water solution will lead finally to production of hydrous (azeotropic) ethanol (theoretical maximum achievable 95.5% wt. ethanol and 4.5% water). To remove the remaining water, special processes are required to reach anhydrous ethanol, that include: chemical dehydration process, dehydration by vacuum distillation process, azeotropic distillation process, extractive distillation processes, membrane processes, adsorption processes, and diffusion distillation process [22]. The evaluation of the energy balance of bioethanol production reveals that most of the energy is required for the distillation, also because of the low concentration of ethanol in the fermented broth [5]. Energy consumption can be reduced during distillation if lower heating is necessary; this can be reached using residual thermal energy from other processes to warm the fermented broth. Moreover, distillations carried out under vacuum enable good ethanol yields even at lower temperatures. The aim of this study was to evaluate the exploitation of the residual thermal energy of a cogeneration plant, producing residual hot water from plant chilling systems, in downstream phases of the fermentation process, by warming the fermented broth with the hot water of the cogeneration plant; vacuum distillation was also tested at temperatures lower than that usually employed in traditional simple batch distillation plants. The possible role of fruit blanching pretreatment, carried out with the hot water from the cogeneration plant—mainly aimed at facilitating the grinding phase and reducing contaminations during alcoholic fermentation—was also tested. Moreover, to further reduce the costs of the process, agricultural wastes were checked as feedstock for fermentations: unmarketable residues of apple, kiwifruit, peaches, and corn threshing residue (CTR). *Saccharomyces bayanus*, characterized by high ethanol tolerance and largely used in cider fermentation, has been chosen as starter. The results obtained support the feasibility of a fermentation process, coupled to a cogeneration plant, fed throughout the year with apple wastes, when available, and with CTR when apple wastes are unavailable, where residual thermal energy from a cogeneration plant is used in fruit blanching and in distillation phases.

2. Materials and Methods

2.1. Strain and Culture Media

Fermentation medium was inoculated with a *Saccharomyces bayanus* commercial strain (Zymoferm Bayanus, Chimica FRANKE, Susa, Torino, Italy). The starter cultures were prepared by inoculating the yeast in 100 mL of YEPD (yeast extract 1%, bacto-peptone 2%, glucose 2%) and incubating it at 25 °C under static conditions in an incubation chamber for 24 h.

The unmarketable residues of apples, kiwifruits, and peaches were cut and ground separately for 2 min (Tables S1 and S2: raw material characterization). The mash obtained from each fruit was stored at −20 °C. Samples of the same fruits were blanched in boiling water: the apples and the peaches for 15 min, the kiwi for 5 min, then the fruits were cut and ground for 2 min and the mash obtained from each fruit stored at −20 °C. The concentrations of sugars (glucose and fructose) were determined in raw materials to check the homogeneity of starting conditions, and throughout the fermentations, to check the trend of fermentation and sugar consumption (Table S2), by D-Fructose/D-Glucose Assay Kit and Ethanol Assay Kit (Megazyme, Bray, Co., Wicklow, Ireland).

The CTR was milled and water was added to the flour in order to obtain a mash 30% *w/v* powder/water ratio. The gelatinization was conducted for 4 h at 85 °C until the complete water absorption. For the liquefaction the Liquozyme® SC DS (Novozymes, Dittingen, Switzerland) was added (0.020% weight of enzyme/dried weight of CTR). The saccharification was carried out at the same time as the fermentation by adding Spirizyme® Ultra (Novozymes, Dittingen, Switzerland) (0.030% weight of enzyme/dried weight of RTC). Starch concentration was analyzed by Kit Total Starch (Megazyme, Bray, Co., Wicklow, Ireland) (Table S3: CTR characterization).

2.2. Fermentations

Lab-scale batch fermentations were carried out on 600 g mash of each kind of fruit and of liquefied CTR, that were inoculated with the starter yeast culture (1×10^6 cell/g of substrate), in 1 liter flask, under static conditions, at 28 and 35 °C. Samples were taken at 24, 48, 72, 96, and 168 h to quantify ethanol production.

Batch and semi-continuous fermentations were also carried out with liquefied CTR and with an apple:kiwifruit 1:1 mix, the latter in a 2 L flask filled with 1200 g mashed fruits, inoculated with the starter yeast culture (1×10^6 cell/g of substrate) and incubated under static conditions at 35 °C. In semi-continuous fermentations, at 120 h, 400 g fermented substrate was withdrawn and substituted with fresh substrate; the same procedure was repeated every 48 h. Samples were withdrawn once a week after one, two, three, and four weeks to determine ethanol and residual sugars concentrations. The concentrations of ethanol were determined by Ethanol Assay Kit (Megazyme, Bray, Co., Wicklow, Ireland). All the assays were carried out in triplicate.

2.3. Downstream

In order to evaluate the amount of ethanol that can be recovered by evaporation/distillation under vacuum at low temperatures from the fermented fruit biomasses heated with the residual hot water from the plant chilling systems of a cogeneration plant, samples of fermented broth were distilled combining different temperatures to different vacuum levels. Temperatures were chosen lower than the temperature of the chilling water of cogeneration plant (83–85 °C). Different water bath temperatures and vacuum levels were obtained in a rotary evaporation system (Laborota 4000, Heidolph Instruments GmbH & Co, Schwabach, Germany). The combinations of temperature and pressure tested are shown in Table 1.

Table 1. Assays of simple batch distillation of fermented broth in a rotary evaporation system: different temperatures and vacuum levels combinations and % ethanol recovered.

Temperature °C	Pressure Mbar	Ethanol % (vol.)
80	1013.25	43.40
80	400	89.59
80	200	93.35
60	175	45.06

2.4. Scale-Up

Scale-up was done for apple and kiwi fruits, fresh or blanched, in a 1000 L thermostated fermenter, loaded with about 600 kg mashed fruits each time. Batch fermentations were carried out for five days under controlled temperature, lower than 35 °C. Exterior air-lock of the bioreactor was loaded with cold water during the fermentation, to avoid excessive warming, and with hot water (83–85 °C), coming from the cooling of the cogeneration plant, during the distillation phase. The same hot water from cogeneration plant was also used to blanch fruits during pretreatment, blanching at 80 °C for 5 min.

3. Results and Discussion

Lab scale fermentations were carried out in order to check the best conditions to reach high ethanol yields; as fruit wastes are available only seasonally, CTR was also evaluated. CTR is an agricultural residue that is easy to store and available throughout the whole year, then usable to supply the fermentation plant when fruit wastes are unavailable.

The ethanol concentrations checked throughout batch fermentations are reported in Figure 1.

Among fruits, the highest ethanol concentrations were reached with blanched fruits, at 35 °C for apple and peach, at 28 °C for kiwifruit and fruit mix. The apple showed, among fruits, the maximum ethanol concentration: $8.71 \pm 0.83\%$ (v/v). The kiwifruits produced $7.97 \pm 0.39\%$ (v/v) and peaches produced $4.26 \pm 0.27\%$ (v/v). The ethanol concentration produced from CTR was $10.22 \pm 0.70\%$ (v/v). Apples were shown to give the best ethanol production among the tested fruits, but kiwifruit also gave good results, while the amount of ethanol obtained from peach was low. CTR showed to be very suitable as substrate of fermentation for bioethanol production.

The ethanol productivity, expressed as g/L/h, is reported in Table 2.

The optimal duration of fermentation was three days for fruits, two days for CTR. A summary of the best selected batch fermentation conditions are reported in Table 3. The short fermentation duration contributes to the economical sustainability of the process.

Semi-continuous fermentations can be sometimes preferred in industrial fermentations, due to several advantages, such as shorter induction times due to the suppression of the lag phase, better control of contaminations and higher yields; however, in our assays, semi-continuous fermentations gave lower productions of ethanol than batch fermentations (Table 4). Then, due to short duration and high yield, batch fermentations should be preferred to semi-continuous fermentations; batch fermentations are also suitable, taking into account that the fermentation plant should be fed with seasonally different substrates.

Scale-up fermentations were carried out with the best conditions selected from the lab-scale fermentations, with blanched apple and kiwifruit. As fruits were ground after blanching at 80 °C, it was necessary to cool down the temperature at the start and throughout the whole fermentation to maintain it lower than 35 °C. An average ethanol concentration of 6.58% and an ethanol productivity of 0.30 g/L/h were obtained with apple (three replications), while with kiwifruit no ethanol production was detected, even if sugars (glucose and fructose) were completely exhausted at the end of the fermentation; a possible explanation could be that an aerobic respiratory catabolism from contaminating microorganisms predominated, preventing yeast growth.

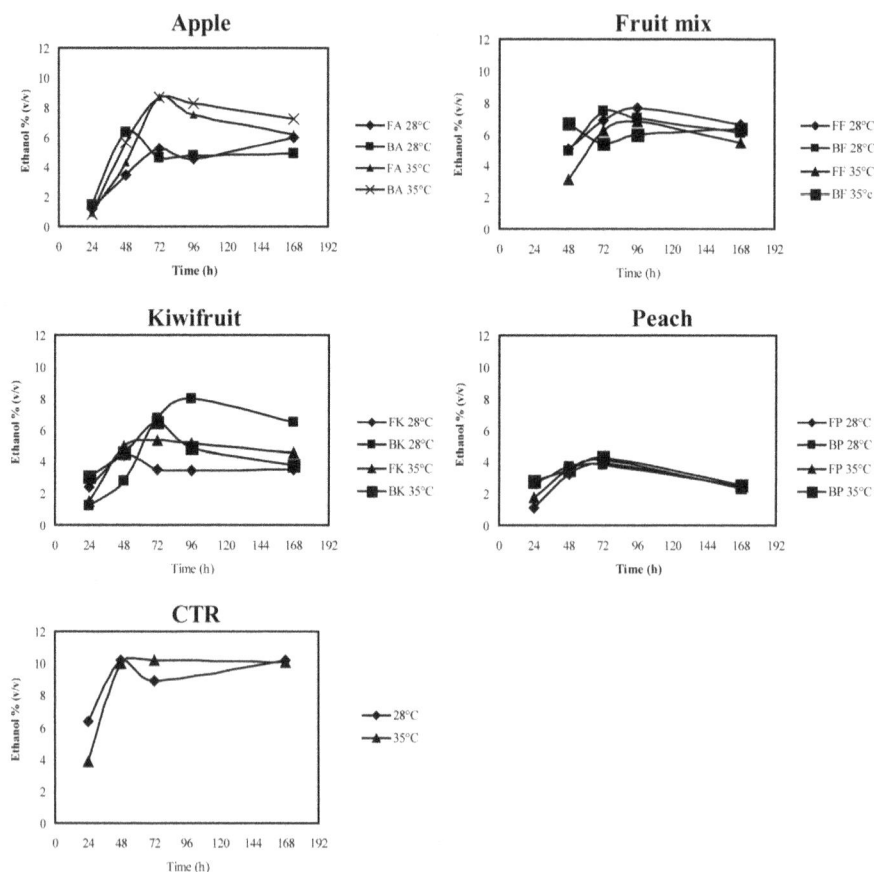

Figure 1. Lab-scale batch fermentation trends. FA: fresh apple; BA: blanched apple; FK: fresh kiwifruit; BK: blanched kiwifruit; FP: fresh peach; BP: blanched peach; FF: fresh fruit mix; BF: blanched fruit mix; CTR: corn threshing residue.

The product of a batch apple fermentation (626 kg, 6.44% ethanol concentration at the end of fermentation) was used to test the ethanol yield that can be reached with distillations at low temperatures. The results of the distillation tests carried out under controlled conditions, at atmospheric pressure and under vacuum, are reported in Table 1. Temperatures of heating water bath ≤80 °C were tested to reproduce the conditions of warming obtained with the hot water coming from the chilling system of a cogeneration plant. The ethanol yielded at 80 °C at atmospheric pressure was only 43.4%, while the yield augmented to 89.59% under vacuum at 400 mbar; a further increase to 93.35% was obtained, raising the vacuum until 200 mbar. A strong pressure decrease (175 mbar) was not sufficient to produce an improvement of ethanol recovery at 60 °C.

Table 2. Ethanol productivity (g/L/h) in batch fermentations. FA: fresh apple; BA: blanched apple; FK: fresh kiwifruit; BK: blanched kiwifruit; FP: fresh peach; BP: blanched peach; FF: fresh fruit mix; BF: blanched fruit mix; CTR: corn threshing residue.

Time (h)	FA 28 °C	BA 28 °C	FA 35 °C	BA 35 °C	FK 28 °C	BK 28 °C	FK 35 °C	BK 35 °C	FP 28 °C
24	0.41	0.47	0.30	0.26	0.79	0.41	0.49	0.99	0.37
48	0.56	0.98	0.71	0.93	0.73	0.46	0.81	0.74	0.53
72	0.57	0.51	0.95	0.95	0.38	0.74	0.58	0.71	0.43
96	0.38	0.39	0.62	0.68	0.27	0.66	0.42	0.40	ND
168	0.28	0.23	0.29	0.34	0.17	0.30	0.21	0.18	0.11

Time (h)	BP 28 °C	FP 35 °C	BP 35 °C	FF 28 °C	BF 28 °C	FF 35 °C	BF 35 °C	RTC 28 °C	RTC 35 °C
24	0.86	0.59	0.91						
48	0.60	0.59	0.58	0.83	0.81	0.51	1.10	2.09	1.27
72	0.42	0.46	0.47	0.75	0.82	0.69	0.59	1.68	1.65
96	ND	ND	ND	0.60	0.58	0.56	0.49	0.98	1.12
168	0.12	0.11	0.12	0.31	0.29	0.26	0.21	0.48	0.47

Table 3. Selection of best results from batch fermentations.

Raw Materials	Temperature (°C)	Time (h)	Ethanol Concentration (% vol.)	Productivity (g/L/h)
Fresh/blanched apple	35	72	8.71 ±0.83	0..5
Blanched kiwifruit	28	96	7.97 ±0.39	0.66
Blanched peach	35	72	4.26 ± 0.27	0.47
CTR	28	48	10.22 ±0.70	1.68

Table 4. Ethanol concentration (% *v/v*) and productivity (g/L/h) in semi-continuous fermentations carried out at 35 °C. BF: blanched fruit mix (apple:kiwifruit 1:1); CTR: residue of threshing of corn.

Sampling	BF		CTR	
Time (Weeks)	Ethanol Concentration (% *v/v*)	Etanol Productivity (g/L/h)	Ethanol Concentration (% vol.)	Etanol Productivity (g/L/h)
1	5.23	0.25	6.91	0.32
2	4.67	0.11	7.30	0.17
3	5.58	0.09	7.45	0.12
4	4.55	0.05	7.76	0.09

In scale-up assay, 47.53% of the produced ethanol was recovered (19.16 L distilled from 40.31 L produced) with distillation at air pressure, warming the whole tank at the end of fermentation with 85 °C hot water. The use of residual hot water from the cogeneration plant allows an energy saving, in each batch fermentation, of 28,000 kcal for warming 500 kg fermented biomass from 35 to 85 °C, and of 6000 kcal to maintain the same biomass at 85 °C during a 12 h distillation; this corresponds to 1775 kcal saved per liter of ethanol produced. If distillation is applied under vacuum, distilled ethanol recovered would rise to 37.49 L; energy consumption for vacuum production would be 2438 kcal per fermentation, then the global energy saving would decrease from 34,000 to 31,562 kcal, but considering the higher distilled ethanol recovery, the energy saved per liter of ethanol would be 842 kcal.

Fermentation **2017**, *3*, 24

4. Conclusions

Several aspects of the production of bioethanol from fruit wastes using the residual heat from chilling water of a cogeneration plant make this process highly suitable for its environmental sustainability. In fact, an alternative to disposal of residual biomasses with high carbon load valorizes them, transforming agricultural wastes in co-products; the use of hot chilling water of a cogeneration plant allows to reduce the energy consumption needed for distillation, contemporarily reducing the environmental impact due to heat dispersal from the cogeneration plant; bioethanol, that can be used as a biofuel, is a renewable energy source.

Among the tested fruits, apple showed the best performances. The possibility to feed the fermentation plant with CTR when fruit wastes are unavailable guarantees functionality throughout the whole year and the economical sustainability of the plant.

Supplementary Materials: The following are available online at www.mdpi.com/2311-5637/3/2/24/s1.

Acknowledgments: This project was financed by the Piedmont Region (Italy) under the FEASR Program (Assessorato Agricoltura, Foreste, Caccia e Pesca, Direzione Agricoltura, PSR 2007-2013, Misura 124 Azione 1). We acknowledge Marco Sola (TAS s.r.l, Ceresole d'Alba (CN) Italy) for kind cooperation during scale-up assays, and Giancarlo Bourlot (Piedmont Region) for helpfulness during the whole project steps.

Conflicts of Interest: The authors declare no conflict of interest.

References

1. Licht's, F.O. *World Ethanol and Biofuels Report*; Informa PLC: London, UK, 2016.
2. European Commission. *Final report of the Biofuel Research Advisory Council, 2006: Biofuels in the European Union. A Vision for 2030 and Beyond*; EUR 22066; Office for Official Publications of the European Communities: Brussels, Belgium, 2006; ISBN 92-79-01748-9. ISSN: 1018-5593.
3. Passoth, V. Molecular Mechanisms in Yeast Carbon Metabolism: Bioethanol and Other Biofuels. In *Molecular Mechanisms in Yeast Carbon Metabolism*; Piškur, J., Compagno, C., Eds.; Springer-Verlag: Berlin/Heidelberg, Germany, 2014; 43p.
4. Zoppellari, F.; Bardi, L. Production of bioethanol from effluents of the dairy industry by *Kluyveromyces marxianus*. *New Biotechnol.* **2013**, *30*, 607–613. [CrossRef] [PubMed]
5. Zabed, H.; Faruq, G.; Sahu, J.N.; Azirun, M.S.; Hashim, R.; Boyce, A.N. Bioethanol production from fermentable sugar juice. *Sci. World J.* **2014**, *12*, 957102. [CrossRef] [PubMed]
6. Elfasakhany, A. Investigations on the effects of ethanol–methanol–gasoline blends in a spark-ignition engine: Performance and emissions analysis. *Eng. Sci. Technol.* **2015**, *18*, 713–719. [CrossRef]
7. Minteer, S. *Alcoholic Fuels*; Taylor & Francis: New York, NY, USA, 2006; 265p.
8. Cheng, J.J.; Timilsina, G.R. Status and barriers of advanced biofuel technologies: A review. *Renew. Energy* **2011**, *36*, 3541–3549. [CrossRef]
9. Banerjee, S.; Mudliar, S.; Sen, R.; Giri, L.; Satpute, D.; Chakrabarti, T.; Pandey, R.A. Commercializing lignocellulosic bioethanol: Technology bottlenecks. *Biofuels Bioprod. Biorefining* **2010**, *4*, 77–93. [CrossRef]
10. Ucuncu, C.; Tari, C.; Demir, H.; Oguz Buyukkileci, A.; Ozen, B. Dilute-Acid Hydrolysis of Apple, Orange, Apricot and Peach Pomaces as Potential Candidates for Bioethanol Production. *J. Biobased Mater. Bioenergy* **2013**, *7*, 1–14. [CrossRef]
11. Edwards, M.C.; Williams, T.; Pattathil, S.; Hahn, M.G.; Doran-Peterson, J. Fermentation, Cell Culture and Bioengineering Replacing a suite of commercial pectinases with a single enzyme, pectate lyase B, in *Saccharomyces cerevisiae* fermentations of cull peaches. *J. Ind. Microbiol. Biotechnol.* **2014**. [CrossRef] [PubMed]
12. Verardi, A.; Blasi, A.; Molino, A.; Albo, L.; Calabrò, V. Improving the Enzymatic Hydrolysis of *Saccharum Officinarum* L. Bagasse by Optimizing Mixing in a Stirred Tank Reactor: Quantitative Analysis of Biomass Conversion. *Fuel Process. Technol.* **2016**, *149*, 15–22. [CrossRef]
13. Amelio, A.; van der Bruggen, B.; Lopresto, C.; Verardi, A.; Calabrò, V. Pervaporation Membrane Reactors: Biomass conversion into alcohols. In *Membrane Technologies for Biorefining*; Figoli, A., Cassano, A., Basile, A., Eds.; Elsevier Ltd.: Duxford, UK, 2016; pp. 331–381. ISBN 978-0-08-100451-7.

14. Xu, Z.; Huang, F. Pretreatment Methods for Bioethanol Production. *Appl. Biochem. Biotechnol.* **2014**, *174*, 43–62. [CrossRef] [PubMed]

15. Bothast, R.J.; Schlicher, M.A. Biotechnological processes for conversion of corn into ethanol. *Appl. Microbiol. Biotechnol.* **2005**, *67*, 19–25. [CrossRef] [PubMed]

16. Lee, S. Ethanol from corn. In *Handbook of Alternative Fuel Technologies*; Lee, S., Speight, J.G., Loyalka, S.K., Eds.; CRC Press: Boca Raton, FL, USA, 2007; pp. 323–341.

17. López-Malo, M.; Querol, A.; Guillamon, J.M. Metabolomic comparison of *Saccharomyces cerevisiae* and the cryotolerant species *S. bayanus* var. uvarum and *S. kudriavzevii* during wine fermentation at low temperature. *PLoS ONE* **2013**, *8*, e60135. [CrossRef] [PubMed]

18. Publicover, K.; Caldwell, T.; Harcum, S.W. Biofuel ethanol production using *Saccharomyces bayanus*, the champagne yeast. In Proceedings of the 32nd Symposium on Biotechnology for Fuels and Chemicals, Clearwater Beach, FL, USA, 19–22 April 2010; pp. 11–34.

19. Verardi, A.; De Bari, I.; Ricca, E.; Calabrò, V. *Hydrolysis of Lignocellulosic Biomass: Current Status of Processes and Technologies and Future Perspectives*; Lima, M.A.P., Ed.; InTech: Rijeka, Croatia, 2012; pp. 95–122.

20. Rose, A.H.; Harrison, J.S. *The yeasts. Volume III: Metabolism and Physiology of Yeasts*, 2nd ed.; Academic Press Limited: London, UK, 1989.

21. Mallouchos, A.; Komaitis, M.; Koutinas, A.; Kanellaki, M. Wine fermentations by immobilized and free cells at different temperatures: Effect of immobilization and temperature on volatile by-products. *Food Chem.* **2003**, *80*, 109–113. [CrossRef]

22. Kumar, S.; Singh, N.; Prasad, R. Anhydrous ethanol: A renewable source of energy. *Renew. Sustain. Energy* **2010**, *14*, 1830–1844. [CrossRef]

fermentation

MDPI

Review

Syngas Fermentation: A Microbial Conversion Process of Gaseous Substrates to Various Products

John R. Phillips, Raymond L. Huhnke and Hasan K. Atiyeh *

Biosystems and Agricultural Engineering Department, Oklahoma State University, 214 Ag Hall, Stillwater, OK 74078, USA; rphi@ostatemail.okstate.edu (J.R.P.); raymond.huhnke@okstate.edu (R.L.H.)
* Correspondence: hasan.atiyeh@okstate.edu; Tel.: +1-405-744-8397; Fax: +1-405-744-6059

Received: 27 April 2017; Accepted: 12 June 2017; Published: 16 June 2017

Abstract: Biomass and other carbonaceous materials can be gasified to produce syngas with high concentrations of CO and H_2. Feedstock materials include wood, dedicated energy crops, grain wastes, manufacturing or municipal wastes, natural gas, petroleum and chemical wastes, lignin, coal and tires. Syngas fermentation converts CO and H_2 to alcohols and organic acids and uses concepts applicable in fermentation of gas phase substrates. The growth of chemoautotrophic microbes produces a wide range of chemicals from the enzyme platform of native organisms. In this review paper, the Wood–Ljungdahl biochemical pathway used by chemoautotrophs is described including balanced reactions, reaction sites physically located within the cell and cell mechanisms for energy conservation that govern production. Important concepts discussed include gas solubility, mass transfer, thermodynamics of enzyme-catalyzed reactions, electrochemistry and cellular electron carriers and fermentation kinetics. Potential applications of these concepts include acid and alcohol production, hydrogen generation and conversion of methane to liquids or hydrogen.

Keywords: syngas fermentation; acetyl-CoA pathway; acetogen; biofuel; gasification

1. Introduction to Syngas Fermentation

Syngas fermentation is a hybrid thermochemical/biochemical platform that takes advantage of the simplicity of the gasification process and the specificity of the fermentation process to deliver ethanol and potentially other chemicals. Biomass is converted to ethanol through the thermochemical platform, i.e., gasification and the biological platform, i.e., fermentation in syngas fermentation [1]. Energy-rich biomass and waste materials are converted by gasification to syngas, which consists of CO, H_2 and CO_2. These gases are then converted to ethanol and other chemicals by acetogenic autotrophic microbes [2]. These microorganisms, "possess a very valuable (trait)" have "the ability to grow in strict autotrophy" and "to produce added-value compounds" [3]. After twenty five years of syngas fermentation research for the production of ethanol, this application is now being deployed at a near commercial scale. However, "these studies have yet to define a methodology for generating high ethanol production levels with stable culture." [4]. In this paper, we present a review of feedstocks, syngas production, metabolic pathway, bioreactor design, mass transfer, thermodynamics, electrochemistry and microbial kinetics of the syngas fermentation process and propose a conceptual model to describe the syngas fermentation.

A process flow diagram for the conversion of switchgrass, a dedicated perennial energy crop, to ethanol is shown in Figure 1. Switchgrass is first converted to syngas in gasification with O_2 and/or steam. CO and H_2 from the cooled syngas is utilized by the bacterial culture in fermentation for cell growth and product synthesis. Beer from the fermentation is then distilled to recover ethanol, and the bottom stream from distillation is returned to the fermentation. Recovered ethanol, taken from overhead of the distillation column, is processed using a molecular sieve to achieve the final product specification.

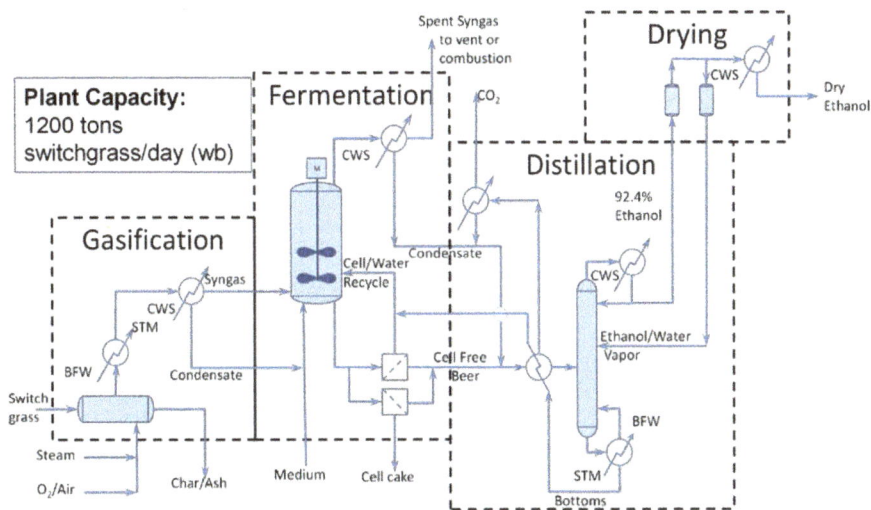

Figure 1. Process flow diagram for the gasification of switchgrass followed by syngas fermentation to produce fuel ethanol; BFW: boiler feedwater system; STM: steam; Syngas: synthesis gas; CWS: circulating water system; wb: wet basis.

A key consideration in any fuel process is preserving the energy content of the feedstock in the final product. Energy is expended in each production step; after solid biomass is heated to a high temperature for the gasification step, energy is recovered from the syngas as steam, and heat is lost to the environment. Energy diverted to cell growth, heat lost from the fermenter, unconverted syngas and unrecovered acetic acid represent energy diverted from the ethanol product. The economy of the fermentation process is enhanced through improvements in efficiency that conserve energy and increase product yield. Energy efficiency represented by retaining the higher heating value from the products, through gasification [5] and as increased product yield from fermentation [6], and the use of energy efficient separation technologies, such as membrane separation, are very important to achieve a profitable commercial process for fuels or chemicals.

1.1. Energy Demand

World energy demand is expected to grow from 553 exajoules (1 EJ = 10^{18} J = 0.948 quadrillion BTU = 0.948 Quad) in 2012 to 865 EJ in 2040 per the U.S. Energy Information Administration's International Energy Outlook for 2013 [7,8]. The world demand for transportation fuels is also projected to rise from 182 EJ in 2010 to 249 EJ in 2040 (from 182 to 229 trillion liters of petroleum). Consumption of liquid biofuels will increase from 2.78 EJ in 2010 (77.2 trillion liters of gasoline equivalent) to 6.21 EJ projected in 2040 (172.4 trillion liters gasoline equivalent).

1.2. Potential Resources

Balan [9] compiled an extensive list of projects for the development of lignocellulosic biofuels supported by the governments of the U.S. and the EU that have achieved a range of success. Syngas fermentation projects have advanced to commercial scale [10,11]. The accounting of biofuels by the EIA includes biomass-to-liquids (BTL) and biodiesel; while ethanol remains the most prominent liquid fuel from biomass. BTL also includes pyrolysis oil and Fischer–Tropsch liquids that share similar thermochemical processes with gasification. Additional sources of syngas include gasification of coal and steam reforming of natural gas.

1.2.1. Biomass

The major portion of projected U.S. biofuels consumption of 1.84 EJ in 2012, rising to 2.34 EJ in 2040 [12], is from ethanol with consumption of 1.71 EJ in 2012 almost exclusively supplied from corn, rising to 1.99 EJ in 2040 with less than 1% projected to derive from cellulosic feedstock. These projections reflect lowered expectations due to slow progress in the technical and economic competence of cellulosic fuel production, coupled with the increased reserves for U.S. oil and gas production. However, global capacity to produce biomass for energy production is projected as 11 to 28 billion tonnes by 2050 [13], which represents 200 to 500 EJ annually.

1.2.2. *Wastes*

In addition to forest and agriculture wastes, municipal solids wastes (MSW) can be an energy resource. Combustible material discarded in the U.S. municipal waste stream is estimated at about 117 million tonnes [14] representing about 2 EJ potential for energy production with only small increase expected through 2030. Use of MSW combined with chemical and petroleum wastes in syngas production may be important as an environmentally-sound management practice. Paper mill wood wastes and black liquor can supply syngas for energy production [15]. These materials are a gathered resource with a negative value, incurring cost for disposal, and represent an opportunity for energy production with environmental benefit.

1.3. *Syngas Production*

Gasification of biomass to produce syngas provides the simple precursors CO and H_2 for fermentation. Atsonios et al. [15] published a process flow diagram for similar production of syngas (followed by catalytic mixed alcohol synthesis). When gasification is coupled with fermentation of the syngas, the robustness and adaptability of the acetogenic bacteria reduce the requirements for gas cleaning and adjustment by the water gas shift reaction required for catalytic conversion of syngas.

Liew et al. [11] discussed a fixed-bed, a circulating fluidized bed (CFB) and entrained flow gasifiers, with preference for CFB for biomass and entrained flow for liquids and solids that are easily pulverized. For syngas fermentation, low pressure and high temperature in the gasification chamber promote CO and H_2 formation and reduce higher molecular weight hydrocarbons or "tar" in the syngas produced, and an atmospheric indirect heated CFB is preferred [5].

Biomass and waste materials contain nitrogen, sulfur, chlorine and other constituent elements, in addition to complex hydrocarbon structures, such as aromatics that decompose slowly in gasification. These compounds remain in the syngas product as N_2 and other minor components, such as ammonia (NH_3), hydrogen sulfide (H_2S) and tars [16,17]. Residual hydrocarbon tars can foul equipment surfaces and orifices and can be inhibitory in fermentation along with chemical species produced in combustion like hydrogen cyanide (HCN). Woolcock and Brown [18] presented an extensive review of syngas contaminants, gas specifications for particular applications and technologies used for gas cleanup. Some chemical species that poison chemical catalysts such as NH_3, carbonyl sulfide (COS) and H_2S can be used as nutrient components for the growth of the acetogenic bacteria when present at low levels. Fermentation uses syngas with a composition dependent on the type of gasifier used and its operating conditions. Fermentation can simplify the process flow diagram for syngas cleaning and emissions treatment, lowering capital requirements compared to catalytic processes, such as Fischer–Tropsch.

1.4. *Microbial Conversion of CO and H_2*

Acetogenic bacteria convert CO, H_2 and CO_2 derived from biomass or waste materials into acetic acid [2]. It is theorized that the acetogenic pathway is as old as life on the Earth [19] and has been optimized by evolution to ensure the survival of species that produce acetyl coenzyme A (acetyl-CoA) from small molecules in natural environments. Acetyl-CoA is an intermediate metabolite that is converted to synthesize cell mass and complex chemicals and yields organic acids and alcohols, most easily acetic acid and ethanol. Production of acetic acid supplies energy for synthesis of cell mass, including lipids, proteins and other complex cell components from the simple inorganic gas substrates (CO, H_2 and CO_2). The ability of some acetogens to reduce organic acids to alcohols, particularly acetic acid to ethanol, is the basis for biofuel production. Knowledge of the acetogenic mechanisms supports successful process design for energy conservation in biofuels' production.

The use of dedicated biomass energy crops, waste biomass and municipal and industrial wastes as feedstock for energy and chemical synthesis promotes reuse and recycling of materials consumed in our society. This can establish a true cycle of renewable, carbon-neutral, energy and chemical production.

2. Chemoautotrophic Microbes

The bacteria used in syngas fermentation belong to a group of prokaryotic single cell organisms termed "acetogens", which are defined by the use of the acetyl-CoA pathway for reductive synthesis of acetyl-CoA from CO_2, energy conservation for growth and assimilation of carbon from CO and CO_2 into biomass [2]. The cellular mechanisms of acetogenesis are present and used by bacteria, archaea and eukaryotes alike. Acetogens inhabit a wide range of ecosystems and have diverse capacities for substrate utilization and product formation, dependent on the growth environment.

Acetogenesis was recognized in 1932 when the production of acetic acid from H_2 and CO_2 by sewage sludge was reported [20]. Subsequently, Klass Wieringa [21] isolated *Clostridium aceticum*, demonstrating synthesis of acetic acid from H_2/CO_2 by this pure culture. The type culture for acetogenesis, *Clostridium thermoaceticum*, reclassified as *Moorella thermoacetica* [22], was isolated by Francis Fontaine [23]. Harland Wood and Lars Ljungdahl studied the acetyl-CoA pathway, providing the definition of the incorporation of CO and of the tetrahydrofolate (THF)-dependent reduction of CO_2 to a methyl group, in the formation of acetyl-CoA. The acetyl-CoA pathway is also referred to as the Wood–Ljungdahl pathway of autotrophic growth. A detailed description of the history of the discovery of acetogenesis is given in a review [2], and the enzymology is reviewed by Ragsdale [24].

2.1. Species and Habitat

Drake et al. [2] cited 100 species of acetogens, from 22 genera in his review. These acetogens were of various morphologies (rods, cocci and spirochetes) with a wide range of temperature optima from 5 to 62 °C. Acetogens were isolated from a wide variety of habitats including soil, sewage sludge, feces, rumen fluid, sediments and industrial wastes. The pH conditions ranged from alkaline to acidic, and most habitats were not strictly anoxic.

The first acetogen reported to produce ethanol from syngas was *Clostridium ljungdahlii* [25,26]. Shortly thereafter, *Butyribacterium methylotrophicum* was reported to produce butanol and ethanol from CO [27]. Other prominent species of acetogenic alcohol producers are *C. autoethanogenum* [28], *C. carboxidivorans*, which has also been shown to synthesize butanol and hexanol [29,30], and *C. ragsdalei* [31]. New species continue to be discovered, including moderately alkaliphilic acetogens that produce ethanol; for example, *Alkalibaculum bacchi* represents yet another new genus and species [32,33].

Mixed culture syngas fermentations for the production of ethanol and acetic acid and conversion of organic acids to their respective alcohols were also reported [34,35]. Enrichment of acetogens in chicken manure in India [4] and cow manure in China [36] shows the potential of natural inocula. However, *Clostridium difficile* and *C. sordellii*, acetogenic human pathogens, were detected in the enriched fermentation [36], which suggests that extreme caution should be exercised in the selection and use of syngas fermenting microorganisms.

The diversity and habitat of acetogens show the potential for additional species to be discovered. A range of fermentative capabilities may be expected from this diverse population, promising new products from syngas fermentation. Successful production using acetogens will likely use the conditions to which the strain has adapted through evolution. The natural environment has a limited source of CO, and acetogens have developed mechanisms that scavenge H_2 to fix CO_2 very effectively via the autotrophic pathways. Nutrients essential to the growth of functional cell mass are the object of competition between a consortium of bacteria and other organisms. Efficient mechanisms for nutrient uptake are required for the bacteria to thrive especially in the environment with very low nutrient concentrations. Isolation of acetogens typically uses a medium enriched with yeast extract with pH stabilized using a Good's buffer, like 2-(N-morpholino)ethanesulfonic Acid (MES) [37]. Several studies have substituted other complex medium components for yeast extract; for example, corn steep liquor [38,39] or cotton seed extract [40]. A defined medium without complex nutrients was used with *C. ljungdahlii* to achieve 48 g/L of ethanol [41].

Culture methods were modified to control mass transfer for successful growth of *C. carboxidivorans* in defined medium and produce butanol and hexanol [30].

2.2. Structure

Acetogens are found as rods, cocci and spirochetes and can be either Gram-positive or Gram-negative [2]. The typical ethanol producing acetogen is a rod-shaped Gram-positive motile bacterium that can form spores. *C. carboxidivorans*, also known as strain P7, is described as "Gram-positive, motile rods (0.5 × 3 µm) occurring singly and in pairs. Cells rarely sporulate, but spores are subterminal to terminal with slight cell swelling. Obligate anaerobe with an optimum growth temperature of 38 °C and an optimum pH of 6.2. Grows autotrophically with H_2/CO_2 or CO and chemoorganotrophically" [29].

The cell membrane is a phospholipid bilayer embedded with proteins, which divides the cytoplasm from the external environment and mediates cell function [42]. Fifteen enzymes closely associated with the acetyl-CoA pathway [24] are listed in Table 1.

Table 1. Enzymes of the acetyl-CoA (Wood–Ljungdahl) pathway.

Enzyme	Reaction	Reference
Carbon Monoxide dehydrogenase	$CO + H_2O \rightarrow CO_2 + 2\,H^+ + 2\,e^-$	[43]
Hydrogenase	$H_2 \rightarrow 2\,H^+ + 2\,e^-$	[44]
Ferredoxin oxidoreductase	$Fd_{Rd} \rightarrow Fd_{Ox} + 2\,e^-$	[45]
Formate dehydrogenase	$CO_2 + NADPH \rightarrow HCOO^- + NADP^+$	[46]
Formate kinase	$HCOO^- + ATP^{4-} + H^+ \rightarrow HCOOPO_3^- + ADP^{3-}$	[47]
Formyl THF synthetase [1]	$HCOOPO_3^- + THF \rightarrow HCOTHF + HPO_4^{2-} + H^+$	[48]
Methenyl THF cyclohydrolase	$HCOTHF + H^+ \rightarrow HC^+THF + H_2O$	[44]
Methylene THF dehydrogenase	$HC^+THF + NADPH \rightarrow H_2CTHF + NADP^+$	[49]
Methylene THF reductase	$H_2CTHF + 2H^+ + 2e^- \rightarrow H_3CTHF$	[50]
Methyl transferase	$H_3CTHF + H^+ + [Co^+]E^{2+} \rightarrow THF + H_3C\left[Co^{3+}\right]E^+$	[44]
Corrinoid-Iron-Sulfur protein	$[Co^+]E^{2+}$	[51]
Acetyl-CoA synthase	$H_3C[Co^{3+}]E^+ + CO + CoASH \rightarrow CH_3COSCoA + [Co^+]E^{2+} + H^+$	[24]
Phosphotransacetylase	$CH_3COSCoA + HPO_4^{2-} + H^+ \rightarrow CH_3COOHPO_3^- + CoASH$	[52]
Acetate kinase	$CH_3COOHPO_3^- + ADP^{3-} \rightarrow CH_3COO^- + ATP^{4-} + H^+$	[52]
Aldehyde dehydrogenase	$CH_3COO^- + NADPH + 2H^+ \rightarrow CH_3CHO + NADP^+ + H_2O$	[53]
Alcohol dehydrogenase	$CH_3CHO + NADPH + H^+ \rightarrow CH_3CH_2OH + NADP^+$	[54]

[1] THF–tetrahydrofolate.

2.3. Pathway

The production of acetic acid and ethanol from syngas, CO, H_2 and CO_2, follows a sequenced set of elementary chemical reactions as seen in Figure 2 [11,55,56]. Each reaction proceeds with an associated enzyme in a specific location within a cell, either free in the cytoplasm, tethered to the surface of the cell membrane or embedded in the membrane. Each cell acts independently, but the combined action of all cells sets conditions in the fermentation bulk liquid. Reactions inside cells are mediated by enzymes (Table 1); each binds specific reactants and converts them to specific products, and these enzymatic reactions are typically reversible. The reactions occur at local conditions of pH and chemical concentrations inside the cell, conditions that determine the activity of the enzymes and the form and availability of reactants. The simple inorganic chemical substrates, CO, H_2 and CO_2, are transformed, step by step, first to acetyl-CoA and then to organic products, such as acetic acid and ethanol. Some acetyl-CoA is diverted to form complex organic cell components, carbohydrates, proteins and lipids. However, the majority of gas consumed provides energy for cell function, resulting in the accumulation of acetic acid and ethanol.

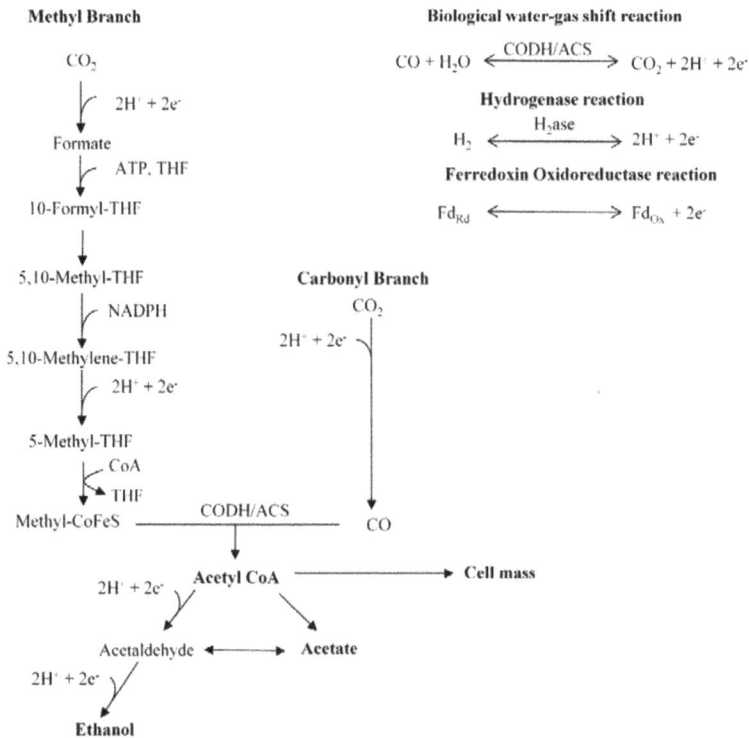

Figure 2. The Wood–Ljungdahl pathway for the production of ethanol and acetic acid; THF: tetrahydrofolate; ACS: acetyl CoA synthase; CODH: carbon monoxide dehydrogenase; H_2ase: hydrogenase; NADPH: reduced nicotinamide adenine dinucleotide phosphate; adapted from [1,57].

2.3.1. Stoichiometry

The production of acetic acid and ethanol from syngas is represented in the literature by the stoichiometry for a single reductant, production from either CO or H_2 with CO_2 [2,25,55,58–60]. The pure component stoichiometry and associated Gibbs free energy, $\Delta G°$, are given in Table 2, Equations (1) and (5) for the production of acetic acid and Equations (6) and (12) for the production of ethanol. The similar stoichiometry from CO and H_2 to form products, 4 moles per mole of acetic acid and 6 moles per mole of ethanol, reinforce that CO and H_2 both act as reductants, providing indistinguishable electrons for the subsequent production reactions. $\Delta G°$ provides an insight into the direction of a reaction and whether or not it is a spontaneous or not. A negative $\Delta G°$ for a reaction means it is spontaneous in the forward direction to make products. A positive $\Delta G°$ for a reaction means it is nonspontaneous in the forward direction. When $\Delta G°$ equals zero, a reaction is at equilibrium. $\Delta G°$ is not correlated with the speed of reaction. Kinetics governs the speed of reactions and how fast a product is formed. The $\Delta G°$ values for all reactions in Table 2 are negative and favorable in the formation of acetic acid and ethanol. A more negative $\Delta G°$ for a reaction makes that reaction more favorable thermodynamically. For example, ethanol production in Equation (6) from only CO is thermodynamically more favorable than in Equations (7) to (12) from both CO and H_2 or from both H_2 and CO_2. In addition, the higher the molar ratios of H_2:CO, the greater the efficiency of incorporating carbon from CO into acetic acid or ethanol.

Table 2. Stoichiometry of acetic acid and ethanol production from syngas and change in Gibbs free energy at 298 °K and 100 kPa.

Products	Reaction		ΔG° kJ/mol
Acetic Acid	$4\,CO + 2\,H_2O \rightarrow CH_3COOH + 2\,CO_2$	(1)	-154.6
	$3\,CO + H_2 + H_2O \rightarrow CH_3COOH + CO_2$	(2)	-134.5
	$2\,CO + 2\,H_2 \rightarrow CH_3COOH$	(3)	-114.5
	$CO + 3\,H_2 + CO_2 \rightarrow CH_3COOH + H_2O$	(4)	-94.4
	$4\,H_2 + 2\,CO_2 \rightarrow CH_3COOH + 2\,H_2O$	(5)	-74.3
Ethanol	$6\,CO + 3\,H_2O \rightarrow CH_3CH_2OH + 4\,CO_2$	(6)	-217.4
	$5\,CO + H_2 + 2\,H_2O \rightarrow CH_3CH_2OH + 3\,CO_2$	(7)	-197.3
	$4\,CO + 2\,H_2 + H_2O \rightarrow CH_3CH_2OH + 2\,CO_2$	(8)	-177.3
	$3\,CO + 3\,H_2 \rightarrow CH_3CH_2OH + CO_2$	(9)	-157.2
	$2\,CO + 4\,H_2 \rightarrow CH_3CH_2OH + H_2O$	(10)	-137.1
	$CO + 5\,H_2 + CO_2 \rightarrow CH_3CH_2OH + 2\,H_2O$	(11)	-117.1
	$6\,H_2 + 2\,CO_2 \rightarrow CH_3CH_2OH + 3\,H_2O$	(12)	-97.0
Acetic Acid	$CO + CO_2 + 6H^+ + 6e^- \rightarrow CH_3COOH + H_2O$	(13)	-94.4
Ethanol	$CO + CO_2 + 10H^+ + 10\,e^- \rightarrow CH_3CH_2OH + 2H_2O$	(14)	-117.1

Production can proceed using either CO or H_2 [55], but fermentation with syngas, containing CO, CO_2 and H_2, typically shows simultaneous uptake of both CO and H_2 [38,41]. Intermediate stoichiometry can be written beginning with production from pure CO and substituting one H_2 for one CO and reducing consumption of H_2O and production of CO_2 by one for each step. In this way, five balanced equations are obtained showing the "quantum" or molecular production of acetic acid from any combination of four, CO plus H_2, with two carbons fixed in acetic acid. Similarly, seven balanced equations are obtained showing the "quantum" production of ethanol from a combination of six, CO plus H_2, with two carbons fixed in ethanol. The overall stoichiometry observed in fermentation will be the average of the "quantum" stoichiometry; for example, 4.3 moles CO plus 1.7 moles H_2 can produce 1 mole of ethanol.

The substitution of H_2 for CO as reductant in fixing two carbons in the product, either acetic acid or ethanol, suggests a general stoichiometry independent of the origin of electrons, whether from CO or H_2. Reducing equivalents of H_2 ($2\,H^+ + 2\,e^-$) are provided by either CO or H_2, while carbon comes from CO and CO_2. The methyl group of acetic acid is formed from CO_2, and the carbonyl is formed from CO. The general stoichiometry of acetic acid formation is given in Equation (13) and the general stoichiometry of ethanol formation in Equation (14). Consumption of four reductants (including CO and H_2) and two carbons (CO or CO_2 including the CO used as reductant) will produce acetic acid. Consuming six reductants per two carbons will produce ethanol. The energy, H^+ and e^- are supplied by oxidation of CO or H_2, and Equations (15) and (16) describe acetic acid and ethanol production.

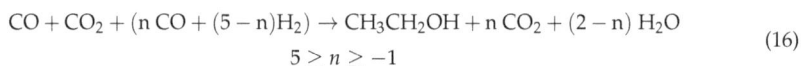

$$CO + CO_2 + (n\,CO + (3 - n)H_2) \rightarrow CH_3COOH + n\,CO_2 + (1 - n)\,H_2O$$
$$3 \geq n \geq -1 \tag{15}$$

$$CO + CO_2 + (n\,CO + (5 - n)H_2) \rightarrow CH_3CH_2OH + n\,CO_2 + (2 - n)\,H_2O$$
$$5 \geq n \geq -1 \tag{16}$$

A negative coefficient for CO_2 or H_2O as a product indicates that the species is added as a reactant. For reactions on a molecular level, n is an integer, and Equations (15) and (16) represent the quantum stoichiometry. However, on a molar level of reaction, n is not restricted to integer values, and Equations (15) and (16) represent the average stoichiometry.

2.3.2. Production Reactions

The acetyl-CoA pathway has been defined over 70 years of research [24,56] and is shown in Figure 2. Energy and carbon from syngas are used to produce acetyl-CoA. CO_2 is converted to a methyl group in the tetrahydrofolate cycle, through a series of reactions that consume one adenosine triphosphate (ATP) and three reducing equivalents of hydrogen ($2 H^+ + 2 e^-$ derived from CO or H_2). Acetic acid can be released from the cell into the bulk liquid (by diffusion or facilitated diffusion) [61] or reduced through acetaldehyde to ethanol consuming another two reducing equivalents [62].

Carbon

Carbon enters the acetyl-CoA pathway reactions as CO_2 or CO. CO_2 is required for the formation of formate [46], which is bound to tetrahydrofolate and reduced to form the methyl group of acetyl-CoA. A methyl cation is transferred to acetyl-CoA synthase (ACS) via an enzyme that contains cobalt (and is called the corrinoid iron-sulfur protein or CoFeSP). The carbonyl of acetyl-CoA is derived from CO bound to carbon monoxide dehydrogenase (CODH), transferred within the bi-functional enzyme to the acetyl-CoA synthase (ACS) active site [63] and condensed with the methyl group and coenzyme A to form acetyl-CoA (Figure 2). Acetyl-CoA is either incorporated in cell components or converted to acetic acid inside the cell, and the conversion of acetyl-CoA to acetic acid via acetyl-phosphate replaces the ATP used to convert CO_2 to the methyl cation.

Acetic acid is released by the acetate kinase enzyme to the cytoplasm (inside the cell) and is reduced to ethanol by carboxylic acid reductase [53] and alcohol dehydrogenase using reduced electron carriers like the reduced nicotinamide adenine dinucleotide (NADH) that are not strongly associated with the membrane. The enzymes acetaldehyde dehydrogenase and alcohol dehydrogenase, which are required to reduce acetic acid to ethanol, have been isolated from *Moorella thermoacetica* (formerly *Clostridium thermoaceticum*) and *Clostridium formicoaceticum* [53,54]. Based on commercial acetone-butanol-ethanol (ABE) fermentation, acyl-CoA (acetyl- or butyryl-CoA) is reduced to aldehyde and then to alcohol using low potential electron carriers, at the expense of one ATP. This loss of ATP in equimolar ratio to ethanol production would make ethanol production from syngas impossible. Fraisse [54] and White [53] found that acetic acid is converted directly to acetaldehyde; acetyl-CoA is not the direct precursor of acetaldehyde and ethanol; and ATP is not lost when alcohol is produced. Aldehyde dehydrogenase and alcohol dehydrogenase were also found to be functional for the reduction of other carboxylic acids to their respective alcohols, including reduction of butyric acid to butanol [30,53,64]. Ethanol production is affected by the internal electrochemical potential and internal pH of the cell, which are determined by the concentration of accumulated CO and H_2.

Energy

During syngas fermentation, energy flows by the transfer of electrons. One pair of electrons ($2 e^-$) is supplied for reaction in the cell by each CO oxidized on CODH or H_2 oxidized on hydrogenase (H_2ase); a pair of protons ($2 H^+$) is released to the cytoplasm for each oxidized CO [43] or H_2 [44]. The electrons are distributed to reaction sites within the cell by electron carriers like ferredoxin and NAD(P)H. Electrons from CODH and H_2ase are first transferred to the membrane-associated clostridial ferredoxin [65] and then are transferred to other electron carriers like NAD(P)H for use in pathway reactions and other cell function. Electrons are transferred via enzymes and cofactors coded in the acetogenic genome in the *Rnf* (for *Rhodobacter* nitrogen fixing) operon. The *Rnf* operon produces a membrane-bound protein complex that is critical to electron transfer and translocation of protons across the cell membrane [66–68].

The *Rnf* complex is proposed to mediate "electron bifurcation", wherein electrons from H_2 are bound on ferredoxin at very low potential using energy supplied from electrons transferred at higher potential to NADH. This "bifurcation" is proposed to translocate protons across the cell membrane through the *Rnf* complex proteins. However, electrons from H_2 are likely transferred via ferredoxin to the *Rnf* complex, then distributed to NAD(P)H and membrane-integral flavins, equalizing the intracellular potential of all electron carriers. The reduced flavin nucleotides carry protons across the membrane, and the associated electrons, still near the potential of H_2, reduce methylene-THF to methyl-THF in a critical reaction of the production pathway. To develop the bifurcation concept, Schuchmann and Muller [69] reported methylene-THF reductase to be neither membrane associated, nor membrane attached. However, Hugenholtz et al. [70] found this critical enzyme was membrane bound, but easily displaced by disruption of the cells for isolation of proteins. The cell membrane serves to insulate the low potential electrons transferred to the flavin and, thus, establish the proton gradient, membrane potential and the proton-motive force used to produce ATP for cell growth.

A single crossover integration in *C. ljungdahlii* was reported to block the production of "a membrane associated polyferredoxin accepting electrons from ferredoxin and transferring them to membrane domains of the *Rnf* complex" [68]. Autotrophic growth on H_2 and CO_2 was blocked by this mutation, and the "proton gradient, membrane potential and protonmotive force collapsed". The reduction of ferredoxin well above its midpoint potential ($E^{\circ\prime}= -420\,mV$) by H_2 drives the function of the *Rnf* complex, but the imposed mutation blocked the transfer of electrons from H_2 into the *Rnf* complex circuit, and without a supply of energy, the chemical potential across the membrane was dissipated. Presumably, autotrophic growth of the *C. ljungdahlii* mutant on CO would also have been blocked, although this was not reported by [68]. Insulation of the electron current by the cell membrane is critical to the function of the *Rnf* electron transfer chain that transports H^+ across the membrane and terminates in reduction of methylene-THF to methyl-THF.

2.3.3. Key Oxidation/Reduction Reactions in the Acetyl-CoA Pathway

Electrochemical reactions involve the transfer of electrons and protons. Electrons are transferred from a reduced chemical as it is oxidized to a less reduced (or oxidized) chemical [42,71]. The reduced and oxidized forms of both the electron donor and electron acceptor are called the redox couple. The reaction releasing the oxidized form and electrons, from its reduced form, is called a "half-cell reaction". Important half-cell reactions of the acetyl-CoA pathway are shown in Table 3. Each half-cell reaction (and redox couple) has a mid-point potential (expressed in mV) at which the concentrations of the reduced and oxidized forms are equal. CO_2/CO is a redox couple, and the CO_2/CO half-cell can be paired with the H_2 half-cell; ferredoxin mediates this electron transfer in acetogens [45]. The combined half reactions comprise the water-gas shift reaction, Equation (17); CO and H_2O are converted to H_2 and CO_2 in this reversible reaction.

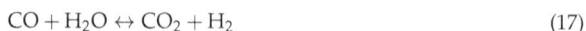

$$CO + H_2O \leftrightarrow CO_2 + H_2 \qquad (17)$$

The elementary reactions of the pathway are balanced for charge and conserve elemental species when written as in Figure 2. The chemical equations can be analyzed using pH and chemical concentrations at the enzymes to establish thermodynamic relationships. Several important reactions of the pathway are characterized by paired electrochemical half-cell reactions. The electrons are supplied by the hydrogen (H^+/H_2) and CO_2/CO couples. Electrons are distributed to electron carriers such as ferredoxin (Fd_{Ox}/Fd_{Rd}) and nicotinamide adenine dinucleotide ($NAD^+/NADH$). In the terminal redox couple (CH_3COOH/CH_3CH_2OH), acetic acid is reduced to ethanol.

Table 3. Selected half-cell reactions of the acetyl-CoA pathway. $\Delta G_r{}^\circ$ and E° indicate the standard Gibbs free energy change and midpoint potential of the half-cell reaction at pH = 0, while $\Delta G_r{}^{\circ\prime}$ and $E^{\circ\prime}$ are at pH = 7.0, n_e and Δm_H are the numbers of electrons transferred and protons consumed, respectively, and Π_{prod}/Π_{react} is the form of the mass action ratio.

Half Cell Reduction	$\Delta G_r{}^\circ$ (kJ/mol)	E° (mV)	n_e	Δm_H	$\Delta G_r{}^{\circ\prime}$ (kJ/mol)	$E^{\circ\prime}$ (mV)	Π_{prod}/Π_{react}
$2H^+ + 2e^- \leftrightarrow H_{2(g)}$	0	0	2	−2	79.90	−414	P_{H2}
$CO_2 + 2H^+ + 2e^- \leftrightarrow CO_{(g)} + H_2O$	20.03	−104	2	−2	99.93	−518	P_{CO}/P_{CO2}
$CH_3COOH + 2H^+ + 2e^- \leftrightarrow CH_3CHO + H_2O$	−7.67	40	2	−2	72.23	−374	C_{Ald}/C_{HA}
$CH_3CHO + 2H^+ + 2e^- \leftrightarrow CH_3CH_2OH$	−41.85	217	2	−2	38.05	−197	C_{Et}/C_{Ald}
$NAD^+ + H^+ + 2e^- \leftrightarrow NADH$	21.80	−113	2	−1	61.75	−320	C_{NADH}/C_{NAD+}
$Fd_{Ox} + 2e^- \leftrightarrow Fd_{Rd}$	81.05	−420	2	0	81.05	−420	C_{Fdr}/C_{Fdo}

2.4. ATP and Cell Growth

Autotrophic growth and production are dependent on the transport of protons and electrons across the cell membrane to generate the proton-motive force that drives synthesis of ATP [72]. The proton-motive force consists of a pH differential plus a difference in electrochemical potential as shown in Equation (18) [42].

$$\Delta p = \Delta\varphi - \frac{2.3RT}{F}\Delta pH \qquad (18)$$

where Δp is the proton-motive force (mV) driving transfer of protons across the membrane, $\Delta\varphi$ is the potential difference across the membrane (mV), ΔpH is the pH differential across the membrane, R is the gas constant (8.314 J/mol K), T is the temperature (K) and F is the Faraday constant (96.485 J/mV mol e$^-$).

Protons released into the cytoplasm are consumed in the formation of acetyl-CoA, acetic acid and ethanol, maintaining the charge balance, while one pair of protons (2 H$^+$) is expelled from the cell for each acetyl-CoA formed. This proton pair is carried across the membrane by a reduced flavin electron carrier, while the electrons are used in the reduction of methylene-THF to methyl-THF [49]. The removal of H$^+$ from the interior of the cell develops a differential of pH and electrochemical potential across the membrane. Protons, as positively-charged particles, are attracted to the more negatively-charged interior of the cell and driven by the higher concentration of protons outside the cell [73]. This proton-motive force pulls protons through an ATP synthase, driving rotation in the ATPase structure that mechanically forms and releases ATP from three binding sites for ATP/ADP + P$_i$ on the enzyme [74,75]. One ATP is consumed in converting formate to formyl phosphate, and one ATP is recovered in the conversion of acetyl phosphate to acetate. Product formation via the acetyl-CoA pathway yields no net ATP; in syngas fermentation, ATP is obtained only from the chemiosmotic mechanism of the ATP synthase [73,76].

3. Microbial Conversion of Gas Phase Substrates

The conversion of CO, H$_2$ and CO$_2$ by acetogenic bacteria to acetic acid and ethanol via the acetyl-CoA pathway is affected by the conditions inside and outside the cell, as depicted in Figure 3. These include pH, temperature and concentrations of nutrients, CO, H$_2$ and CO$_2$, and products like acetic acid and ethanol. Mass transfer also affects the availability of CO, H$_2$ and CO$_2$ inside the cells, and each intermediate reaction, in vivo, will depend on the concentration of its particular reactants and products. The concentrations of intermediate metabolites define the individual reactions and connect the chain of reactions that constitutes the overall stoichiometry of production. The rate of each reaction in the acetyl-CoA pathway is determined by the concentrations of the metabolites involved and the enzyme kinetics supported in the cell.

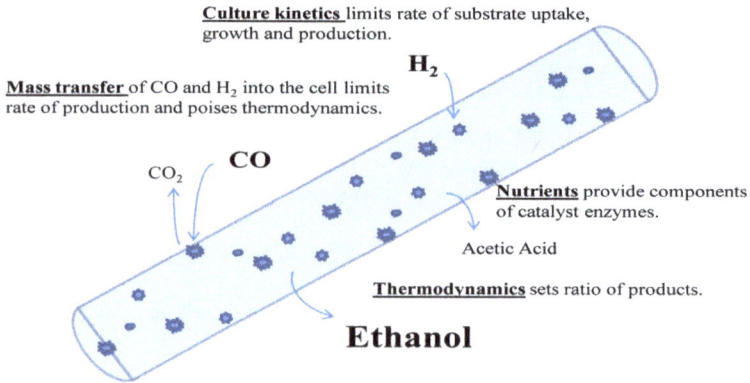

Figure 3. Depiction of an acetogenic bacterial cell showing the supply of CO and H_2 into the cell and efflux of CO_2 by mass transfer, the reaction on enzymes dependent on nutrients taken from the medium to support culture kinetics of growth and production and the thermodynamic determination of products in syngas fermentation.

3.1. Gas Solubility

CO and H_2 are sparingly soluble in water, and their solubility depends on the partial pressure of the individual species according to Henry's law. As an example, for CO:

$$C_{CO} = y_{CO}\, P_T / H_{CO} \tag{19}$$

where C_{CO} is the liquid phase concentration of CO, y_{CO} is the gas phase mol fraction of CO, P_T is the total pressure and H_{CO} is the Henry's law constant for CO. The Henry's law constants for CO, H_2 and CO_2 at 37 °C are given in Table 4. Saturated concentration of either CO or H_2 in water under 100 kPa of pure gas is less than 10^{-3} mol/L. CO and H_2 must be continuously replenished in the liquid medium to support active fermentation. The lowest concentrations of CO and H_2 are inside the cell where the enzymes that catalyze oxidation reside. In contrast, CO_2 is produced in fermentation that consumes CO, and in that case, CO_2 is transferred from inside the cell through the liquid phase to the gas phase. The concentration of CO_2 will be highest inside the cell.

Table 4. Henry's law constants and diffusivities for gases in water at 37 °C [a].

Gas	H (kPa L/mol)	$D_{i,W}$ (m²/s)
CO	121,561	2.50×10^{-9}
H_2	140,262	6.24×10^{-9}
CO_2	4240	2.69×10^{-9}
O_2	101,300	3.25×10^{-9}

[a] Adapted from [77].

3.2. Transport Theory

The low solubility of CO and H_2, gases that provide the energy for syngas fermentation and energy conserved in ethanol product, requires these gases to be continually replenished in the fermentation broth to sustain production. The rate of mass transfer of substrate gas from the bulk gas through the gas-liquid interface and the bulk liquid into the cell, depicted in Figure 4, can be described by film theory [78].

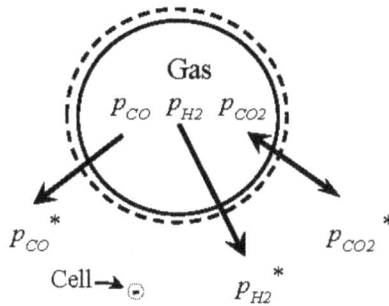

Figure 4. Schematic of gas to liquid mass transfer in the fermentation broth. Partial pressure in the gas phase, p_{CO}, p_{H2}, p_{CO2} and in the equilibrium bulk liquid phase pressure, p_{CO}^*, p_{H2}^*, p_{CO2}^*. The bubble boundary is indicated by the solid line and the liquid film by the dashed line; a single cell is indicated inside the small circle; the scale of the bubble is 1 mm in diameter; the size of the cell is 0.5 μm in diameter by 3 μm long; there are more than 10^{10} cells per liter of fermentation broth.

Diffusion of gas components within the bulk gas is very fast relative to the consumption rate, and the concentration of each species is uniform throughout the gas phase. The concentration of each species in the liquid at the interface is at equilibrium with the bulk gas partial pressure as predicted by Henry's law. The liquid at the interface is part of a stagnant film of fluid through which dissolved gas must transfer by diffusion to the bulk liquid, and since diffusion is driven by the concentration difference, the transfer rate is dependent on the gas diffusivity through water and the thickness of the stagnant film. Outside of the stagnant film, the liquid is assumed to be mobile and turbulent [79], and dissolved gas transfer within the bulk liquid is by bulk flow at rates far exceeding diffusion. The bulk liquid is assumed to be well mixed and homogeneous.

Gas is transferred into the cell by a diffusion process through the cell membrane, which is 6 to 9 nm thick [80]. *C. ragsdalei* cells are rod shaped with typical dimension of 0.5 μm by 3 μm, and even at low cell density (0.02 g cells/L), there are more than 10^{10} cells/L of bulk liquid. The surface area of these cells will exceed the area of the gas-liquid interface by up to three orders of magnitude in a typical fermentation, and resistance to gas transfer across the membrane will be negligible. The gas to liquid mass transfer rate is controlled by diffusion through the film of stagnant liquid at the gas-liquid interface, and the rate of molar gas transfer is proportional to the difference in concentration from the surface of the liquid to the bulk liquid.

The partial pressure of each component in the gas phase is the product of its mole fraction and the total pressure. The partial pressure for CO is calculated by Equation (20).

$$p_{CO} = y_{CO}\, P_T \tag{20}$$

The liquid film mass transfer of CO is represented by Equation (21).

$$-\frac{1}{V_L}\frac{dn_{CO}}{dt} = \frac{k_{L,CO}a}{V_L}(c_{CO}^* - c_{CO,L}) = \frac{\left(\frac{k_{L,CO}a}{V_L}\right)}{H_{CO}}(p_{CO} - p_{CO}^*) \tag{21}$$

where c_{CO}^* is the concentration of CO at the gas-liquid interface in equilibrium by Henry's law; $c_{CO,L}$ is the concentration of CO in the bulk liquid; p_{CO}^* is the CO partial pressure (kPa) in equilibrium by Henry's law with the concentration of CO dissolved in the bulk liquid; p_{CO} is partial pressure in the gas bubble; H_{CO} is the Henry's law constant for CO (kPa L/mol); and V_L is the volume (L) of liquid into which gas is transferred. The molar rate of CO transfer is $-dn_{CO}/dt$ (mol CO/h), where the negative sign denotes consumption from n_{CO} moles of CO in the bulk gas. The constant of proportionality is $k_{L,CO}a/V_L$, which is the overall liquid film mass transfer coefficient (often denoted simply as k_La in the literature) for CO with units of reciprocal time (h^{-1}). The area of the gas/liquid interface is a (m^2). The term $k_{L,CO}$ is the liquid film mass transfer coefficient for CO (L/m^2 h), which includes the effects of turbulence in the liquid, hydrodynamic conditions like viscosity that affect film thickness and gas diffusivity in the aqueous phase. When CO is mass transfer limited, p_{CO}^* is arithmetically zero, and $k_{L,CO}a/V_L$ can be calculated from dn_{CO}/dt and p_{CO} using Equation (21).

The volumetric mass transfer coefficients for H_2 ($k_{L,H2}a/V_L$) and CO_2 ($k_{L,CO2}a/V_L$) differ from $k_{L,CO}a/V_L$, but are proportional. The area of the gas/liquid interface and the liquid volume are the same for all gases, as is the intensity of turbulence in the liquid. The coefficients for these gases will differ due to their diffusivity in the fermentation broth ($D_{i,W}$) through the liquid film, and the measured $k_{L,CO}a/V_L$ from Equation (21) is used to predict values of $k_{L,H2}a/V_L$ and $k_{L,CO2}a/V_L$ based on the surface renewal theory for film transfer [81].

$$\frac{k_{L,CO}a}{V_L} = \sqrt{\frac{D_{CO,W}}{D_{H2,W}}} \left(\frac{k_{L,H2}a}{V_L} \right) = \sqrt{\frac{D_{CO,W}}{D_{CO2,W}}} \left(\frac{k_{L,CO2}a}{V_L} \right) \tag{22}$$

The actual capacity of the fermenter to transfer H_2 and CO_2 is represented in $k_{L,H2}a/V_L$ and $k_{L,CO2}a/V_L$ determined from Equation (22). This capacity can remain unused, in which case H_2 or CO_2 will accumulate in the bulk liquid and in the cell up to saturation of the dissolved gas. The fermentation broth was assumed to be like water, which is 98% of the medium.

The attainment of higher mass transfer represented in the volumetric mass transfer coefficients, $k_{L,CO}a/V_L$ and $k_{L,H2}a/V_L$, is of primary concern in most discussion of syngas fermentation [82–84]. A model of syngas fermentation in the continuously-stirred tank reactor (CSTR) was developed to assess the potential for the production of acetate [85], and mass transfer has been studied in various configurations of fermenters [82,83,86,87]. Klasson et al. [83], however, notes that the rate of mass transfer will not exceed the rate of reaction of the slightly soluble substrates and that the applied mass transfer should balance the supply and consumption of CO and H_2.

3.3. Enzyme Catalyzed Reactions

Conversion of CO and H_2 to acetic acid, ethanol and cell mass is performed on a platform of enzymes contained in the cells (Figure 3 and Table 1). The cell membrane separates the cytoplasm from the bulk liquid fermentation broth, and enzymes are either suspended in the cytoplasm or associated with or embedded in the membrane. Intracellular conditions of pH, oxidation reduction potential (ORP) and chemical composition are related to the bulk liquid by diffusion and membrane transport and can differ in significant ways that are essential to cell function [42,88]. The concentrations of dissolved CO, H_2 and CO_2 inside the cells are nearly the same (within 5%) as the bulk liquid, since the transfer of gas into the cells occurs along a short mass transfer path through a very thin membrane (6 to 9 nm) with a large total surface area. The observed rates of consumption of gas and the formation of products in the defined stoichiometry of the production pathway reveal the mass flux of carbon, protons and electrons through the pathway reactions. However, in analogy to the catalytic conversion of syngas to ethanol, production is, "impacted by kinetic and thermodynamic constraints." as previously reported [15]. The dissolved gas concentrations set the thermodynamics of reactions, set the concentrations of intermediate metabolites and determine the kinetic rates. Fermentation occurs in this intracellular environment, and the mass flux through the biological pathways can be quantified and controlled to achieve targeted results on the macroscopic scale.

3.4. Thermodynamics

Syngas fermentation thermodynamics have been examined [89] using transformed thermodynamics, and it was concluded that CO was always preferred over H_2 as a substrate for fermentation. CO inhibition of hydrogenase or thermodynamic disfavor was suggested as the reason for low and delayed uptake of H_2 in syngas fermentation. These thermodynamic calculations assumed bulk liquid concentration saturated from the gas phase partial pressures of H_2, CO and CO_2. While acetogenic fermentation of gas containing both CO and H_2 can exhibit periods of exclusive CO uptake, CO and H_2 are typically consumed together [41], and the concentrations of dissolved CO and H_2 are changed significantly to effect mass transfer.

The ordered chemical reactions in the acetyl-CoA pathway occur in sequence to produce acetyl-CoA, acetic acid and ethanol from CO_2, CO and H_2. Each reaction is mediated by an enzyme that catalyzes the reaction, and each reaction proceeds in the direction of favored thermodynamics, for which $\Delta G_r < 0$. The thermodynamics of biological reactions are addressed in biochemistry texts [88,90] and reviews [66,71]; these treatments discuss the criteria for a reaction to proceed, $\Delta G_r < 0$, and for thermodynamic equilibrium, $\Delta G_r = 0$, and the dependence of ΔG_r on concentration of reactants and products through the mass action ratio [88]. The effect of pH on ΔG is not discussed extensively, although Lehninger (1982) stated, "Biochemical reactions take place near pH 7.0 and often involve H^+" to introduce the standard free energy at pH 7.0, $\Delta G_r^{\circ\prime}$. The dependence of ΔG_r on pH and the application in redox reactions in the cell are discussed in Cramer and Knaff [42]. Thermodynamic Cramer and Knaff data for reactions and compounds of interest in biological systems are available in the appendix of Thauer et al. [71], and these data can be used to define the thermodynamic position of the reactions of the acetyl-CoA pathway. Cramer and Knaff [42] emphasized the division of the intracellular space, where the enzymes reside, from the bulk liquid in fermentation. The production reactions occur inside the cell, while measurements like pH and ORP are taken in the bulk liquid. Fermentation thermodynamics are characterized by parameters that cannot be measured directly, and such parameters must then be inferred by calculation from available measurements. These calculations require assumptions informed by the biochemical mechanisms to build the equations for data analysis and a predictive model of fermentation.

3.5. Electrochemistry

Many reactions in the acetyl-CoA pathway are oxidation-reduction reactions, in which electrons are transferred from one molecule to another. Electron donors are oxidized, and the electron acceptors are reduced. In the water-gas shift reaction in Equation (17), CO is oxidized to CO_2, and H^+ is reduced to H_2. The two half-reactions are shown in Equations (23) and (24).

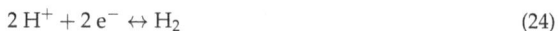

$$CO + H_2O \leftrightarrow CO_2 + 2\,H^+ + 2\,e^- \tag{23}$$

$$2\,H^+ + 2\,e^- \leftrightarrow H_2 \tag{24}$$

CO donates $2\,e^-$ that are used to produce H_2. The water-gas shift reaction is reversible, and H_2 can be oxidized to produce CO from CO_2. Reaction proceeds in the direction for which $\Delta G_r < 0$. The reaction is in equilibrium when $\Delta G_r = 0$.

The oxidized and reduced forms of a chemical comprise a redox couple, for example H_2/H^+ and CO/CO_2. The oxidized form accepts electrons (and sometimes H^+) and becomes reduced. When the half-reaction is set at the standard conditions of 1.0 mol/L reactants and products, the redox couple exhibits a characteristic tendency or potential to donate electrons. This potential, measured in volts, with equal concentrations of the oxidized and reduced forms, is the midpoint potential. This is referred to as E° at pH 0. E° for a half-cell reaction can be calculated from ΔG_r° as in Equation (25) [71,88].

$$E^\circ = -\Delta G_r^0 / n_e F \tag{25}$$

where n_e is the number of electrons transferred and F is the Faraday constant (0.0965 kJ/mV mol e^-). Note that this potential is a characteristic of the half-cell reaction, not a differential. The Gibbs free energy change for a half-cell reaction, ΔG_r, changes with concentrations of products and reactants;

the electrochemical potential of the half-cell changes, as well. The potential (E) is given by the Nernst equation [80,88].

$$E = -\frac{\Delta G_r}{n_e F} = E^0 - \frac{RT}{n_e F} ln\left(\prod C_{(Products)} / \prod C_{(Reactants)}\right) + 2.302 \frac{RT}{n_e F} \Delta m_H pH \tag{26}$$

The notation ($\prod C_{Products} / \prod C_{Reactants}$) represents the mass action ratio for the reaction [88], and Δm_H is the number of protons produced in the reaction. E is the potential of the redox couple to donate electrons under the actual conditions, and each redox couple exhibits its characteristic potential under those conditions. A redox couple at lower potential (more negative) donates electrons (is oxidized) to couples at higher potential. Two half-cell reactions like Equations (23) and (24) are combined, an oxidation with a reduction, in a balanced reaction, as shown in the water-gas shift reaction in Equation (17). When the reaction reaches equilibrium, $\Delta G_r = 0$, and both redox couples are at the same potential E. The degree of reduction of each couple is reflected in the mass action ratio that gives E for the half-cell in Equation (26).

3.6. Electron Carriers

Bar-Even [91] asks, "Does acetogenesis require especially low reduction potential?" and applies similar thermodynamic analysis under the bifurcation concept. Bar-Even acknowledges lower concentrations of reactants in syngas fermentation reactions, but does not recognize the very low dissolved gas concentrations (particularly for CO) that produce mass transfer driving force in syngas fermentation. When mass transfer is rate limiting, the transfer of gas to the cell is the slowest process in the fermentation, and all reaction steps in the production pathway are fast relative to the rate of gas supply. Under mass transfer limitation, the reactions of the production pathway approach thermodynamic equilibrium, and all electrochemical half-cell reactions inside the cell approach the same potential, E_{Cell}. The assumption of thermodynamic equilibrium at one intracellular potential sets a boundary condition that defines the thermodynamic state of the pathway reactions. The approach to this assumed thermodynamic reaction state provides a convenient method to describe the reaction set for study and modeling of syngas fermentation.

The potential of the oxidation-reduction reactions of the pathway can be estimated by Equation (26) using E^0 calculated from Equation (25). Then, Equation (26) can be rearranged to calculate the mass action ratio as in Equation (27).

$$\left(\prod C_{(Products)} / \prod C_{(Reactants)}\right) = \exp\left[-\frac{\Delta G_r^o}{RT} - \frac{n_e FE}{RT} + 2.302 \Delta m_H pH\right] \tag{27}$$

Mass action ratios for selected half-cell reactions from the Wood–Ljungdahl pathway are presented in Table 3. Note that the ratios of products to reactants are the ratios of concentrations or partial pressures, except for the partial pressure of H_2. The half-cells are typically two electron reductions, $n_e = 2$, and that most reductions consume two protons, $\Delta m_H = -2$, except NADH/NAD$^+$ and Fd_r/Fd_o consume one proton and no protons, respectively. The values of $\Delta G^{o\prime}$ and $E^{o\prime}$ given in Table 3 are calculated at pH 7.0 and match values given by Thauer et al. [71].

The electrochemical couples are defined by the mass action ratio of products to reactants in the half-cells at a given pH. The CO/CO_2 half-cell is defined by $p_{CO}{}^*/p_{CO2}{}^*$, while the H_2 half-cell is defined by $p_{H2}{}^*$ alone. The calculated $p_{H2}{}^*$ defines the potential at a given pH and is the best measure of the internal electrochemical potential, E_{Cell}, that sets the ratio of ethanol to acetic acid attained. Equation (27) correlates the concentrations of chemicals inside the cell to the intracellular pH (pH$_{ic}$) and E_{Cell}.

3.7. Kinetics

Thermodynamics control the direction and possible extent of the reactions in the production pathway, while kinetics describes the rates of reactions and the overall rates of CO and H_2 consumption, acetic acid and ethanol accumulation, as well as cell growth. The overall rates are expected to be proportional to cell mass (XV_L) in the fermenter, with the coefficient of proportionality being the specific growth rate (μ) for growth and the specific uptake rate (q_{CO} for CO and q_{H2} for H_2). Individual reaction rates are related to the concentrations of the reactants and products using a kinetic model, such as Michaelis–Menten for enzyme-mediated reactions [92]. The specific growth and specific uptake rates are likewise correlated to the concentration of substrates, like CO and H_2 inside the cell, in a kinetic model, such as the Monod equation [92,93]. The concentrations of substrates and products that are important in syngas fermentation are the CO, H_2, CO_2, acetic acid and ethanol dissolved in the bulk liquid. These concentrations are likewise thermodynamic quantities that can be measured or predicted. The dissolved CO, H_2 and CO_2 are represented by the dissolved partial pressures, $p_{CO}{}^*$, $p_{H2}{}^*$ and $p_{CO2}{}^*$, and these can be calculated from mass transfer analysis of the experimentally-observed uptake.

Description of fermentation kinetics incorporates time differentials of measured parameters that describe the cell culture. The specific growth rate is the production of cell mass per unit of cell mass per time, g_x/g_xh or in h^{-1}, and calculated as:

$$\mu = \frac{1}{X}\frac{dX}{dt} \tag{28}$$

where X is the cell mass concentration, in g/L. The specific uptake of CO (q_{CO}) or H_2 (q_{H2}) is the consumption of the gas per unit cell mass per time, $mol/g_x.h$, which is estimated as:

$$q_{CO} = \frac{1}{XV_L}\frac{dn_{CO}}{dt} \tag{29}$$

$$q_{H2} = \frac{1}{XV_L}\frac{dn_{H2}}{dt} \tag{30}$$

$$q_{CO+H2} = \frac{1}{XV_L}\frac{dn_{CO+H2}}{dt} \tag{31}$$

Syngas fermentation by autotrophic acetogens produces complex chemicals including proteins, sugars, nucleic acids and lipids from CO and H_2. This progression from small to complex must occur through a reversal of reactions typical in sugar fermentation; acetyl-CoA to pyruvate to retrace the glycolytic pathways and branches that produce amino acids [94]. Most reactions of the autotrophic pathway operate near thermodynamic equilibrium. Reaction rates depend on the dissolved concentrations of CO, H_2 and CO_2, and prominent redox reactions used in the pathway dispose syngas fermentation to inhibitions and competition of substrates for enzyme binding sites. Moreover, the production reactions are reversible, and the production rate depends on product concentrations. An effective model of syngas fermentation should include the prediction of reaction rates using the same intracellular potential, pH and concentrations of CO, H_2 and CO_2 that define the thermodynamics.

3.8. Conceptual Model of Fermentation

The initial description of syngas fermentation borrowed from the phenomenological description of the ABE (acetone-butanol-ethanol) fermentation that was commercially prominent in the last century [59,95,96]. This concept of alcohol production persists in the basis of ongoing research [3,11] and is the basis for organism development through genetic modification [97]. Ramió-Pujol et al. [3] notes, "successful production of alcohols in clostridia relies on the metabolic shift from acido-genesis (production of acids) to solventogenesis (production of alcohols). The mechanisms governing this shift have been extensively investigated, especially in acetone–butanol–ethanol (ABE) fermenting clostridia." However, "little is known about the regulatory circuits and molecular mechanisms for the transition to the solventogenesis". It is appropriate that the study of syngas fermentation might lend knowledge to better understand the ABE fermentation.

Syngas fermentation has been modeled by correlating cell growth and productivity with the partial pressure of CO in the gas phase [58,98]. However, the isolated focus on the CO concentration in the supply gas ignores both the presence of H_2 and CO_2 in the fermentation reactions and the difference in concentration imposed by the transfer of each gas into the liquid phase. Growth of *C. ljungdahlii* on H_2/CO_2 shows H_2 to be a competent source of energy for growth and production in syngas fermentation [55]. The requirement for CO_2 as carbon entering the methyl branch of the acetyl-CoA pathway in Figure 2 shows the importance of the CO_2 concentration in the production of acetyl-CoA and subsequent synthesis of acetic acid, ethanol and cell mass. Further, CO and H_2 are used together in syngas fermentation, and both provide electrons to the fermentation reactions [41]. A single parameter model of syngas fermentation using CO partial pressure in the bulk gas is not adequate. Chen et al. [93] prepared an ambitious model to describe syngas fermentation through the space of a bubble column fermenter. However, appropriate data to populate the model constants were lacking. The model does not apply the chemical engineering unit operations with appropriate assumptions to derive rigorous thermodynamic and kinetic parameters for the equations. The model utility can be improved by applying these engineering techniques.

We propose a new conceptual model of syngas fermentation that includes the growth of acetogens with concurrent ethanol production and high conversion of CO and H_2, reduced dissolved concentration of sparingly soluble CO and H_2 resulting from high rate of gas transfer to the intracellular enzymes, less inhibition of the hydrogenase enzyme at very low dissolved concentration of CO and de facto mass transfer limitation for CO in active syngas fermentation. Further, concurrent uptake of CO and H_2 with electron flow from both species to reduce ferredoxin establishes the thermodynamic equilibrium of the water-gas shift within the cell, and the reduction of acetic acid to ethanol in redox reactions, coupled to oxidation of CO and H_2 via cellular electron carriers, suggests a single intracellular redox potential (E_{Cell}) and pH (pH_{ic}). The redox reactions of the acetyl-CoA pathway shown in Figure 2 operate near thermodynamic equilibrium at E_{Cell} and pH_{ic}. A mathematical model constructed with equations conforming to this novel conceptual model describes observed fermentation behaviors and has proven useful in fermentation analysis and control [99].

3.9. Reactor Design

Serum bottles are useful for culture maintenance, but the inherent batch operation is marked by transient conditions of substrate supply and cell and product concentrations. The baffled CSTR (continuously-stirred tank reactor) fermenter equipped with gas dispersion impellers can be operated in semi-batch mode with batch liquid and continuous gas feed, with fed-batch liquid, or with both continuous gas and liquid feed. Two-stage CSTR fermenters have been operated with the first CSTR configured to promote growth of the acetogenic culture with acid production and the second CSTR operated at low pH under nutrient limitation and low gas conversion to achieve high ethanol concentration [100]. Column fermenters that show promise include a bubble column with a ceramic monolith to support biofilm [84], a trickle bed with biofilm [87,101,102] and a biofilm supported on a hollow fiber membrane for gas dispersion [86,87]. Biofilms retain cells, but long-term mass transfer and fouling may limit application. Ethanol productivity was reported to increase in a two-stage CSTR and bubble column with gas and cell recycling because more cells can be accumulated and more gas can be processed in two-stage bioreactors [103]. Chen et al. [93] developed differential equations to describe syngas fermentation through a bubble column, but as yet lack appropriate data for modeling and validation. Column fermenters can be modeled as a series of CSTR, and each CSTR stage can be characterized and designed to deliver mass transfer appropriate to meet a portion of the goals for overall fermentation.

Fermenter equipment can be designed using computer simulation models to meet the requirements for commercial fermentation. Syngas fermentation should be performed with continuous feed of syngas and liquid medium and removal of product for uninterrupted production. Continuous operation must provide high conservation of energy from the syngas into ethanol, a high concentration of ethanol and stable operation without shutdown over long periods. High energy conservation is only achieved through high conversion of both CO and H_2, as well as high specificity for ethanol as the exclusive product. High concentrations of CO and H_2 promote a high ethanol concentration relative to acetic acid. Stable operation that maintains a steady state marked by high activity of the bacterial culture for CO and H_2 uptake is promoted by tight process control and equipment designed for mechanical reliability, redundancy and ease of maintenance.

The typical laboratory CSTR operates with plug flow characteristic for gas conversion, but achieves a single aggregate state of the fermentation parameters in the well-mixed liquid. Since the liquid parameters define the thermodynamics and kinetics of the fermentation, all goals of syngas fermentation cannot be achieved in a single CSTR stage. The laboratory CSTR fermenter is essential in defining the parameters of successful syngas fermentation, but efficient and economical commercial syngas fermentation for biofuel production can be realized in carefully-designed packed column fermenters that provide multistage gas contact during fermentation. Rich syngas contact with high liquid volume at the column bottom will promote reduction of acid to alcohol; partially converted syngas will promote culture growth with low inhibition in the middle; and high mass transfer could convert residual CO and H_2 before the spent gas exits the column at the top. These characteristics ensure product specificity, productivity and efficient energy conservation, which are all essential to process economy.

4. Potential Products

Growth of acetogens in syngas fermentation using a mineral-defined medium shows production of complex cell components from single carbon substrates, CO and CO_2, with energy derived from CO and H_2 [30]. This implies a reversal of glycolytic pathways to form pyruvate and then sugars that compose the membranes from the syngas components. Energy from ATP and reduced electron carriers sufficient to supply fermentation reactions that branch from the glycolytic pathway to form amino acids, nucleic acids and lipids is available in syngas fermentation through the chemiosmotic mechanisms that drive the membrane-bound ATPase and electron transfers. The accumulation of butyric acid, hexanoic acid, butanol and hexanol has been demonstrated for *C. carboxidivorans* [30], and a broad range of potential products awaits techniques developed to enhance accumulation.

Ethanol and acetic acid are products derived directly from acetyl-CoA without the expense of ATP. Ethanol that can be recovered by distillation is the most prominent product. Acetic acid requires more elaborate recovery, such as extraction, but is a high volume chemical and potentially could be produced by oxidation of ethanol. Ethylene, globally one of the highest selling chemicals, could be formed by dehydration of ethanol [104].

An additional ATP is expended by the cells to condense two acetyl-CoA to butyryl-CoA, which is converted to butyric acid and then to butanol in steps similar to ethanol production. Butanol is sought as a "drop-in" biofuel for use in existing petroleum infrastructure, as a solvent and as precursor for subsequent synthesis. Propionic acid, propanol, hexanoic acid, hexanol, acetone, isobutanol, butanediol, amino and fatty acids are other potential products proposed from syngas fermentation [34,35,105–107]. A biological water-gas shift is proposed to produce H_2 [108], and syngas can be biologically converted to methane [109] so that syngas energy and subsequent products might be obtained from biological conversion of natural gas.

5. Techno-Economic Analysis

The annual ethanol production in the United States increased from 52 billion liters in 2010 to 59 billion liters in 2017 [110]. The global annual ethanol production increased from 50 billion liters in 2007 to 101 billion liters in 2016 [111]. These statistics show huge demand for ethanol worldwide. Ethanol produced globally is mostly made from grains and sugar cane. Corn ethanol and gasoline prices in the United States in April 2017 are about $0.43 per liter and $0.44 per liter, respectively [112]. Current corn ethanol prices are similar to prices reported in 1982 (Figure 5). However, current gasoline price is about 60% higher than in 1982. For lignocellulosic ethanol to compete in the fuel market, its selling price should be comparable to corn ethanol prices on an energy basis.

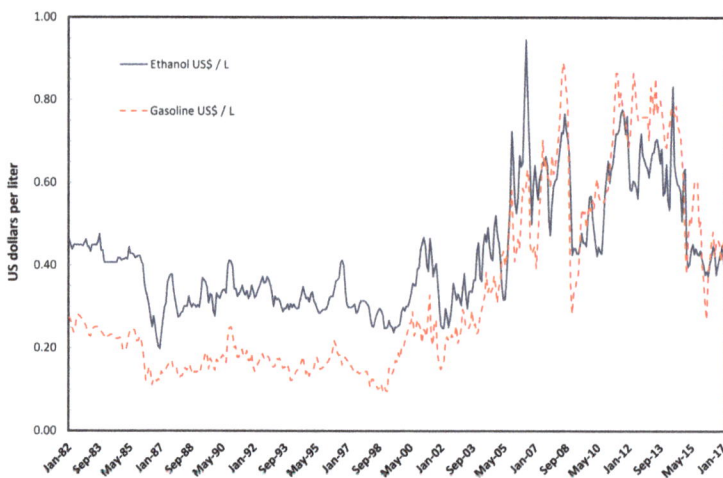

Figure 5. Historical prices of corn ethanol and gasoline in the United States [113].

Lignocellulosic biofuel producers experience delayed plans for commercialization due to difficulty in technology scale-up and securing financing with low petroleum and natural gas costs. Technological and economic challenges in commercialization of lignocellulosic biofuels must be solved to address increased world energy demand, concerns of climate change and to build a sustainable biofuel industry.

Techno-economic analysis (TEA) provides assessments of cost-competitiveness and market penetration potential of alternative biofuel production technologies to researchers, engineers, investors and policy makers [114]. TEA can also facilitate sensitivity analyses of key process parameters to improve feasibility and provide future directions for biofuels research. TEAs are typically based on process and plant design assumptions including experimentally-derived or assumed parameters to estimate process performance, biofuel cost and yield and capital and operating costs. The results obtained from TEA are strongly dependent on the models used and the assumptions made.

Several studies report TEA for the conversion of lignocellulosic biomass to ethanol using the enzymatic hydrolysis fermentation (EHF) process [115–118] and the gasification-mixed alcohol catalytic conversion (GMA) platform [5,119]. However, few studies were found on TEA for ethanol production through the hybrid gasification-syngas fermentation (GF) process [117,120]. TEAs of various thermochemical technologies for cellulosic biofuels have been recently reviewed [121].

Spath and Dayton [120] reported a minimum ethanol selling price (MESP) of $0.44/L for the GF process of 2206 metric tonnes per day (MTPD) with a feedstock cost of $38.70 per metric tonnes (mT), as shown in Table 5. Piccolo and Bezzo [117] estimated that the cost of ethanol production via GF was about 30% higher than for EHF with an assumption of ethanol concentration in the fermentation beer three-fold higher in EHF. The higher ethanol cost using the GF process was due to additional cost for energy required in the distillation of beer containing 24 g/L ethanol compared to 70 g/L ethanol in EHF. The MESP for EHF was 5% lower than for GF when ethanol concentration in the beer in the GF was assumed to be 50 g/L due to higher ethanol yield with the EHF process [117]. However, MESP for EHF with ammonia fiber explosion pretreatment [122] was 4% higher than for GF with 50 g/L ethanol [117]. For the GF process, the cost of biomass feedstock, ethanol concentration and ethanol yield were identified as the main contributors to the MESP.

Typically, total capital investment (TCI) of the GF process is higher than the EHF process due to the additional cost of the gasification system (Table 5). However, the GF process has the potential to achieve high ethanol yields (440 L/Mg) compared to 340 L/Mg for the EHF process [123]. This is due to utilization of all components of the biomass, including lignin during gasification to produce syngas converted into ethanol. The type of gasifier used in thermochemical conversion technologies and pretreatment methods in the biochemical conversion platform greatly affect the production cost of biofuel [118,122].

Table 5. Techno-economic analysis (TEA) of GF, GMA and EHF processes; all values in 2015 dollars.

Process [a]	Plant Size (MTPD) [b]	Feedstock Cost ($/mT)	Ethanol Yield (L/mT)	TCI [c] (M$)	MESP [d] ($/L)	Reference
GF	2206	38.70	289	NR [e]	0.44	[120]
	2030	80.13	204	575	1.32 [f]	[117]
	2030	80.13	282	NR	1.07 [g]	
GMA	2140	88.74	236	578	0.86	[5]
	2000	78.06	350	593	0.62	[119]
EHF	2000	74.17	330	509	0.65	[115]
	2030	80.13	310	301	1.01 [h]	[117]
	2000	95.45	289	432	1.03 [i]	[122]
	2000	95.45	250	444	1.11 [j]	

[a] GF: gasification-syngas fermentation; EHF: enzymatic hydrolysis fermentation; GMA: gasification-mixed alcohol catalytic conversion; [b] MTPD: metric tonnes per day; [c] TCI: total capital investment; [d] MESP: minimum ethanol selling price; [e] NR: not reported; [f] ethanol concentration in the beer is 24 g/L; [g] ethanol concentration in the beer is 50 g/L; [h] ethanol concentration in the beer is 70 g/L; [i] diluted acid pretreatment; [j] AFEX: ammonia fiber explosion.

Current TEA studies are based on technical data and assumptions for first generation biorefineries. Further technology advancements will provide stable, controlled and efficient biofuel conversion processes, which are expected to make future biorefineries feasible.

6. Conclusions

Thermochemical gasification of biomass and wastes combined with the simple robust conversion of CO and H_2 by autotrophic acetogenic bacteria to various products provides a versatile and potentially economical process for the production of fuels and chemicals. Biomass and wastes are wide-spread feedstock resources that represent untapped economic opportunity and, often, environmental disposal problems. Production of fuels and chemicals from biomass will reduce economic reliance on fossil carbon and emission of greenhouse gases, approaching sustainable energy derived from solar input. The autotrophic bacteria that mediate syngas fermentation build the complex chemicals that comprise cell mass from the simple molecules CO, CO_2 and H_2. This synthetic capability presents a staggering number of potential products from the enzyme platform of the native organisms. A more rigorous analysis of syngas mass transfer within an improved concept of the fermentation mechanisms allows the determination of thermodynamic and kinetic parameters inside the bacterial cells. These parameters can be incorporated in a mathematical model to advance process design and control for commercial use of syngas fermentation. The review of techno-economic analysis of gasification-syngas fermentation showed a competitive advantage of the hybrid gasification-syngas fermentation technology to make biofuels compared to gasification-mixed alcohol catalytic conversion and enzymatic hydrolysis fermentation processes. Further advancements in fundamental and applied research areas are essential to make biological gas conversion processes feasible for the production of new products in support of the chemical, petrochemical, agricultural, environmental and pharmaceutical industries.

Acknowledgments: This research was supported by the Sun Grant Program—South Center No. DOTS59-07-G-00053, USDA-NIFA Project No. OKL03005 and the Oklahoma Agricultural Experiment Station.

Author Contributions: John R. Phillips, Raymond L. Huhnke and Hasan K. Atiyeh performed the analysis of the literature and wrote the paper.

Conflicts of Interest: The authors declare no conflict of interest.

References

1. Wilkins, M.R.; Atiyeh, H.K. Microbial production of ethanol from carbon monoxide. *Curr. Opin. Biotechnol.* **2011**, *22*, 326–330. [CrossRef] [PubMed]
2. Drake, H.L.; Gossner, A.S.; Daniel, S.L. Old acetogens, new light. *Ann. N. Y. Acad. Sci.* **2008**, *1125*, 100–128. [CrossRef] [PubMed]
3. Ramió-Pujol, S.; Ganigué, R.; Bañeras, L.; Colprim, J. How can alcohol production be improved in carboxydotrophic clostridia? *Process Biochem.* **2015**, *50*, 1047–1055. [CrossRef]
4. Singla, A.; Verma, D.; Lal, B.; Sarma, P.M. Enrichment and optimization of anaerobic bacterial mixed culture for conversion of syngas to ethanol. *Bioresour. Technol.* **2014**, *172*, 41–49. [CrossRef] [PubMed]
5. Valle, C.R.; Perales, A.L.V.; Vidal-Barrero, F.; Gomez-Barea, A. Techno-economic assessment of biomass-to-ethanol by indirect fluidized bed gasification: Impact of reforming technologies and comparison with entrained flow gasification. *Appl. Energy* **2013**, *109*, 254–266. [CrossRef]
6. Griffin, D.W.; Schultz, M.A. Fuel and chemical products from biomass syngas: A comparison of gas fermentation to thermochemical conversion routes. *Environ. Progress Sustain. Energy* **2012**, 219–224. [CrossRef]
7. Energy Information Administration (EIA). *Annual Energy Outlook*; Department of Energy: Washington, DC, USA, 2012.
8. Energy Information Administration (EIA). *International Energy Outlook*; Department of Energy: Washington, DC, USA, 2013.
9. Balan, V.; Chiaramonti, D.; Kumar, S. Review of US and EU initiatives toward development, demonstration, and commercialization of lignocellulosic biofuels. *Biofuels Bioprod. Biorefin.* **2013**, *7*, 732–759. [CrossRef]
10. Energy Information Administration (EIA). *Biofuels Issues and Trends*; Department of Energy: Washington, DC, USA, 2012.
11. Liew, F.M.; Köpke, M.; Simpson, S.D. Gas fermentation for commercial biofuels production. In *Liquid, Gaseous and Solid Biofuels—Conversion Techniques*; Fang, Z., Ed.; InTech: Rijeka, Croatia, 2013; pp. 125–173.
12. Energy Information Administration (EIA). *Annual Energy Outlook (US) with Projections to 2040*; Department of Energy: Washington, DC, USA, 2015.
13. Perlack, R.D.; Stokes, B.J. *U.S. Billion-Ton Update: Biomass Supply for a Bioenergy and Bioproducts Industry*; Department of Energy: Washington, DC, USA, 2011.
14. Environmental Protection Agency. *Municipal Solid Waste Generation, Recycling, and Disposal in the United States: Facts and Figures for 2012*; Environmental Protection Agency: Washington, DC, USA, 2012.
15. Atsonios, K.; Christodoulou, C.; Koytsoumpa, E.I.; Panopoulos, K.D.; Kakaras, E. Plant design aspects of catalytic biosyngas conversion to higher alcohols. *Biomass Bioenergy* **2013**, *53*, 54–64. [CrossRef]
16. Ahmed, A.; Cateni, B.G.; Huhnke, R.L.; Lewis, R.S. Effects of biomass-generated producer gas constituents on cell growth, product distribution and hydrogenase activity of *Clostridium carboxidivorans* P7(T). *Biomass Bioenergy* **2006**, *30*, 665–672. [CrossRef]
17. Xu, D.; Tree, D.R.; Lewis, R.S. The effects of syngas impurities on syngas fermentation to liquid fuels. *Biomass Bioenergy* **2011**, *35*, 2690–2696. [CrossRef]
18. Woolcock, P.J.; Brown, R.C. A review of cleaning technologies for biomass-derived syngas. *Biomass Bioenergy* **2013**, *52*, 54–84. [CrossRef]
19. Ferry, J.G.; House, C.H. The stepwise evolution of early life driven by energy conservation. *Mol. Biol. Evol.* **2006**, *23*, 1286–1292. [CrossRef] [PubMed]

20. Fischer, F.; Lieske, R.; Winzer, K. Biological gas reactions II concerning the formation of acetic acid in the biological conversion of carbon oxide and carbonic acid with hydrogen to methane. *Biochem. Z.* **1932**, *245*, 2–12.

21. Wieringa, K.T. Over het verwijnenvan waterstofen koolzuur onder anaerobe voorwaarden. *Antonie Leeuwenhoek* **1936**, *3*, 263–273. [CrossRef]

22. Collins, M.D.; Lawson, P.A.; Willems, A.; Cordoba, J.J.; Fernandezgarayzabal, J.; Garcia, P.; Cai, J.; Hippe, H.; Farrow, J.A.E. Thephylogeny of the genus *Clostridium*—Proposal of five new genera and eleven new species combinations. *Int. J. Syst. Bacteriol.* **1994**, *44*, 812–826. [CrossRef] [PubMed]

23. Fontaine, F.E. A new type of glucose fermentation by *Clostridium thermoaceticum* n. sp. *J. Bacteriol.* **1942**, *43*, 700–715.

24. Ragsdale, S.W. Enzymology of the Wood-Ljungdahl pathway of acetogenesis. *Ann. N. Y. Acad. Sci.* **2008**, *1125*, 129–136. [CrossRef] [PubMed]

25. Barik, S.; Prieto, S.; Harrison, S.B.; Clausen, E.C.; Gaddy, J.L. Biological Production of Alcohols from Coal through Indirect Liquefaction. *Appl. Biochem. Biotechnol.* **1988**, *18*, 363–378. [CrossRef]

26. Tanner, R.S.; Miller, L.M.; Yang, D. *Clostridium ljungdahlii* sp. nov., an Acetogenic Species in Clostridial rRNA Homology Group I. *Int. J. Syste. Bacteriol.* **1993**, *43*, 232–236. [CrossRef] [PubMed]

27. Worden, R.M.; Grethlein, A.J.; Jain, M.K.; Datta, R. Production of butanol and ethanol from synthesis gas via fermentation. *Fuel* **1991**, *70*, 615–619. [CrossRef]

28. Abrini, J.; Naveau, H.; Nyns, E.-J. *Clostridium autoethanogenum*, sp. nov., an anaerobic bacterium that produces ethanol from carbon monoxide. *Arch. Microbiol.* **1994**, *161*, 345–351. [CrossRef]

29. Liou, J.S.-C.; Balkwill, D.L.; Drake, G.R.; Tanner, R.S. *Clostridium carboxidivorans* sp. nov., a solvent-producing clostridium isolated from an agricultural settling lagoon, and reclassification of the acetogen *Clostridium scatologenes* strain SL1 as *Clostridium drakei* sp. nov. *Int. J. Syst. Evol. Microbiol.* **2005**, *55*, 2085–2091. [CrossRef] [PubMed]

30. Phillips, J.R.; Atiyeh, H.K.; Tanner, R.S.; Torres, J.R.; Saxena, J.; Wilkins, M.R.; Huhnke, R.L. Butanol and hexanol production in *Clostridium carboxidivorans* syngas fermentation: Medium development and culture techniques. *Bioresour. Technol.* **2015**, *190*, 114–121. [CrossRef] [PubMed]

31. Saxena, J. Development of an optimized and cost-effective medium for ethanol production by *Clostridium* strain P11. In *Botany and Microbiology*; University of Oklahoma: Norman, OK, USA, 2008; p. 131.

32. Allen, T.D.; Caldwell, M.E.; Lawson, P.A.; Huhnke, R.L.; Tanner, R.S. *Alkalibaculum bacchi* gen. nov., sp. nov., a CO-oxidizing, ethanol-producing acetogen isolated from livestock-impacted soil. *Int. J. Syst. Evol. Microbiol.* **2010**, *60*, 2483–2489. [CrossRef] [PubMed]

33. Liu, K.; Atiyeh, H.K.; Tanner, R.S.; Wilkins, M.R.; Huhnke, R.L. Fermentative production of ethanol from syngas using novel moderately alkaliphilic strains of *Alkalibaculum bacchi*. *Bioresour. Technol.* **2012**, *104*, 336–341. [CrossRef] [PubMed]

34. Liu, K.; Atiyeh, H.K.; Stevenson, B.S.; Tanner, R.S.; Wilkins, M.R.; Huhnke, R.L. Continuous syngas fermentation for the production of ethanol, n-propanol and n-butanol. *Bioresour. Technol.* **2014**, *151*, 69–77. [CrossRef] [PubMed]

35. Liu, K.; Atiyeh, H.K.; Stevenson, B.S.; Tanner, R.S.; Wilkins, M.R.; Huhnke, R.L. Mixed culture syngas fermentation and conversion of carboxylic acids into alcohols. *Bioresour. Technol.* **2014**, *152*, 337–346. [CrossRef] [PubMed]

36. Xu, S.; Fu, B.; Zhang, L.; Liu, H. Bioconversion of H_2/CO_2 by acetogen enriched cultures for acetate and ethanol production: The impact of pH. *World J. Microbiol. Biotechnol.* **2015**, *31*, 941–950. [CrossRef] [PubMed]

37. Tanner, R.S. Cultivation of Bacteria and Fungi. In *Manual of Environmental Microbiology*, 3rd ed.; ASM Press: Washington, DC, USA, 2007; pp. 69–78.

38. Maddipati, P.; Atiyeh, H.K.; Bellmer, D.D.; Huhnke, R.L. Ethanol production from syngas by *Clostridium* strain P11 using corn steep liquor as a nutrient replacement to yeast extract. *Bioresour. Technol.* **2011**, *102*, 6494–6501. [CrossRef] [PubMed]

39. Saxena, J.; Tanner, R.S. Optimization of a corn steep medium for production of ethanol from synthesis gas fermentation by *Clostridium ragsdalei*. *World J. Microbiol. Biotechnol.* **2012**, *28*, 1553–1561. [CrossRef] [PubMed]

40. Kundiyana, D.K.; Huhnke, R.L.; Maddipati, P.; Atiyeh, H.K.; Wilkins, M.R. Feasibility of incorporating cotton seed extract in *Clostridium* strain P11 fermentation medium during synthesis gas fermentation. *Bioresour. Technol.* **2010**, *101*, 9673–9680. [CrossRef] [PubMed]

41. Phillips, J.R.; Klasson, K.T.; Clausen, E.C.; Gaddy, J.L. Biological production of ethanol from coal synthesis gas - Medium development studies. *Appl. Biochem. Biotechnol.* **1993**, *39*, 559–571. [CrossRef]

42. Cramer, W.A.; Knaff, D.B. Energy transduction in biological membranes: A textbook of bioenergetics. In *Springer Advanced Texts in Chemistry*; edn Springer Study; Springer-Verlag: New York, NY, USA, 1991.

43. Roberts, J.R.; Lu, W.P.; Ragsdale, S.W. Acetyl-coenzyme A synthesis from methyltetrahydrofolate, CO, and coenzyme A by enzymes purified from *Clostridium thermoaceticum*: Attainment of in vivo rates and identification of rate-limiting steps. *J. Bacteriol.* **1992**, *174*, 4667–4676. [CrossRef] [PubMed]

44. Ljungdahl, L.G. The Autotrophic Pathway of Acetate Synthesis in Acetogenic Bacteria. *Annu. Rev. Microbiol.* **1986**, *40*, 415–450. [CrossRef] [PubMed]

45. Ragsdale, S.W.; Lindahl, P.A.; Münck, E. Mössbauer, EPR, and optical studies of the corrinoid/iron-sulfur protein involved in the synthesis of acetyl coenzyme A by *Clostridium thermoaceticum*. *J. Biol. Chem.* **1987**, *262*, 14289–14297. [PubMed]

46. Yamamoto, I.; Saiki, T.; Liu, S.M.; Ljungdahl, L.G. Purification and properties of NADP-dependent formate dehydrogenase from *Clostridium thermoaceticum*, a tungsten-selenium-iron protein. *J. Biol. Chem.* **1983**, *258*, 1826–1832. [PubMed]

47. Mejillano, M.R.; Jahansouz, H.; Matsunaga, T.O.; Kenyon, G.L.; Himes, R.H. Formation and utilization of formyl phosphate by N10-formyltetrahydrofolate synthetase: Evidence for formyl phosphate as an intermediate in the reaction. *Biochemistry* **1989**, *28*, 5136–5145. [CrossRef] [PubMed]

48. Sun, A.Y.; Ljungdahl, L.; Wood, H.G. Total synthesis of acetate from CO_2 II. Purification and properties of formyltetrahydrofolate synthetase from *Clostridium thermoaceticum*. *J. Bacteriol.* **1969**, *98*, 842–844. [PubMed]

49. Ljungdahl, L.G.; O'Brien, W.E.; Moore, M.R.; Liu, M.-T. Methylenetetrahydrofolate dehydrogenase from *Clostridium formicoaceticum* and methylenetetrahydrofolate dehydrogenase, methenyltetrahydrofolate cyclohydrolase (combined) from *Clostridium thermoaceticum*. In *Methods in Enzymology*; Donald, B., McCormick, L.D.W., Eds.; Academic Press: Orlando, FL, USA, 1980; pp. 599–609.

50. Park, E.Y.; Clark, J.E.; de Vartanian, D.V.; Ljungdahl, L.G. 5,10-methylenetetrahydrofolate Reductases: Iron-sulfur-zinc flavoproteins. In *Chemistry and Biochemistry of flavoenzymes*; Müller, F., Ed.; CRC Press: Boca Raton, FL, USA, 1991; pp. 389–400.

51. Lu, W.P.; Harder, S.R.; Ragsdale, S.W. Controlled potential enzymology of methyl transfer reactions involved in acetyl-CoA synthesis by CO dehydrogenase and the corrinoid/iron-sulfur protein from *Clostridium thermoaceticum*. *J. Biol. Chem.* **1990**, *265*, 3124–3133. [PubMed]

52. Drake, H.L.; Hu, S.I.; Wood, H.G. Purification of five components from *Clostridium thermoaceticum* which catalyze synthesis of acetate from pyruvate and methyltetrahydrofolate. Properties of phosphotransacetylase. *J. Biol. Chem.* **1981**, *256*, 11137–11144. [PubMed]

53. White, H.; Strobl, G.; Feicht, R.; Simon, H. Carboxylic acid reductase: A new tungsten enzyme catalyses the reduction of non-activated carboxylic acids to aldehydes. *Eur. J. Biochem.* **1989**, *184*, 89–96. [CrossRef] [PubMed]

54. Fraisse, L.; Simon, H. Observations on the reduction of non-activated carboxylates by *Clostridium formicoaceticum* with carbon monoxide or formate and the Influence of various viologens. *Arch. Microbiol.* **1988**, *150*, 381–386. [CrossRef]

55. Phillips, J.R.; Clausen, E.C.; Gaddy, J.L. Synthesis gas as substrate for the biological production of fuels and chemicals. *Appl. Biochem. Biotechnol.* **1994**, *45*, 145–157. [CrossRef]

56. Latif, H.; Zeidan, A.A.; Nielsen, A.T.; Zengler, K. Trash to treasure: Production of biofuels and commodity chemicals via syngas fermenting microorganisms. *Curr. Opin. Biotechnol.* **2014**, *27*, 79–87. [CrossRef] [PubMed]

57. Liew, F.; Martin, M.E.; Tappel, R.C.; Heijstra, B.D.; Mihalcea, C.; Köpke, M. Gas fermentation—A flexible platform for commercial scale production of low-carbon-fuels and chemicals from waste and renewable feedstocks. *Front. Microbiol.* **2016**, *7*, 1–28. [CrossRef] [PubMed]

58. Hurst, K.M.; Lewis, R.S. Carbon monoxide partial pressure effects on the metabolic process of syngas fermentation. *Biochem. Eng. J.* **2010**, *48*, 159–165. [CrossRef]

59. Vega, J.L.; Prieto, S.; Elmore, B.B.; Clausen, E.C.; Gaddy, J.L. The Biological Production of Ethanol from Synthesis Gas. *Appl. Biochem. Biotechnol.* **1989**, *20*, 781–797. [CrossRef]

60. Ukpong, M.N.; Atiyeh, H.K.; de Lorme, M.J.M.; Liu, K.; Zhu, X.; Tanner, R.S.; Wilkins, M.R.; Stevenson, B.S. Physiological response of *Clostridium carboxidivorans* during conversion of synthesis gas to solvents in a gas-fed bioreactor. *Biotechnol. Bioeng.* **2012**, *109*, 2720–2728. [CrossRef] [PubMed]

61. Ezeji, T.; Milne, C.; Price, N.D.; Blaschek, H.P. Achievements and perspectives to overcome the poor solvent resistance in acetone and butanol-producing microorganisms. *Appl. Microbiol. Biotechnol.* **2010**, *85*, 1697–1712. [CrossRef] [PubMed]

62. Köpke, M.; Mihalcea, C.; Bromley, J.C.; Simpson, S.D. Fermentative production of ethanol from carbon monoxide. *Curr. Opin. Biotechnol.* **2011**, *22*, 320–325. [CrossRef] [PubMed]

63. Seravalli, J.; Ragsdale, S.W. Channeling of carbon monoxide during anaerobic carbon dioxide fixation. *Biochemistry* **2000**, *39*, 1274–1277. [CrossRef] [PubMed]

64. Isom, C.E.; Nanny, M.A.; Tanner, R.S. Improved conversion efficiencies for n-fatty acid reduction to primary alcohols by the solventogenic acetogen "*Clostridium ragsdalei*". *J. Ind. Microbiol. Biotechnol.* **2015**, *42*, 29–38. [CrossRef] [PubMed]

65. Shanmugasundaram, T.; Wood, H.G. Interaction of ferredoxin with carbon monoxide dehydrogenase from *Clostridium thermoaceticum*. *J. Biol. Chem.* **1992**, *267*, 897–900. [PubMed]

66. Buckel, W.; Thauer, R.K. Energy conservation via electron bifurcating ferredoxin reduction and proton/Na+ translocating ferredoxin oxidation. *Biochim. Biophys. Acta-Bioenerg.* **2013**, *1827*, 94–113. [CrossRef] [PubMed]

67. Poehlein, A.; Schmidt, S.; Kaster, A.K.; Goenrich, M.; Vollmers, J.; Thurmer, A.; Bertsch, J.; Schuchmann, K.; Voigt, B.; Hecker, M.; et al. An ancient pathway combining carbon dioxide fixation with the generation and utilization of a sodium ion gradient for ATP synthesis. *PLoS ONE* **2012**, *7*, e33439. [CrossRef] [PubMed]

68. Tremblay, P.L.; Zhang, T.; Dar, S.A.; Leang, C.; Lovley, D.R. The Rnf complex of *Clostridium ljungdahlii* is a proton-translocating ferredoxin:NAD+ oxidoreductase essential for autotrophic growth. *mBio* **2013**, *4*, 1–12. [CrossRef] [PubMed]

69. Schuchmann, K.; Muller, V. Autotrophy at the thermodynamic limit of life: A model for energy conservation in acetogenic bacteria. *Nat. Rev. Microbiol.* **2014**, *12*, 809–821. [CrossRef] [PubMed]

70. Hugenholtz, J.; Ivey, D.M.; Ljungdahl, L.G. Carbon monoxide-driven electron transport in *Clostridium thermoautotrophicum* membranes. *J. Bacteriol.* **1987**, *169*, 5845–5847. [CrossRef] [PubMed]

71. Thauer, R.K.; Jungermann, K.; Decker, K. Energy conservation in chemotrophic anaerobic bacteria. *Microbiol. Mol. Biol. Rev.* **1977**, *41*, 100–180.

72. Das, A.; Ljungdahl, L.G. Composition and primary structure of the F1F0 ATP synthase from the obligately anaerobic bacterium *Clostridium thermoaceticum*. *J. Bacteriol.* **1997**, *179*, 3746–3755. [CrossRef] [PubMed]

73. Ivey, D.M.; Ljungdahl, L.G. Purification and characterization of the F1-ATPase from *Clostridium thermoaceticum*. *J. Bacteriol.* **1986**, *165*, 252–257. [CrossRef] [PubMed]

74. Nakamoto, R.K.; Scanlon, J.A.B.; Al-Shawi, M.K. The rotary mechanism of the ATP synthase. *Arch. Biochem. Biophys.* **2008**, *476*, 43–50. [CrossRef] [PubMed]

75. Von Ballmoos, C.; Cook, G.M.; Dimroth, P. Unique rotary ATP synthase and its biological diversity. *Annu. Rev. Biophys.* **2008**, *37*, 43–64. [CrossRef] [PubMed]

76. Ragsdale, S.W. Enzymology of the acetyl-CoA pathway of CO_2 fixation. *Critical Rev. Biochem. Mol. Biol.* **1991**, *26*, 261–300. [CrossRef] [PubMed]

77. Hougen, O.A.; Watson, K.M.; Ragatz, R.A. *Chemical Process Principles*; Wiley: New York, NY, USA, 1954.

78. Bird, R.B.; Stewart, W.E.; Lightfoot, E.N. *Transport Phenomena*, 2nd ed.; Wiley: New York, NY, USA, 2002.

79. Charpentier, J.-C. Mass-transfer rates in gas-liquid absorbers and reactors. In *Advances in Chemical Engineering*; Drew, T.B., Cokelet, G.R., Hooper, J.W., Vermeulen, T, Eds.; Academic Press: Orlando, FL, USA, 1981; pp. 1–133.

80. Bailey, J.E.; Ollis, D.F. *Biochemical Engineering Fundamentals*; McGraw-Hill: New York, NY, USA, 1986.

81. McCabe, W.L.; Smith, J.C. *Unit Operations of Chemical Engineering*; McGraw-Hill: New York, NY, USA, 1976.

82. Munasinghe, P.C.; Khanal, S.K. Syngas fermentation to biofuel: Evaluation of carbon monoxide mass transfer coefficient (k_La) in different reactor configurations. *Biotechnol. Prog.* **2010**, *26*, 1616–1621. [CrossRef] [PubMed]

83. Klasson, K.T.; Ackerson, M.D.; Clausen, E.C.; Gaddy, J.L. Mass transport in bioreactors for coal synthesis gas fermentation. *Abstr. Pap. Am. Chem. Soc.* **1992**, 1924–1930.

84. Shen, Y.W.; Brown, R.; Wen, Z.Y. Enhancing mass transfer and ethanol production in syngas fermentation of *Clostridium carboxidivorans* P7 through a monolithic biofilm reactor. *Appl. Energy* **2014**, *136*, 68–76. [CrossRef]

85. Vega, J.L.; Antorrena, G.M.; Clausen, E.C.; Gaddy, J.L. Study of gaseous substrate fermentations: Carbon monoxide conversion to acetate. 2. continuous culture. *Biotechnol. Bioeng.* **1989**, *34*, 785–793. [CrossRef] [PubMed]

86. Shen, Y.; Brown, R.; Wen, Z. Syngas fermentation of *Clostridium carboxidivoran* P7 in a hollow fiber membrane biofilm reactor: evaluating the mass transfer coefficient and ethanol production performance. *Biochem. Eng. J.* **2014**, *85*, 21–29. [CrossRef]

87. Orgill, J.J.; Atiyeh, H.K.; Devarapalli, M.; Phillips, J.R.; Lewis, R.S.; Hunke, R.L. A comparison of mass transfer coefficients between trickle bed, hollow fiber membrane and stirred tank reactors. *Bioresour. Technol.* **2013**, *133*, 340–346. [CrossRef] [PubMed]

88. Nicholls, D.G.; Ferguson, S.J. *Bioenergetics*; Academic Press: Olrando, FL, USA, 2002.

89. Hu, P.; Bowen, S.H.; Lewis, R.S. A thermodynamic analysis of electron production during syngas fermentation. *Bioresour. Technol.* **2011**, *102*, 8071–8076. [CrossRef] [PubMed]

90. Lehninger, A.L. *Principles of Biochemistry*; Worth Publishers: New York, NY, USA, 1982.

91. Bar-Even, A. Does acetogenesis really require especially low reduction potential? *Biochim. Biophys. Acta* **2013**, *1827*, 395–400. [CrossRef] [PubMed]

92. Shuler, M.L.; Kargi, F. *Bioprocess Engineering Basic Concepts*; Prentice Hall PTR: Upper Saddle River, NJ, USA, 2002.

93. Chen, J.; Gomez, J.A.; Hoffner, K.; Barton, P.I.; Henson, M.A. Metabolic modeling of synthesis gas fermentation in bubble column reactors. *Biotechnol. Biofuels* **2015**, *8*, 1–12. [CrossRef]

94. Krivoruchko, A.; Zhang, Y.; Siewers, V.; Chen, Y.; Nielsen, J. Microbial acetyl-CoA metabolism and metabolic engineering. *Metab. Eng.* **2015**, *28*, 28–42. [CrossRef] [PubMed]

95. Bredwell, M.D.; Srivastava, P.; Worden, R.M. Reactor design issues for synthesis-gas fermentations. *Biotechnol. Prog.* **1999**, *15*, 834–844. [CrossRef] [PubMed]

96. Rogers, P.; Chen, J.-S.; Zidwick, M.J. Organic acid and solvent production. In *The Prokaryotes*; Dworkin, M., Falkow, S., Rosenberg, E., Schleifer, K.-H., Stackebrandt, E., Eds.; Springer: New York, NY, USA, 2006; pp. 511–755.

97. Straub, M.; Demler, M.; Weuster-Botz, D.; Durre, P. Selective enhancement of autotrophic acetate production with genetically modified *Acetobacterium woodii*. *J. Biotechnol.* **2014**, *178*, 67–72. [CrossRef] [PubMed]

98. Klasson, K.T.; Ackerson, C.M.D.; Clausen, E.C.; Gaddy, J.L. Biological conversion of synthesis gas into fuels. *Int. J. Hydrogen Energy* **1992**, *17*, 281–288. [CrossRef]

99. Atiyeh, H.K.; Phillips, J.R.; Huhnke, R.L. Fermentation Control for Optimization of Syngas Utilization. World Intellectual Property Organization: Geneva, Switzerland; WO2016077778 A1, 2016.

100. Kundiyana, D.K.; Huhnke, R.L.; Wilkins, M.R. Effect of nutrient limitation and two-stage continuous fermentor design on productivities during "*Clostridium ragsdalei*" syngas fermentation. *Bioresour. Technol.* **2011**, *102*, 6058–6064. [CrossRef] [PubMed]

101. Devarapalli, M.; Atiyeh, H.K.; Phillips, J.R.; Lewis, R.S.; Huhnke, R.L. Ethanol production during semi-continuous syngas fermentation in a trickle bed reactor using *Clostridium ragsdalei*. *Bioresour. Technol.* **2016**, *209*, 56–65. [CrossRef] [PubMed]

102. Devarapalli, M.; Lewis, R.S.; Atiyeh, H.K. Continuous ethanol production from synthesis gas by *Clostridium ragsdalei* in a trickle-bed reactor. *Fermentation* **2017**, *3*, 1–13. [CrossRef]

103. Richter, H.; Martin, M.E.; Angenent, L.T. A two-stage continuous fermentation system for conversion of syngas into ethanol. *Energies* **2013**, *6*, 3987–4000. [CrossRef]

104. Nguyen, T.T.N.; Belliere-Baca, V.; Rey, P.; Millet, J.M.M. Efficient catalysts for simultaneous dehydration of light alcohols in gas phase. *Catal. Sci. Technol.* **2015**, *5*, 3576–3584. [CrossRef]

105. Kopke, M.; Mihalcea, C.; Liew, F.M.; Tizard, J.H.; Ali, M.S.; Conolly, J.J.; Al-Sinawi, B.; Simpson, S.D. 2,3-Butanediol production by acetogenic bacteria, an alternative route to chemical synthesis, using industrial waste gas. *Appl. Environ. Microbiol.* **2011**, *77*, 5467–5475. [CrossRef] [PubMed]

106. Dürre, P. Butanol formation from gaseous substrates. *FEMS Microbiol. Lett.* **2016**, *363*, 1–7. [CrossRef] [PubMed]

107. Hu, P.; Chakraborty, S.; Kumar, A.; Woolston, B.; Liu, H.; Emerson, D.; Stephanopoulos, G. Integrated bioprocess for conversion of gaseous substrates to liquids. *Proc. Natl. Acad. Sci. USA* **2016**, *113*, 3773–3778. [CrossRef] [PubMed]

108. Pakpour, F.; Najafpour, G.; Tabatabaei, M.; Tohidfar, M.; Younesi, H. Biohydrogen production from CO-rich syngas via a locally isolated Rhodopseudomonas palustris PT. *Bioprocess Biosyst. Eng.* **2014**, *37*, 923–930. [CrossRef] [PubMed]
109. Youngsukkasem, S.; Chandolias, K.; Taherzadeh, M.J. Rapid bio-methanation of syngas in a reverse membrane bioreactor: Membrane encased microorganisms. *Bioresour. Technol.* **2015**, *178*, 334–340. [CrossRef] [PubMed]
110. EIA: 4-Week average U.S. *Oxygenate Plant Production of Fuel Tthanol (Thousand Barrels Per Day)*; Energy Department of Energy Information Administration: Washington, DC, USA, 2017.
111. Renewable Fuels Association (RFA). *World Fuel Ethanol Production*; Renewable Fuels Association: Washington, DC, USA, 2016.
112. Nasdaq. *Ethanol Futures.* 2017. Available online: http://www.nasdaq.com/markets/ethanol.aspx (accessed on 27 April 2017).
113. USDA NASS, Quick Stats Database. *USDA ERS—U.S. Bioenergy Statistics: Fuel Ethanol, Corn and Gasoline Prices*; Service UER. USDA Economic Research Service: Washington, DC, USA, 2017.
114. Klein-Marcuschamer, D.; Oleskowicz-Popiel, P.; Simmons, B.A.; Blanch, H.W. Technoeconomic analysis of biofuels: A wiki-based platform for lignocellulosic biorefineries. *Biomass Bioenergy* **2010**, *34*, 1914–1921. [CrossRef]
115. Humbird, D.; Davis, R.; Tao, L.; Kinchin, C.; Hsu, D.; Aden, A.; Schoen, P.; Lukas, J.; Olthof, B.; Worley, M. *Process Design and Economics for Biochemical Conversion of Lignocellulosic Biomass to Ethanol: Dilute-Acid Pretreatment and Enzymatic Hydrolysis of Corn Stover*; National Renewable Energy Laboratory (NREL): Golden, CO, USA, 2011.
116. Dwivedi, P.; Alavalapati, J.R.; Lal, P. Cellulosic ethanol production in the United States: Conversion technologies, current production status, economics, and emerging developments. *Energy Sustain. Dev.* **2009**, *13*, 174–182. [CrossRef]
117. Piccolo, C.; Bezzo, F. A techno-economic comparison between two technologies for bioethanol production from lignocellulose. *Biomass Bioenergy* **2009**, *33*, 478–491. [CrossRef]
118. Anex, R.P.; Aden, A.; Kazi, F.K.; Fortman, J.; Swanson, R.M.; Wright, M.M.; Satrio, J.A.; Brown, R.C.; Daugaard, D.E.; Platon, A. Techno-economic comparison of biomass-to-transportation fuels via pyrolysis, gasification, and biochemical pathways. *Fuel* **2010**, *89*, S29–S35. [CrossRef]
119. Dutta, A.; Talmadge, M.; Hensley, J.; Worley, M.; Dudgeon, D.; Barton, D.; Groenendijk, P.; Ferrari, D.; Stears, B.; Searcy, E. Techno-economics for conversion of lignocellulosic biomass to ethanol by indirect gasification and mixed alcohol synthesis. *Environ. Progress Sustain. Energy* **2012**, *31*, 182–190. [CrossRef]
120. Spath, P.L.; Dayton, D.C. Preliminary Screening-Technical and Economic Assessment of Synthesis Gas to Fuels and Chemicals with Emphasis on the Potential for Biomass-Derived Syngas. 2003. Available online: http://oai.dtic.mil/oai/oai?verb=getRecord&metadataPrefix=html&identifier=ADA436529 (accessed on 27 April 2017).
121. Brown, T.R. A techno-economic review of thermochemical cellulosic biofuel pathways. *Bioresour. Technol.* **2015**, *178*, 166–176. [CrossRef] [PubMed]
122. Kazi, F.K.; Fortman, J.A.; Anex, R.P.; Hsu, D.D.; Aden, A.; Dutta, A.; Kothandaraman, G. Techno-economic comparison of process technologies for biochemical ethanol production from corn stover. *Fuel* **2010**, *89*, S20–S28. [CrossRef]
123. Tanner, R.S. Production of ethanol from synthesis gas. In *Bioenergy*; Wall, J.D., Harwood, C.S., Demain, A.L., Eds.; American Society of Microbiology: Washington, DC, USA, 2008; pp. 147–151.

fermentation

MDPI

Article

Phenols Removal from Hemicelluloses Pre-Hydrolysate by Laccase to Improve Butanol Production

Rosalie Allard-Massicotte [1,2], Hassan Chadjaa [1] and Mariya Marinova [2,3,*

[1] Centre National en Électrochimie et en Technologies Environnementales, Shawinigan, QC G9N 6V8, Canada; rosalie.allard-massicotte@polymtl.ca (R.A.-M.); hchadjaa@collegeshawinigan.qc.ca (H.C.)
[2] Department of Chemical Engineering, Polytechnique Montréal, Montréal, QC H3T 1J4, Canada
[3] Department of Wood and Forest Sciences, Université Laval, 2405 rue de la Terrasse, Québec, QC G1V 0A6, Canada
* Correspondence: mariya.marinova.1@ulaval.ca

Received: 9 May 2017; Accepted: 26 June 2017; Published: 30 June 2017

Abstract: Phenolic compounds are important inhibitors of the microorganisms used in the Acetone-Butanol-Ethanol (ABE) fermentation. The degradation of phenolic compounds in a wood pre-hydrolysate, a potential substrate for the production of ABE, was studied in this article. First, physicochemical methods for detoxification such as nanofiltration and flocculation were applied and the best combination was selected. With a flocculated sample, the concentration of phenolic compounds decreases from 1.20 to 0.28 g/L with the addition of a solid laccase at optimum conditions, which is below the phenolic compounds limit of inhibition. This results in an increase in butanol production, more than double, compared to a pre-hydrolysate non-treated with laccase enzymes.

Keywords: phenolic compounds; detoxification; butanol; pre-hydrolysate; laccase enzymes

---

1. Introduction

The acetone-butanol-ethanol (ABE) fermentation was developed in the late 19th and the early 20th century and is one of the first fermentation processes adapted to a large scale. The main reason for this was the growing need of acetone for munitions factories during the two World Wars [1]. Today, the production of butanol and its derivatives is attractive for international markets [2]. In addition, butanol is a high-energy fuel, less corrosive than ethanol, and its derivatives are used for the production of latex, plasticizers and coatings [3]. Although ABE fermentation with *Clostridium acetobutylicum* has been well known, improvements are still being proposed to decrease the cost of substrate and the subsequent operating costs. In a previous work conducted by Ajao et al. [4,5], the detoxification of a wood pre-hydrolysate, obtained from a dissolving Kraft pulp mill, was conducted by filtration and flocculation, prior to ethanol production by fermentation.

In the current study, the pre-hydrolysate was used as a substrate to grow *C. acetobutylicum* and to produce butanol. During the pre-hydrolysis step, phenolic compounds which are lignin residues are generated [6,7]. They can damage the structure of the cell membrane and the ability of microorganisms to absorb sugars, which makes them very toxic for the ABE fermentation [8]. The reduction of the phenolic compounds concentration by flocculation was significant [4]; however, it was not sufficient to improve fermentation yield. The conventional treatment of the pre-hydrolysate cannot reduce the concentration of phenolic compounds below the inhibition limit, as set by Mechmech et al. [9].

In this work, the degradation of the phenolic compounds using solid and liquid laccase enzymes was investigated to complement the physicochemical treatment. Laccases are copper-containing enzymes and can oxidize several types of phenolic compounds [10–12]. The enzymatic degradation of phenolic compounds was investigated on aqueous solutions inspired by wastewater treatment [13–15]. These studies have been done with laccases or peroxydases, oxidative enzymes similar to the laccases. The method can significantly improve the degradation of phenolic compounds, thus reaching the thresholds that do not allow inhibition. There are two important aspects to understand the relevance of this work: the world population is concerned about the environment protection and the revalorization of resources can be an additional income for pulp and paper mills. The detoxification of lignocellulosic biomass by laccases was investigated to increase the quality of the substrate in ethanol production [16]. In the current study, it is shown that the degradation of the phenolic compounds prior to fermentation using laccases can increase the production of butanol from a wood pre-hydrolysate and improve the cost-effectiveness of the process. In North America, where the pulp and paper industry is under transformation, looking to implement innovative processes, it is important to maximize the reuse of resources in a biorefinery context [17].

2. Materials and Methods

2.1. Microorganism, Culture Maintenance and Inoculum Preparation

The culture preparation was performed as described by Mechmech et al. [9]. *C. acetobutylicum* ATCC 824 was obtained from the American Type Culture Collection (ATCC) and was cultured in sterilized Reinforced Clostridium Medium (RCM), its composition being (in g/L): tryptose 10, beef extract 10, glucose 5, yeast extract 3, soluble starch 0.5, sodium chloride 5, l-cysteine-HCl 0.5. The culture was kept under anaerobic conditions at 37 °C for 18–22 h, with shaking at 110 rpm, until an Optical Density (OD_{600}) of 1.9–2 was obtained. Glycerol 50% *v/v* was added to the bacterial culture to obtain a final concentration of 25% and the stock culture was immediately frozen at -80 °C until use. For complete anaerobic conditions, a drop of sodium sulfide nonahydrate was added in a cryogenic tube.

The inoculum for ABE fermentation was prepared in a RCM medium. The medium was first boiled and then purged with a gas mixture of 80% N2 and 20% CO2 to remove oxygen from the culture media. The culture was inoculated to a proportion of 1/2000 with stock culture. Culture conditions were identical to those of the stock culture. Inoculum was ready to be inoculated at an OD600 between 0.6 and 0.9.

2.2. Fermentation Medium

The fermentations experiments were carried out in a complex fermentation medium. The medium was composed of 60 g/L of xylose and resazurin 0.001% base. Before sterilization, the base was first boiled and then purged with a gas mixture of 80% N_2 and 20% CO_2 during 5 min. After sterilization, a purged sterile solution of 200 g/L of yeast extract was added in a ratio of 1/40. A filtered mixture with KH_2PO_4 50 g/L, K_2HPO_4 50 g/L, ammonium acetate 220 g/L, para-aminobenzoic acid 0.1 g/L, thiamin 0.1 g/L, biotin 0.001 g/L, $MgSO_4 \cdot 7H_2O$, 20 g/L, $MnSO_4 \cdot H_2O$ 1 g/L, $FeSO_4 \cdot 7H_2O$ 1 g/L and NaCl 1 g/L was added in a ratio of 1/100. In some experiments, a phenolic compound was added to the xylose solution to maintain a specific concentration. Screw capped Schott bottles of 250 mL were filled with 200 mL of complete culture media and used for anaerobic fermentation. The culture media for the hydrolysate test consists of 195 mL of treated hydrolysate and 5 mL of yeast extract. Xylose (60 g/L) was added in order to compare with the control solution. Before inoculation with 10 mL of inoculums, the bottles were slightly open in anaerobic jars containing Gas Pak envelopes (BD Gas Pak™EZ Anaerobe Container System, Franklin Lakes, NJ, USA) with indicators (BD BBL™ Dry Anaerobic Indicator Strips, Franklin Lakes, NJ, USA) for 48 h to create perfect anaerobic conditions in the mixtures. If necessary, a solution containing 5% *p/v* of $Na_2S \cdot 9H_2O$ was added in a ratio of 1/400 to eliminate the traces of oxygen. After inoculation, the cultures were incubated at 37 °C, 110 rpm and with a pH control. During fermentation, 5 mL samples were periodically withdrawn to analyze OD_{600}, residual xylose and alcohols. All fermentation experiments were performed in duplicate.

2.3. Analytical Methods

Spectrometer (Pharmacia Biotech Novaspec®II, Piscataway, NJ, USA) was used to monitor the growth of *C. acetobutylicum* and to determine the total phenols concentration with the Folin–Ciocalteu reagent method [18].

Gas chromatograph (GC 7890A, Agilent Technologies, Santa Clara, CA, USA) w ith an OV 624 capillary column and a flame ionization detector (H_2 flow rate: 30 mL/min; air flow rate: 2.23 mL/min) was used to measure butanol, acetone and ethanol concentrations in the fermentation medium after the inhibition test [9].

High performance liquid chromatography (HPLC Agilent Technology, Germany) was used to determinate the concentration of vanillin, catechol, syringaldehyde gallic acid and simple sugars. To measure phenols concentration, separation was made using a mixture of 15% acetonitrile and 85% phosphoric acid 10 mM on a Nucleosil C18 (150 X 4.6 mm) column with a diode array detector (DAD) at 313 nm and 280 nm. Dilution for phenol analysis was performed in the same solvent to reach a maximum concentration of 500 ppm. To measure the simple sugar concentration, a refractive index detector and an EC Nucleodur RP-NH_2 colum (250 mm × 4.6 mm, 5μm) were used with a mixture of 75% acetonitrile and 25% deionized water as a solvent. A temperature of 40 °C and a flow rate of 1.5 mL/min were applied for better separation [9]. Dilution for simple sugar analysis was performed in 50% acetonitrile and 50% deionized water to reach a maximum concentration of 10 g/L.

2.4. Degradation with Laccase Enzymes

For degradation assays, two types of laccase enzymes provided by industrial partners were used: a solid (or dehydrated) laccase and a liquid laccase. All degradation tests were performed in a test tube (14 mm diameter) with a 5 mL mixture composed of syringaldehyde, vanillin, catechol and gallic acid with the same concentrations. Total phenolic compounds concentrations of 2 g/L, 4 g/L and 6 g/L were used. After pH adjustment with 4M NaOH, a fixed dose of laccase was added and the test tube was incubated for 7h at the appropriate temperature and rotation at 180 rpm. When a large volume of flocculated hydrolysates was detoxified by laccase, an erlenmeyer was used with a ratio erlemeyer volume/flocculated hydrolysate volume of 2.5. A preliminary screening was done to determinate the optimal temperature and pH conditions for the subsequent experiments. The samples were analyzed once; however, occasional duplicate analyses were conducted to check the method used. Since the fermentation and enzyme degradation tests were performed, respectively, in duplicate and in triplicate, irregular results were easily detected and corrected.

2.5. Preparation and Treatment of Hydrolysates

Detoxification approaches described by Mechmech et al. [9] were first applied before the degradation of the phenolic compounds with laccase. To extract the hemicelluloses, a mixture of 60% aspen and 40% maple wood chips was treated with hot water and steam in a pilot digester at FPInnovations (Pointe-Claire, Québec, Canada). That pre-hydrolysate was forwarded to the Centre National en Électrochimie et en Technologies Environnementales (CNETE). Two different methods were applied to detoxify the pre-hydrolysate. First, pre-hydrolysate was filtered through the organic membrane NF270 (Molecular weight cut-off 200–400 Da), then hydrolyzed with 1.5% *w/w* sulfuric acid at 121 °C for 60 min to increase the monomeric sugars concentration. After these treatments, it was coagulated/flocculated. The pH was raised to 6.5 and ferric sulfate ($Fe_2(SO_4)_3$) with a ratio 1 g Fe/1 g phenolic compound was added. Flocculation experiments were carried out in jar tests with an agitation of 150 rpm for 10 min., then 50 rpm for 30 min. In a second step, the same operations were performed, but without applying the nanomembrane filtration.

3. Results and Discussion

3.1. Optimization of the Degradation Conditions

The screening of the solid and liquid laccase enzymes indicates that their efficiency is strongly affected by temperature and pH. The results with a 2 g/L synthetic solution containing syringaldehyde, vanillin, catechol and gallic acid in equal quantity have shown that the degradation of the phenolic compounds by laccases is optimal at a pH 8 and a temperature of 50 °C (Figures 1 and 2).

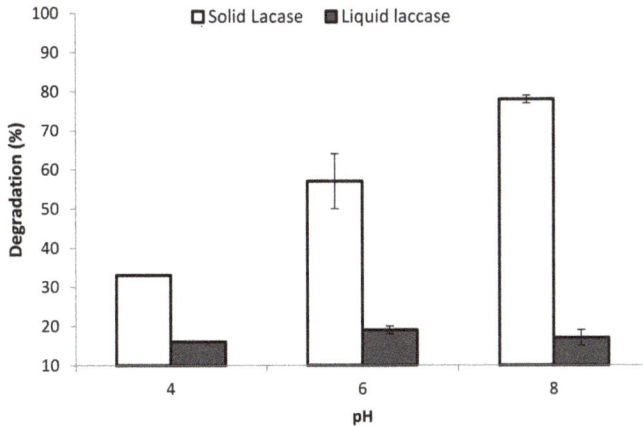

Figure 1. Degradation of phenolic compounds by laccase enzymes at different pH.

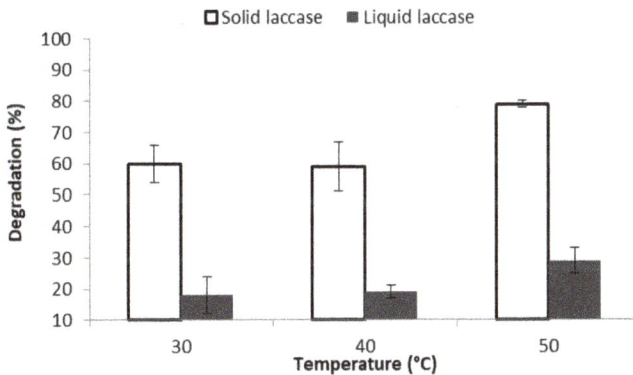

Figure 2. Degradation of phenolic compounds by laccase enzymes at different temperatures.

The optimal dose of the solid laccase is 100 mg of enzyme/g of phenolic compounds. The results are similar when a dosage of 200 mg/g is used; however, for economic reasons, it is preferable to use 100 mg enzyme/g of phenolic compounds (Figure 3). At optimal pH, temperature and enzyme dose, the degradation of the phenolic compounds by the solid laccase is 79%. The most relevant parameters for this type of laccase are the pH, the dose of laccase and the initial concentration of phenolic compounds ($p < 0.01$). Moreover, there is an interaction between the initial concentration of phenolic compounds and the dose of laccase; therefore, the dose of enzyme should be adjusted according to the initial concentration of phenolic compounds. An interaction regression coefficient of -0.24 between the dose of laccase and the initial concentration of phenolic compounds and a correlation value of 0.83 were calculated.

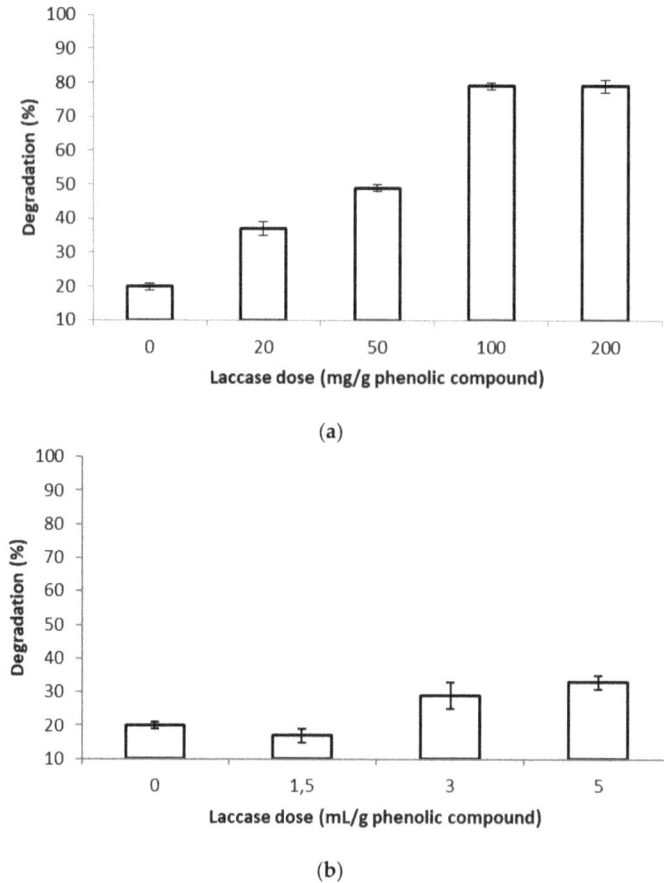

(a)

(b)

Figure 3. Degradation of phenolic compounds with different dose of laccase enzymes: (**a**) solid laccase; (**b**) liquid laccase.

The optimal dose for the liquid laccase is 5 mL enzyme/g of phenolic compounds (Figure 3). At optimal conditions, the degradation of the phenolic compounds by liquid laccase is 33%. For economic reasons, it is suggested to use 3 mL/g of phenolic compounds. When 3 mL of laccase enzyme are used, a degradation percentage of 29% is reached. It is important to mention that for the liquid laccase, the impact of the pH on the degradation efficiency is not significant ($p > 0.05$) and for practical reasons an initial pH of 8 was used. On the other hand, the temperature, the dose of laccase and the initial concentration of phenolic compounds are critical for the efficiency of the liquid laccase. A statistical analysis was performed and the results have shown a non-significant effect of pH and an important effect of the dose of laccase, the initial concentration of the phenolic compounds and the temperature. The regression coefficients for the three parameters were, respectively, 0.56, 0.55 and -0.49, with a p-value < 0.01.

According to the results on the degradation of phenolic compounds with laccases on various stages of the conventional detoxification presented in Section 3.4, the best results were obtained for a sample with pH 8. It is important to point out that there is a link between the degradation time and the added dose of laccase ($p < 0.01$). In fact, the more the laccase dose was high, faster the degradation was. The control sample without laccase demonstrated a degradation of the phenolic compounds of 20% in time (Figure 3). Therefore, a part of the degradation was performed naturally and not by the laccases. The corresponding equation for the naturally occurring phenolic degradation in time was: $y = 0.011x^2 - 0.12x + 1.8$ with a correlation value of 0.77. The degradation mainly occurs during the first hour, after that there was a stabilization of the phenolic compounds concentration.

3.2. Influence of the Individual Phenolic Compounds

The degradation of the phenolics is influenced by the type of individual phenolic compounds in the mixture. When the solid laccase was used, gallic acid was always degraded in the first two hours of incubation, followed by catechol which was entirely degraded too, then syringaldehyde and vanillin which were partially degraded at the end of incubation (Figure 4). Although it is not shown in the figure, it seems that there is a link with the temperature. When the temperature is lower (40 and 30 °C), the degradation of gallic acid and catechol is accelerated, while the other phenolic compounds are not degraded. This implies a change in the conformation of the enzyme to make it suitable for the degradation of syringaldehyde and vanillin at high temperature. The conformation of the enzyme can explain that preference. Usually, when the number of alcohol group increases, the speed of degradation increases too. The laccases directly affect the alcohol groups by oxidation of a reducing substrate and the formation of a free radical [10]. The accessibility of these groups can also influence their oxidation and affect the degradation order. The nature of the individual phenolic compounds is therefore a critical factor for the degradation.

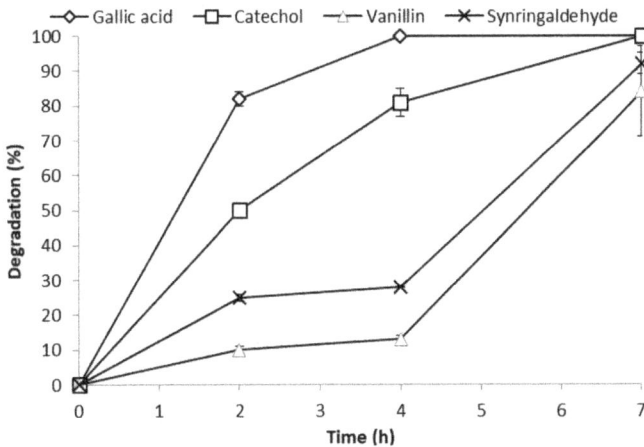

Figure 4. Degradation of the individual phenolic compounds using the solid laccase enzyme as a function of time.

The liquid laccase was efficient for gallic acid and syringladehyde degradation, but had a limited effect on vanillin and catechol, as shown in Figure 5. This confirms the importance of the type of phenolic compounds on the degradation by laccases. Gallic acid is degraded first, then syringaldehyde with 74% and 68% degradation, respectively. Only 35% and 29% of the catechol and vanillin, respectively, are degraded by the liquid laccase. These results are lower, compared to the solid laccase.

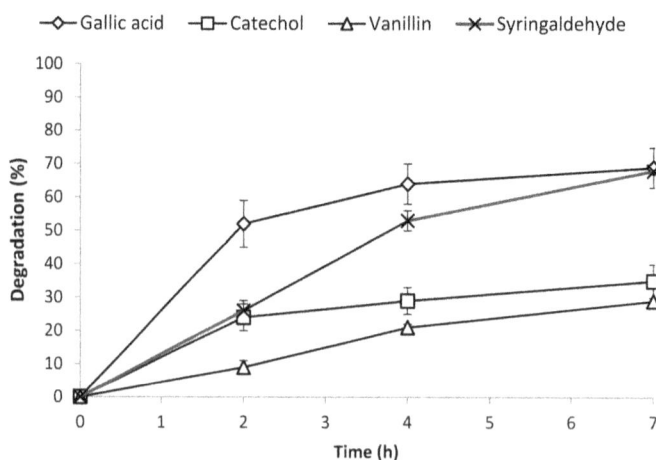

Figure 5. Degradation of the individual phenolic compounds using the liquid laccase enzyme as a function of time.

A sample with a predominance of gallic acid would be degraded faster than a sample rich in vanillin. It would be interesting to test the capacity of degradation on other type of phenolic compounds. On the other side, considering the variety of phenolic compounds, it is almost impossible to determine all individual efficiencies. Therefore, it is suggested to determine the group preference, as it is done with the alcohol group in this work.

3.3. Influence of the Initial Phenolic Compounds Concentration

To determine the capacity of laccase to degrade a concentrated mixture of phenolic compounds, their degradation at different initial concentrations was studied. At the same conditions and proportional dosage, there is a better degradation at the lowest initial concentration (Figure 6). With initial concentration of 2 g/L of phenolic compounds, the degradation is 79%, at 4 g/L it drops to 55% and at 6 g/L only 25% of the phenolic compounds are degraded by the solid laccase. It is important to point out that the increase of the dosage up to 200 mg/g of phenolic compounds can partially offset the increase of the initial concentration. Indeed, with 200 mg/g of phenolic compounds, the degradation reaches 38% with an initial phenolic compounds concentration of 6 g/L. It is 13% more than with only 100 mg/g of phenolic compounds. There is a link between the dosage and the efficiency at high concentrations of phenolic compounds. For the liquid laccase, an initial concentration of 2 g/L of phenolic compounds allows a degradation of 25% (Figure 6). However, with 4 g/L, only 15% of the phenolic compounds are degraded and with 6 g/L, 14%. Regarding the liquid laccase, the difference between the results was due to an increase in the concentration of syringaldehyde, catechol and vanillin. It was difficult to degrade these compounds by the liquid laccase and the residual amount was higher than expected. For the solid laccase, a similar phenomenon occurs. Regardless of the initial concentration (2, 4 or 6 g/L), gallic acid will always be degraded entirely in 2 h. The catechol degradation is around 100%, although it is slightly less effective when the initial concentration of the phenolic compounds increases to 6 g/L. The residual phenolics are vanillin and syringaldehyde. It seems to be difficult for the solid laccase to degrade those phenolic compounds at higher concentrations.

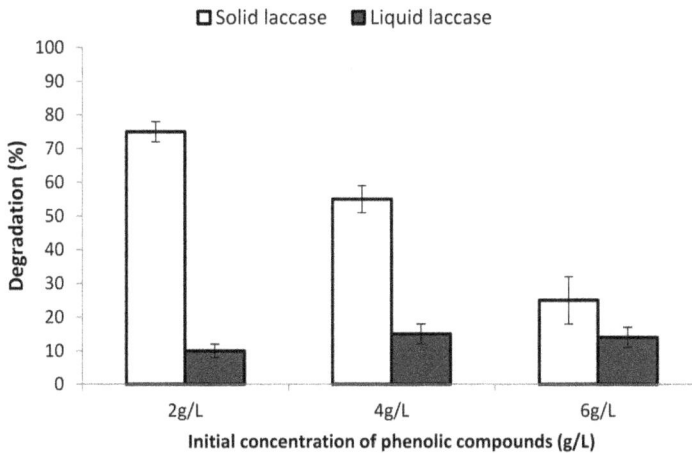

Figure 6. Degradation of the phenolic compounds by laccase enzymes as a function of the initial concentration of phenolic compounds.

3.4. Hydrolysate Detoxification by Laccase

Finally, the efficiency of the laccase enzymes on pre-hydrolysate samples pretreated with different detoxification methods were tested. As shown in Figure 7, the best degradation of the phenolic compounds occurs in the flocculated non-filtrated hydrolysate, and by the solid laccase, at its optimum conditions. In fact, this is the only case where laccase sufficiently degrades phenolic compounds to a concentration below the minimum level of inhibition determined by Mechmech et al. [9]. The level of degradation can reach a leftover of phenolic compounds of 0.28 g/L, under the limit of inhibition of ABE fermentation of 1.1 g/L in flocculated media. However, none of the other samples can reach this limit or the limit of 0.3 g/L for the non-flocculated sample. Theoretically, the percentage of degradation is expected to be similar from one sample to another, but in practice, the composition of each sample may vary significantly. For example, phenolic compounds proportion can be different between flocculated samples and untreated samples. The results show a significant difference in the degradation of the phenolic compounds for each intermediate stage of detoxification. However, there is a link between the enzyme type and the efficiency of detoxification by laccase at each intermediate stage (Figure 7). The liquid laccase seems to be more efficient when used before flocculation, with a degradation percentage of 42% for the untreated hydrolysate and 36%, for the filtered hydrolysate. Two hypotheses are envisaged. First, it is possible that the flocculation removes mainly phenolic compounds that would also be degraded by the liquid laccase, thus reducing the total amount of initial phenolic compounds to be degraded. Then, the residual ferric sulfate used for flocculation can form a complex with the laccase, thus making the enzyme less effective or inactive. The solid laccase is very efficient on flocculated hydrolysate and less efficient on untreated hydrolysate. The hydrolysate composition is complex and may be the source of the variation in the effectiveness of the laccase enzymes.

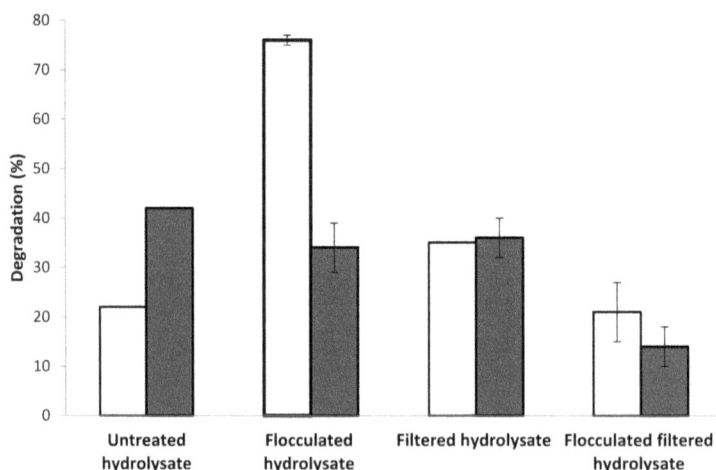

Figure 7. Degradation percentage of phenolic compounds by laccase enzymes in hydrolysates pre-treated by different methods. White: solid laccase; Black: liquid laccase.

3.5. Impact of the Additional Degradation by Laccase on ABE Fermentation

The production of butanol by *C. acetobutylicum* from a detoxified and not detoxified substrate by laccase enzymes was tested. In the case of the substrate detoxified by the solid enzyme, the production of butanol by *C. acetobutylicum* can double, compared with the substrate containing toxic phenolic compounds. The butanol production increases from 1.54 g/L to 4.17 g/L. The results are homogeneous and indicate a clear trend on the effectiveness of the solid laccase. The detoxification by laccase enhances the efficiency of the ABE fermentation, by decreasing the concentration of the phenolic compounds under the limit of inhibition. In the case of the liquid laccase, no butanol production occurs. The reason for this outcome should be investigated, but a negative interaction between the bacteria and the enzyme, or its degradation products is suspected.

4. Conclusions

In this work, it has been shown that laccase enzymes efficiently degrade phenolic compounds in wood hydrolysates. At an optimum temperature of 50 °C, pH of 8 and enzyme dose of 100 mg/g of phenolic compounds, the degradation of the phenolic compounds reaches 77%. The use of laccase for wood hydrolysate detoxification reduces the phenolic compounds concentration to 0.28 g/L, far below the limit of inhibition. The hydrolysate detoxification combining flocculation and laccase enzymes prior to fermentation increases the amount of butanol produced.

Acknowledgments: This work was supported by a grant from the College-University I2I Program of the Natural Sciences and Engineering Research Council of Canada (grant number 437803-12) and BioFuelNet Canada, a Network of Centers of Excellence. The authors wish to express their gratitude to CNETE for hosting the experimental work, especially to Nathalie Martel and Louis Tessier for their help, to FPInnovations for supplying the pre-hydrolysate samples and to Fraunhofer and Metgen Oy for providing the enzyme samples.

Author Contributions: Rosalie Allard-Massicotte, Hassan Chadjaa and Mariya Marinova conceived and designed the experiments; Rosalie Allard-Massicotte performed the experiments; Rosalie Allard-Massicotte and Hassan Chadjaa analyzed the data; Hassan Chadjaa and Mariya Marinova contributed to reagents/materials/analysis tools; Rosalie Allard-Massicotte and Mariya Marinova wrote the paper.

Conflicts of Interest: The authors declare no conflict of interest.

Fermentation **2017**, *3*, 31

References

1. Jones, D.T.; Woods, D.R. Acetone-butanol fermentation revisited. *Microbiol. Rev.* **1986**, *50*, 484–524.
2. Lee, S.Y.; Park, J.H.; Jang, S.H.; Nielsen, L.K.; Kim, J.; Jung, K.S. Fermentative butanol production by clostridia. *Biotechnol. Bioeng.* **2008**, *101*, 209–228. [CrossRef] [PubMed]
3. Kirschner, M. n-Butanol. *Chem. Mark. Report.* **2006**, *269*, 42.
4. Ajao, O.; LeHir, M.; Rahni, M.; Marinova, M.; Chadjaa, H.; Savadogo, O. Concentration and Detoxification of Kraft Prehydrolysate by Combining Nanofiltration with Flocculation. *Ind. Eng. Chem. Res.* **2015**, *54*, 1113–1122. [CrossRef]
5. Ajao, O.; Rahni, M.; Marinova, M.; Chadjaa, H.; Savadogo, O. Retention and flux characteristics of nanofiltration membranes during hemicellulose prehydrolysate concentration. *Chem. Eng. J.* **2015**, *260*, 605–615. [CrossRef]
6. Chandel, A.K.; da Silva, S.S.; Singh, O. V. Detoxification of Lignocellulose Hydrolysates: Biochemical and Metabolic Engineering Toward White Biotechnology. *Bioenergy Res.* **2013**, *6*, 388–401. [CrossRef]
7. Baral, N.R.; Shah, A. Microbial inhibitors: formation and effects on acetone-butanol-ethanol fermentation of lignocellulosic biomass. *Appl. Microbiol. Biotechnol.* **2014**, *98*, 9151–9172. [CrossRef] [PubMed]
8. Palmqvist, E.; Hahn-Hagerdal, B. Fermentation of lignocellulosic hydrolyzates. II: Inhibitors and mechanisms of inhibition. *Bioresour. Technol.* **2000**, *74*, 25–33. [CrossRef]
9. Mechmech, F.; Chadjaa, H.; Rahni, M.; Marinova, M.; Ben Akacha, N.; Gargouri, M. Improvement of butanol production from a hardwood hemicelluloses hydrolysate by combined sugar concentration and phenols removal. *Bioresour. Technol.* **2015**, *192*, 287–295. [CrossRef] [PubMed]
10. Madhavi, V.; Lele, S.S. Laccase: properties and applications. *Bioresources* **2009**, *4*, 1694–1717.
11. Wang, Z.X.; Cai, Y.J.; Liao, X.R.; Tao, G.J.; Li, Y.Y.; Zhang, F.; Zhang, D.B. Purification and characterization of two thermostable laccases with high cold adapted characteristics from Pycnoporus sp. SYBC-L1. *Process Biochem.* **2010**, *45*, 1720–1729. [CrossRef]
12. Sherif, M.; Waung, D.; Korbeci, B.; Mavisakalyan, V.; Flick, R.; Brown, G.; Abou-Zaid, M.; Yakunin, A.F.; Master, E.R. Biochemical studies of the multicopper oxidase (small laccase) from Streptomyces coelicolor using bioactive phytochemicals and site-directed mutagenesis. *Microb. Biotechnol.* **2013**, *6*, 588–597. [CrossRef] [PubMed]
13. Bayramoğlu, G.; Arica, M.Y. Enzymatic removal of phenol and p-chlorophenol in enzyme reactor: Horseradish peroxidase immobilized on magnetic beads. *J. Hazard. Mater.* **2008**, *156*, 148–155. [CrossRef] [PubMed]
14. Dasgupta, S.; Taylor, K.E.; Bewtra, J.K.; Biswas, N. Inactivation of enzyme laccase and role of cosubstrate oxygen in enzymatic removal of phenol from water. *Water Environ. Res.* **2007**, *79*, 858–867. [CrossRef] [PubMed]
15. Kauffmann, C.; Petersen, B.R.; Bjerrum, M.J. Enzymatic removal of phenols from aqueous solutions by Coprinus cinereus peroxidase and hydrogen peroxide. *J. Biotechnol.* **1999**, *73*, 71–74. [CrossRef]
16. Moreno, A.D.; Ibarra, D.; Fernandez, J.L.; Ballesteros, M. Different laccase detoxification strategies for ethanol production from lignocellulosic biomass by the thermotolerant yeast Kluyveromyces marxianus CECT 10875. *Bioresour. Technol.* **2012**, *106*, 101–109. [CrossRef] [PubMed]
17. Marinova, M.; Mateos-Espejel, E.; Jemaa, N.; Paris, J. Addressing the increased energy demand of a Kraft mill biorefinery: The hemicellulose extraction case. *Chem. Eng. Res. Des.* **2009**, *87*, 1269–1275. [CrossRef]
18. Singleton, V.L.; Rossi, J.A.J. Colorimetry of total phenolics with phosphomolybdic-phosphotungstic acid reagents. *Am. J. Enol. Vitic.* **1965**, *16*, 144–158.

MDPI AG

St. Alban-Anlage 66

4052 Basel, Switzerland

Tel. +41 61 683 77 34

Fax +41 61 302 89 18

http://www.mdpi.com

Fermentation Editorial Office

E-mail: fermentation@mdpi.com

http://www.mdpi.com/journal/fermentation